Advances in Soil Dynamics
Vol. 1

ASAE Monograph Number 12
Published by
American Society of Agricultural Engineers

Pamela DeVore-Hansen, Editor
Books & Journals
July 1994

ADVANCES IN SOIL DYNAMICS

Vol. 1

For information, contact the American Society of Agricultural
Engineers, 2950 Niles Road, St. Joseph, Michigan 49085-9659 USA
Phone 616.429.0300 Fax 616.429.3852

Library of Congress Catalog Card Number 94-72088
International Standard Book Number: 0-929355-52-0
Manufactured in the United States of America

Preface

Using the broadest concept of soil dynamics—soils moving under the action of gravitational and mechanical forces—the discipline of soil dynamics covers a wide range of applications: agricultural crop production, earthmoving, forestry, mining, military and space vehicle locomotion, and sea-bed operations. Since its inception, the discipline of soil dynamics has made continuous progress. Many basic tenets of soil dynamics and the history of soil dynamics were reviewed in 1985 during the International Conference on Soil Dynamics held at Auburn, Alabama (Anonymous, 1985a; Gill, 1985a).* This book, *Advances in Soil Dynamics*, represents an attempt to summarize some of the important developments in this field, particularly the progress made in the last 30 years.

Soil dynamics is uniquely an agricultural engineering discipline (Anonymous, 1934; Jones, 1941; Vincent, 1961; ASAE, 1990). History shows that the two pioneers who identified the importance of soil dynamics in the area of machine design and performance were a Russian and an American. Russian academician V. P. Goryachkin recognized the importance of studying combined soil-machine relations in the 1890s. In the 1920s Dr. Mark L. Nichols, without knowledge of Goryachkin's work, similarly surmised that the dynamic properties of soil held the key to the solution of problems associated with soil-machine interactions. They both realized that physical or mechanical forces applied to soil caused the soil to behave or react in manners that appeared to obey basic laws of physics, mathematics, and mechanics. They both:

1. Measured the basic operating parameters of soil engaging machines
2. Identified and measured those dynamic properties that seemed to inherently govern the soil-machine behaviors.
3. Developed theories of specific types of soil manipulations carried out by machines
4. Attempted to combine all of the information to create rational designs of soil engaging machines.

Today, the role of soil dynamics in solving problems and developing technology for things that move across and through soil is recognized as being important around the world. Soil dynamics is an engineering discipline that is studied and practiced worldwide.

---

* References cited are included in the list of references for Chapter 1.

Soil dynamics was developed because a scientific discipline was needed to express the fundamental soil-machine interactions required in the design and use of soil engaging machines and vehicles. Later, soil-plant interactions were included in the discipline. The basic principles of soil dynamics, while initially developed for the study of mineral soils, are universally applied to describe the mechanical behavior of other granular materials such as snow, peat, stones, gravel, rock, food and feed grains, ores, coal, fertilizers, pesticides, and soil amendments. Many applications of soil dynamics are documented in the major fields of application: agriculture, earth construction, forestry, military, mining, and sea-bed and space operations. New developments appearing in major fields of application are creating requirements that cannot be met with existing information.

In the past 50 years, increasing the use of soil-engaging machines and equipment was the most common way to decrease labor input, reduce drudgery, and increase production. However, this action was not without important consequences.

- Many machine and vehicle developments were made with little or no use of known basic principles of soil dynamics.
- The primary performance consideration was often based on limited criteria such as increasing drawbar pull or cubic yards of excavation per hour.
- Increased productivity could be obtained by simply increasing the size and speed of machines. Very large agricultural machines were developed and used without considering possible disruptive effects on plant roots, soil-water relationships, soil aeration, and soil structure.
- Secondary performance considerations of energy, economics, the environment, performance effectiveness and efficiency, larger soil forces, increased traffic, and soil compaction were not emphasized until recently.
- Little credence was given to alternate or innovative systems or operations.

As we approach the 21st century, an awareness of and a concern for the proper association of machines and soil is of universal interest. Soil dynamics is important to future development of soil-machine relationships.

Soil dynamics, when correctly applied, has a proven track record. The success of soil dynamics was clearly demonstrated by the commitment of universities, government laboratories, and private companies, worldwide, to soil dynamics research, education, and manufacturing activities. The progress in the development of the discipline of soil dynamics is well documented in technical literature. The only limitation of soil dynamics is that because it is a young discipline, many of its principles and applications remain to be identified and developed.

Changes in world conditions have created a new environment for science and technology. Massive government research and development projects, many involving soil dynamics, have been terminated. Private companies are down-sizing to gain efficiency, often at the expense of research and development activities. Cost-saving practices often result in research being shifted out of the country to lower-cost areas overseas. Universities also are reducing costs, often at the expense of science and research. Global markets are creating greater competition for the production of agricultural and manufactured products. Global corporations are redistributing research, development, and manufacturing functions around the world in response to the globalization of business.

New crop production methods and manufactured products must increasingly rely on the technical advantages that can be gained from the use of soil dynamics information. Past soil dynamics research and application activities generated information and data that defined basic principles of soil dynamics and provided the basis to improve the design, manufacture, and use of machinery. That information was analyzed, cataloged, and made readily available for users, but new information is needed to meet future requirements. The fact that there have been 124 recent reports of soil dynamics research using soil bins is proof that soil dynamics is playing an important role in current research.

Given the current atmosphere, future requirements for soil dynamics information cannot be met unless greater efforts are made not only to identify additional basic principles, but also to define more accurately the existing basic soil dynamics principles. Soil dynamics education and research projects at universities and research agencies continue to provide facilities, experience, and skills necessary to train the personnel and carry out the required work to meet future needs. In addition, the discipline must be introduced into the cooperative study of additional fields and applications.

One of the most closely associated of these fields is that having to do with the interaction between plant dynamics and soil dynamics. The intricate details of plant roots and seedlings growing through soil have fascinated researchers and those who work with plants for hundreds of years. In the study of these details, much progress has been made in increasing our understanding, which has led to improvements in machines and production practices. It is now commonly possible to observe the beautiful sight of vast fields of a planted crop with thousands of seedlings appearing as though they had emerged within one hour of each other. Yields have been increased through better management of the soil-root environment. This is being accomplished while controlling the problem of soil erosion. Mathematical models have been developed that allow those who work with soils and plants to predict the response of plants to varied conditions. Many excellent books and articles have been published, the information in which can improve our knowledge of why plants respond as they do to different soil and weather conditions.

In recognition of the increasing demands for qualitative and quantitative information on soil-machine interactions and physical soil-plant interactions, the American Society of Agricultural Engineers is publishing a new series of monographs entitled *Advances in Soil Dynamics*. The inclusion of *advances* in the title presumes that readers know the basics of soil dynamics. The reader who desires a review of soil dynamics principles, is referred to *Soil Dynamics in Tillage and Traction* (Gill and Vanden Berg, 1967), *Soil Mechanics Problems of Agricultural Machines* (Sitkei, 1967), *Machines for Moving the Earth* (Zelenin et al, 1975), and *Agricultural Soil Mechanics* (Koolen and Kuipers, 1983).

Advances in Soil Dynamics is primarily intended to concentrate the documentation of the state of the art of the discipline of soil dynamics. The series will provide the means for timely updating of the important aspects of soil dynamics progress. References will identify specific personnel and sources of information that will be of direct assistance to readers. *Advances in Soil Dynamics* will also provide benchmarks upon which to base future research and development of additional basic soil dynamics principles and applications.

Advances in Soil Dynamics will cover state-of-the-art information on: soil bins; soil physical properties; plant-soil dynamics; modeling techniques; tillage mechanics; traction mechanics; soil compaction; and sensors, transducers, and control.

Volume I includes the material on soil bins, soil physical properties and plant-soil dynamics. The remaining topics will be covered in subsequent volumes.

The editors are grateful to the ASAE Soil Dynamics Research Committee, PM-45 and the ASAE Monographs Committee, P-512, for their support for these *advances*. Moreover, the editors greatly appreciate the following organizations for sponsoring this project by providing financial support:

1. CATERPILLAR INC., PEORIA, IL
2. DEERE AND COMPANY, MOLINE, IL
3. FIRESTONE AGRICULTURAL TIRE COMPANY, DES MOINES, IA
4. INGERSOLL PRODUCTS CORPORATION, CHICAGO, IL
5. J I CASE, RACINE, WI
6. VERMEER MANUFACTURING CO.

The financial support received from Agequipment Group, L.P., Lockney Tex. and DMI Inc., Morton, Ill. is also sincerely appreciated

Dedication

These *Advances in Soil Dynamics* are dedicated to Dr. William R. Gill, Director and Soil Scientist (retired), U.S. Department of Agriculture, Agricultural Research Service, National Soil Dynamics Laboratory, Auburn, Alabama and Dr. Glen E. VandenBerg, Agricultural Engineer, U.S. Department of Agriculture, Agricultural Research Service, Beltsville, Maryland (formerly a researcher at the National Soil Dynamics Laboratory) for their lifelong unselfish and dedicated service to the discipline of Soil Dynamics. USDA Handbook 316, *Soil Dynamics in Tillage and Traction*, which was written by these two pioneers, was instrumental in bringing a clear focus to this discipline.

ADVANCES IN SOIL DYNAMICS

Volume 1

Coordinator
William J. Chancellor

Chapter 1: Soil Dynamics and Soil Bins

William R. Gill, U.S. Department of Agriculture, Agricultural Research Service, National Soil Dynamics Laboratory, Auburn, Alabama

Robert L. Schafer, U.S. Department of Agriculture, Agricultural Research Service National Soil Dynamics Laboratory, Auburn, Alabama

Robert D. Wismer, Technical Center, Deere and Co., Moline, Illinois

Chapter 2: Soil Physical Properties

William J. Chancellor, Biological and Agricultural Engineering Department University of California, Davis, California

Chapter 3: Advances in Soil-Plant Dynamics

Henry D. Bowen, Biological and Agricultural Engineering Department, North Carolina State University, Raleigh, North Carolina

Thomas H. Garner, Agricultural and Biological Engineering Department, Clemson University, Clemson, South Carolina

David H. Vaughn, Biological Systems Engineering Department, Virginia Polytechnic Institute and State University, Blacksburg, Virginia

Acknowledgment

All the authors contributing to this volume wish to thank the following persons who reviewed various parts of the manuscript:

Awatif E. Hassan
Department of Forestry
North Carolina State University
Raleigh, North Carolina

Sverker Persson
Agricultural and Biological Engineering Department
Pennsylvania State University
State College, Pennsylvania

Elizabeth D. Schafer
Loachapoka, Alabama

Michael Singer
Department of Land, Air and Water Resources
University of California, Davis
Davis, California

Gajendra Singh
Division of Agricultural and Food Engineering
Asian Institute of Technology
Bangkok, Thailand

Ernest W. Tollner
Biological and Agricultural Engineering Department
University of Georgia
Athens, Georgia

and all members of the editorial board.

The authors were fortunate to be able to draw upon the resources of the library at the USDA-ARS National Soil Dynamics Laboratory, Auburn, Alabama, and particularly the library holdings of Russian article translations done by William R. Gill.

Table of Contents

CHAPTER 1 – SOIL DYNAMICS AND SOIL BINS

CHAPTER 2 – SOIL PHYSICAL PROPERTIES

REFERENCES

Chapter 1

Soil Dynamics and Soil Bins

William R. Gill   Robert L. Schafer   Robert D. Wismer

The key to successful soil dynamics research is to identify the basic laws and principles of behavior of different soils and soil conditions, accurately describe the mechanical reactions of different soils to mechanically applied forces, and apply this knowledge to the design or effective implementation of soil-machine systems.

In the 1920s Dr Mark L. Nichols, a soil dynamics pioneer, developed and used soil bins to study basic soil-machine systems, and to identify and quantify soil behaviors that were of fundamental importance to the solution of soil-machine problems (Gill and Clark, 1985; Gill 1985; SAE, 1990; Schafer, 1993). It was Nichols' experience with small soil bins that led to the proposal for the construction in 1933 of the large soil bin facility that is now the National Soil Dynamics Laboratory, Agricultural Research Service, United States Department of Agriculture (NSDL, ARS, USDA) located at Auburn, Alabama. Other soil bins were built and used around the world.

Because soil bins offer a practical means for controlling soil and machine parameters, knowledge of their design, equipment, and methods of use are important for the future development of soil dynamics. This chapter will provide a broad coverage of soil bins, their advantages, nature, equipment, methods of use, and management.

Soil Bins

History

The history of soil bins is not well known, but their development was most probably a co-development with soil dynamics. Goryachkin (1968/1898) traced the theory of the plow back as far as 1752. By 1889 Goryachkin applied theories of mechanics to soil failure during plowing. His only reference to soil bins appears in association with a redrawing of Nichols' glass-sided box (Goryachkin, 1968/1898 p 24). The same picture appears in a paper that Baver presented at a meeting in Moscow (Nichols and Baver, 1930 p 176; Nichols, 1929 p 4).

The use of soil bins can be traced from Germany to the large soil bins at the NSDL, and perhaps to the soil bin in the Agricultural Engineering Building at the University of Nebraska. In a 1914 German publication, Georg Küehne described a model plow testing apparatus. Soil was compressed into a block 100 cm long, 40 cm wide, and 7 cm thick on the platform of the model soil bin. Model plows were mounted on a plow carriage that was moved longitudinally with a windlass along an elevated guide rail; the plow mount was moved vertically with an adjustment knob. The elevated guide rail, mounted on a gantry, was moved transversely with an adjustment screw. Draft was measured with a recording dynamometer (fig 1.1).

In 1927 Küehne built an indoor soil bin in the Agricultural Machinery Department, Technical University of Munich (Soehne, 1985 pp 106-107) (fig 1.2). Nichols met Küehne in the United States in 1926 and, although there is no record, it is rational to assume that they talked about soil dynamics (Gill, 1990, pp 18-19).

A survey by Wismer (1984) found 90 soil bins to be in use; four were located in Poland. Karczewski (1986) showed the importance of personally dealing with respondents in surveys. As an insider he resurveyed Poland and found there were 11 soil bins. In reality, there may have been 150 soil bins in use around the world. Since then several new soil bins have been built (Onwualu and Watts, 1989; Martin and Buck, 1987; Fielke and Pendry, 1986). The number of soil bins is impressive.

A literature search for references appearing in *Agricultural Engineering Abstracts* (AEA) was conducted to identify the number of citations concerning soil bins during a 3½ year period (1989-92). Soil bins have different names depending upon the country and the language. Names frequently used are: soil bin, soil tank, soil canal, soil channel, soil box, and model box. This search yielded a total of 124 different citations: soil bin, 116: soil tank, 3; soil canal, 1; soil box, 2; soil channel, 1; model box, 1.

Soil Bin Experiences

The published record clearly shows soil bins provide a widespread and important means of studying soil-machine relations. We have many years of experience working with soil bins: The Army Mobility Research Center, US Army Engineers Waterways Experiment Station (WES) (McRae et al, 1965); Deere and Company, Technical Center (Deere) (Wismer and Forth, 1969); Iowa State University (Schafer, 1968); and The National Soil Dynamics Laboratory (Gill, 1990). In this chapter an attempt will be made to glean wisdom and hindsights from those soil bin experiences and to provide documented insights that may be useful for the construction, renovation, or use of soil

Figure 1.1—Soil bin developed by Küehne in Berlin (from Küehne, 1914).

Figure 1.2—Indoor soil bin built in the Agricultural Machinery Department, Technical University of Munich (Soehne, 1985, p. 107).

bins. Obviously, we are only qualified to relate our personal experiences, and we are not implying that our experiences and facilities are better than others around the world. Our goal is to use these personal experiences to communicate general concepts and principles of soil bins and soil dynamics that we think are important.

Advantages of Soil Bins

Advantages of soil bins include the following.

1. Soil bins may be tailored for the available space, resources, and personnel, varying in size from small boxes standing on laboratory benches to large soil bins housed in separate research facilities.
2. A collection of unique soils at a single location provides ready access to soils with different physical-mechanical properties.
3. Many different soil parameters can be studied when bin soils are reused in different soil bin studies. As basic knowledge about each soil's dynamic properties increases, better interpretations can be made of research results.
4. Soils, cleaned of stones and foreign materials, provide a more precise research medium for basic studies in comparison to field test plots.
5. Covered soil bins provide the opportunity to work year-round under all weather conditions.

6. Covered soil bins prevent the long-term contamination of the test soils by atmospheric pollutants and dust storms.
7. Soil bin test carriages, operating on fixed guide rails, provide reliable means of maintaining soil-tool or soil-wheel geometry.
8. Physical models used in association with soil bins may increase the number of variables that may be studied.
9. Modeling techniques permit using reliable data secured from smaller, more economical machinery models to predict performance for larger, more expensive prototypes.
10. Laboratory soil bins provide opportunities for the use of more sophisticated controls and data acquisition systems than would be the case for field test operations.
11. Other professional staff members of the soil bin's parent organization may be ideally suited to broaden soil bin research through cooperative undertakings with the soil bin support staff.
12. Much soil bin expertise is freely available from the world's soil dynamics community.

There are also other factors that should be evaluated when considering the construction or remodeling of soil bins.

1. Construction funds must be followed by adequate recurring funds for personnel, equipment, and operating costs. Without adequate follow-up support, soil bins may soon fall into disuse.
2. Soil bins, being special purpose structures, cannot be used for many other kinds of studies.
3. Soil bins are practically limited to tests using remolded soils.
4. Construction and operation of a small soil bin would provide soil bin experience which would be invaluable in designing and operating a large soil bin facility.
5. If year-round, all-weather testing is required, increased mechanization and instrumentation of field test equipment is a viable alternative.
6. An alternative to developing a soil bin facility is to undertake cooperative research in soil bins owned by other organizations. It is also possible to contract for special research projects to be carried out at a number of private research institutes experienced in soil dynamics research.
7. Extensive contacts and information exchanges among soil bin user groups are important in maximizing the utility of the limited number of facilities available.

Design of Soil Bins

Without specifically defining what constitutes small or large soil bins, we will characterize the conditions that may indicate their differences. Small soil bins tend to be self-contained, and they may share a building with unrelated activities. They will be able to support studies on small test devices. Small soil bins generally have their functional components organized into a single unit, powered and controlled by the soil bin controller. Because all activities are carried out on a single soil bin, the soil bin may be automated. Large soil bins may be widely scattered so they cannot be easily automated as a unit. Large indoor soil bins usually occupy the entire building complex that houses their activities. They will be able to support studies on full-size, soil-engaging machines and test devices. Large soil bins tend to be used for a number of separate and distinct activities that require separate support vehicles, with some requiring human operators.

Soil bins are fixed (stationary) or movable. Generally there is a dynamometer fixture on which the tool or machine is mounted along with appropriate transducers. In the first instance, the soil bin remains in a fixed position, while the dynamometer fixture moves on guide rails. In the second instance, the dynamometer fixture remains in a fixed position, while the soil bin moves on guide rails.

The majority of soil bin facilities are small and owned by universities or industries. The most sophisticated small soil bins in the United States are owned by industries, like Deere and Company and the Caterpillar Corporation (CAT), which use soil bins to develop or improve products. These small bins are designed to study models or small test devices.

Large soil bins are generally owned by government agencies such as the USDA (Agricultural Research Service, NSDL) and the US Army (Waterways Experiment Station and the Land Locomotion Laboratory). The large soil bins 50 to 70 m long and 5 m wide at the NSDL were designed for the purpose of undertaking long-range studies of the behavior and performance of full-size soil-engaging machines and vehicles. Large soil bin facilities frequently also have small soil bins.

Soil bin expertise generally applies equally to large and small facilities. Fortunately, much soil bin expertise is readily available. Most organizations having soil bins have accepted visiting scientists for extended periods of time. Both the hosts and visitors benefit from these visits.

The soil dynamics community, while dispersed, is sufficiently cooperative such that personal contact can be easily established. While developing a small soil bin at the University of Illinois, Martin and Buck (1987) acknowledged the assistance of seven different experts who made contributions to their design.

Soil bins rarely wear out. When idled, they are frequently disassembled into components that can be easily reassembled or placed in remote storage areas. It is not unusual that "retired" soil bins are reactivated for other new projects. There are two known cases where soil bins have changed ownership. An industrial company's outgrown small soil bin and a government agency's excess soil bin were donated to universities for use in research and teaching programs. The reuse, restoration, or transfer of soil bin ownership emphasizes the lasting value of soil bins, especially to educational institutes. When a small soil bin is needed, the possibility of acquiring one from an outside source should not be overlooked.

The following points should be included when considering or planning a new soil bin facility: (1) the goals of the research program – long range or short term; (2) the technical nature and degree of difficulty of carrying out the program – research, development, or testing; (3) the type and volume of program to be carried out; (4) the personnel and funds available to be dedicated to support the program in general, to maintain and adapt test devices to dynamometers, and to redesign or remodel the facility;

(5) potential cooperators, their possible contributions, and their possible demands; (6) the personnel, funds, and equipment available to be dedicated to the planning of the work, analysis, and evaluation of project data, preparation of project reports, and overall management of the program; (7) the space available for the soil bin and its support facilities; (8) the size and weight of the machine or machine components that are to be used in the program; and (9) the operating speeds, power, and force requirements needed to fulfill the research program requirements.

Small Soil Bins

Small soil bins, varying from 4 to 16 m long, are designed to study models or smaller components of full size devices (Wismer and Forth, 1969; Durant et al, 1981; Martin and Buck, 1987). Small soil bins often have glass walls or other special features to expedite the study of specific soil behaviors.

Overhead rail systems have been adopted for use in a high-speed soil bin constructed at the Silsoe Research Institute, Great Britain (Stafford, 1979), and at Salinas, California (Wilkins et al, 1979). Overhead rail systems permit removing the soil bins or soil bin sections for soil preparation or other purposes without encountering interference from the dynamometer carriage.

Wilkins et al (1979) designed a soil bin to evaluate planter units. The planter units were used to sow seeds in soil bins 4 ft long. Then the soil bins were moved into a greenhouse where seedling emergence and plant growth could be studied.

The NSDL constructed and operated a small circular rotating soil bin, having inside and outside diameters of 16 and 18 ft, respectively, (ARS, 1965). Its capability for continuous soil fitting and measurements made the bin useful in wear studies. The power required to rotate the heavily loaded soil bin was essentially constant, ie, unrelated to the added forces required during soil preparation or force measurement activities.

Nedorezov and Mosieenko (1986) built a pressurized soil bin for the study of submerged sea floor soil at Soviet construction institutes in Kiev and Odessa, Ukraine. Pressure applied to the chamber simulated hydrostatic pressures experienced at working depths up to 6000 m.

Not all small soil bins require the same amount of capital investment: inexpensive soil bins, built with available materials, often provide valuable service. Godwin et al (1980) at Silsoe College, Great Britain, built a small bin 13.0 m long, 0.5 m deep, with an adjustable width of 0.8 to 2.0 m, upon a solid concrete floor. Manual soil preparation aided by rollers provided a suitable means for preparing soil. Measuring equipment was of high caliber, and high-quality research was successfully conducted.

Transparent walls of thick tempered glass or plastic, when installed in soil bins, permit the observation of soil movements caused by the action of test devices. Transparent walls can be temporarily erected in the center of large soil bins or built into the walls of any size bin. The glass-sided soil bin was one of the earliest innovations in soil dynamics studies (Nichols, 1929). More recent research reporting the use of this technique in soil boxes and small soil bins included studies of root cutting (Feller et al, 1971); soil deformation during cutting (Gill, 1969); the impact of tools on stones (Studman and Field, 1975); and the injection of fluids into soil (Araya and Kawanishi, 1984).

The primary limitation to soil bins is the possibility of stress-strain distortion due to the close proximity of the sides and bottom, which distortion often requires the use of scale models and associated uncertainty of scaling laws.

Soil Boxes

Soil boxes are a special class of very small soil bins. Soil boxes have been used with great success in low-speed soil dynamics research (Nichols, 1925; Chancellor and Schmidt, 1962; Gill, 1968 and 1969). Soil boxes are normally fixed, but a few have been movable. Because of the small size, they have unique advantages. For example, research can be carried out in small spaces in laboratories because the small volumes of soil needed to fill the boxes simplify storage and handling requirements. Various shapes of boxes can be used, including those with transparent sides. Soil fitting may be accomplished by manual packing or compression with a simple loading device, and the small test devices require only small forces that can be measured with standard measuring equipment. Further, the small size of the operation means that it can be handled by a single person, yet a large number of variables may be studied in a reasonable time, using a reasonable effort. Finally the lack of noise, dirt, and dust associated with soil box studies makes them compatible with general laboratory activities.

Large Soil Bins

Large soil bins are expensive to build, operate, and maintain. They require a long-range investment and must be designed with sufficient flexibility to meet long-term needs. A lot of the flexibility in large soil bins depends on the design of the soil preparation and test equipment that is used on the bins. Large soil bins tend to have adequate space for operation of full-size equipment, replicated treatments, and attainment of realistic operational velocities. The soils are usually homogeneous without any natural soil horizons or inclusions, except those that are part of the system under study.

Some important features of large soil bins are:

1. They are large enough to study full-scale machines or their components.
2. They provide sufficient bin space for the repetition of tests or the evaluation of additional variables and sufficient depth to permit the study of deep soil compaction.
3. Sufficient length is available to permit the attainment of reasonable speeds of soil bin car operation.
4. Soil processing can be either mechanized or automated.
5. Large bins provide the possibility of studying a wide range of machine activities including: tillage, traction, excavation and earth moving, and soil compaction.

The NSDL outdoor soil bins are typical of large soil bins. They measure 75 m long × 6 m wide, with depths varying from 0.6 to 1.5 m, which is sufficient for the study of deep soil compaction by large machinery. Tile drain lines run longitudinally on the bottom of the bins and provide adequate subsoil drainage. A surface drainage system permits catching surface drainage separately. The footings and bin walls were designed to support the maximum foreseeable weight of any soil bin car. Common rails between bins permitted grouping more soil bins into a given area.

The WES small-scale test facility has a fixed soil bin that is assembled from movable soil bin sections (McRae et al, 1965). This segmented soil bin permits extensive soil-fitting activities away from the site where the measurements are made. The bin sections, filled with prepared soil, are rolled on tracks to the measurement area, where they are assembled into a long fixed bin relative to the movable dynamometer carriage. A 50-m-long overhead guide rail permits the attainment of high speed by the dynamometer carriage and preserves freedom of action during the assembly or disassembly of the fixed soil bin.

The South Australia Institute of Technology constructed an outdoor oval-shaped soil bin with a test track length of 257 m. The facility was designed primarily to evaluate designs of chisels and sweeps for use in Australian agriculture (Fielke and Pendry, 1986). The endless test track permits running several laps during measurement runs. The sides of the oval are comprised of two 50-m straight sections that eliminate the lateral accelerations that characteristically occur in circular soil bins.

Covered Soil Bins

The construction of large covered soil bins should be seriously considered when a new facility is being built. Adding covers to at least some of the existing bins of an outdoor soil bin facility will significantly improve the facility's capabilities.

Covered soil bins need certain special features. Internal combustion engines operating inside closed buildings require pollution control equipment. Vents, windows, doors, and fans are needed to purge the pollutant fumes. However, excessive air flow through the building causes differential drying across the soil surface. Roofs require insulation to prevent moisture condensation. Transparent roof panels and auxiliary lighting are required to provide adequate visibility during soil bin studies during all types of weather and at all times of the year.

Soil Bin Equipment

The structure and form of soil bin equipment depends on the size and degree of sophistication of the soil bin facility. Studies in small soil boxes and soil bins that require small volumes of soil and delicate or intricate measuring techniques can be conducted manually.

Studies involving larger volumes of soil must be studied in soil bins in which most of the necessary operations are mechanized. Small soil bins that integrate all equipment into a single unit can be automated to provide a high degree of uniformity and economies of labor and time.

Operations in large indoor or outdoor soil bins with very large volumes of soil must be mechanized. These facilities, although mechanized, are difficult to automate because of their wide dispersion of soil bins. The need to operate on different soil bins, conduct different operations, and use different soil bin cars presents situations in which human operators rather than automation provide the best means of control.

Soil preparation and measurements are the two most time-consuming operations during studies of soil-machine interactions in soil bins. These operations must be mechanized and automated to the greatest extent possible. The system needed to support soil bin operations includes: power sources and drives, soil preparation devices, measurement devices, data acquisition and processing equipment, controls, special equipment, and maintenance and support equipment.

Soil bin equipment must be designed to prevent fuel, oil, and grease from contaminating the soils. Mechanical equipment maintenance involving change of engine and transmission fluids must not be performed over uncovered bin soils.

Power Sources and Drive Systems

The range of speeds and power requirements of the test devices determine the quality and capacity of the power sources and drives. Test program requirements determine the size of the soil bin cars and their general performance needs. The power sources and drive systems may be mounted on individual cars, shared by several cars, or mounted in stationary positions remote from the cars.

High-speed cars powered by cable-pull drives have been designed and used at WES (McRae et al, 1965); in Israel (Feller et al, (1971); and in Great Britain (Stafford, 1979). Mobile dynamometer carriages are designed with low masses to reduce acceleration-deceleration forces. The choice of power source and drive systems determines the distance required to accelerate to a desired speed, pass over the test soil section, then safely decelerate to a safe stop. In one instance, when speed was limited by the length of the soil bin, the bin was lengthened to secure the desired speed (Wismer and Forth, 1969).

Power sources and drive systems don't always have to be complicated. Pendulums which reach high velocities have been used in soil cutting and pulverization studies (Ellen, 1984). Turnage (1970) used a pneumatic-hydraulic loading system for high velocity penetration studies.

Drive systems can be powered by pneumatics, hydraulics, electric motors, and internal combustion engines. Electric, hydraulic, and pneumatic supply lines require frequent coupling-uncoupling when the motors and drives are mounted on separate units. WES suspended, with reasonable safety, electric power cables above its high-speed dynamometer carriage from an overhead trolley. For motors mounted on small soil bins or slowly moving soil bin cars, the power supply lines may be suspended overhead with a system of pulleys and weights, pulled from a rotating reel, or dragged along the ground.

A special series of drives is required to position and orient soil preparation or test devices. The position of every device touching the soil must be accurately set and maintained. On large soil bins, the general positioning of an implement or test device relative to the soil may be accomplished in two steps. The car superstructure may contain a mechanism to make rapid large changes. A subframe and drives can provide for fine adjustments. Vertical and lateral adjustments of soil loosening or packing devices are used during each step of soil preparation operations to ensure proper soil manipulations.

Soil Preparation Equipment

Soil preparation equipment includes devices to add moisture to the soil and an array of soil-processing devices including chisels, rotary tillers, scraper blades, V-wheel packers, and rollers. Soil drying may be accomplished by removing soil from the bin and placing it in a dryer or by repeated stirring of soil exposed to a drying atmosphere. Wismer and Forth (1969) described soil preparation devices for small soil bins, and a general review of soil-processing equipment used in soil bins is presented in Wismer (1984).

Measurement and Control Equipment

Measuring cars are fitted with transducers and other devices that hold the test device in the desired operating position as it is moved on or through the soil. Simultaneously, they may provide control of the transverse position of the test device in the bin, the distance and velocity of travel along the bin, and other variables such as angular displacement and angular velocity. Standardized mounts permit rapid interfacing of test devices with the dynamometer mounting plates and ensure the preplanned orientations of the test devices.

It may be advantageous to use self-contained, separate measuring units. A self-contained automated penetrometer may be used to routinely map the soil condition at different points during soil bin operations without requiring support from the main data processing system.

Instrumentation Packages

Data acquisition and control computers and all of the associated recording and display equipment are required to process data acquired during the conduct of test programs. In addition to coordinating data acquisition, the package provides computer control of the test units.

Special-Purpose Equipment

Special-purpose equipment is needed to conduct operations not directly associated with the research operations. Stafford (1979), McRae et al (1965), and Wilkins et al (1978), developed hybrid soil bin systems in which forklifts, pug mills, or other conventional batch handling equipment served as part of the soil bin system. Hybrid systems reduce design requirements and provide overall efficiencies and economies.

NSDL uses a transport car to move cars to and from the soil bins and storage areas. A platform car is used to support heavy loads on the bin rails. The main use has been to move loaded soil trucks along the bin while filling the bin. Initially, NSDL used cover cars to protect test conditions when rain occurred during periods of testing.

Other equipment which may be necessary includes: portable electric generators; air compressors; portable walkways or platforms to reduce soil compaction when walking on prepared soils; power cables to connect the main power supply directly to the soil bin cars or its on-board components; data communication cables (where telemetering is not used) to link the soil bin cars into the data communication network; radios or other communication equipment to coordinate work activities among soil bin personnel.

Maintenance and Storage Equipment

In large soil bin facilities, storage and maintenance areas can be combined. Car storage should not infringe upon unoccupied soil bins; such storage results in the need to shift cars around in order to secure access to the various soil bins. Space must be provided where the soil-engaging components may be washed free of soil before storage or use in a different type of soil.

Storage areas should be large enough to provide room for a variety of dynamometers and soil preparation tools, in addition to test devices such as tillage tools and tires. The rapid availability of critical items designed specifically for use with soil bins provides a high degree of readiness (flexibility) in the facility.

Maintenance requires cranes and a wide selection of hand and power tools, spare parts, etc, needed for the maintenance of units and all of their subcomponents.

The Bin Soils

General Considerations

The proper selection and mixing of soils for use in soil bins is important to providing a uniform test medium. The volume of soil in large soil bins precludes changing soils on a routine basis to secure different soil physical properties. A soil bin facility used for general applications should have a minimum of three soil types: sand, loam, and clay. Soils differ markedly, and the proper selection of soil is so important that securing knowledgeable assistance in making the selection is mandatory. The properties of prospective soils must be measured to confirm that they possess the desired properties.

It is possible to create artificial soils by mixing natural soils or by using mixtures of bentonite or other clay minerals and glycol, oil, or other fluid. Unique properties can be achieved in such mixtures (Moechnig and Hoag, 1979; Batchelder et al, 1970; Reaves, 1966).

Soil Selection

The process described here was developed for the collection of large volumes of soil at the NSDL. The principles apply equally to selection and collection of smaller volumes of soil.

The soil selection process should focus on soils with specific soil parameters known to govern the soil-machine relations of particular interest. These parameters include: stress-strain relations, shear resistance, soil-metal sliding friction, internal and external friction, abrasion, soil adhesion, soil cohesion, compression resistance, and plasticity. Certain static soil properties are important for describing soil conditions that are created by soil manipulation: bulk density, soil moisture characteristics, clod size distribution, porosity, mechanical composition, mineral composition.

Not all types of soil may be found where the soil bin facility is located. The proper choice of soil should not be compromised because of the distance to the collection site. The NSDL received shipments of soil by rail from sites spread around the southeastern United States and a number of substantial supplies of soil by sea and rail from Hawaii. The NSDL has provided a large supply of soil to WES for special studies; a substantial supply of soil to Deere for use in a small soil bin; and smaller lots to various US locations.

The selection process includes the following:

1. Identify the exact location of the test soil in the field soil profile (the desired soil may not be on the surface).
2. Expose the layer of test soil.
3. Scrape the test soil into windrows with a grader blade.
4. Pile the windrows into a series of separate piles and blend (re-stir) with a front-end loader.
5. Combine adjacent piles into fewer piles and blend.
6. Load into transporting trucks from piles in different parts of the field.
7. Spread soil from each truck along the length of the soil bin.

When filling deep soil bins it is important to ensure that the soil is not overly compacted at depths that cannot be loosened from the surface of the bin. The NSDL bins were filled from dump trucks supported on a platform car operating on the bin rails. The procedure permitted filling the bins without excessively compacting the soils in the bins. Lighter equipment was used inside the bin to spread, mix, and firm the soil sufficiently to prevent subsequent settling.

Soil-Fitting Equipment

Soil fitting is defined as the process used to prepare the bin soils to provide the desired test conditions. Once the bin is filled, the soil should be mixed to the maximum possible depth for homogeneity. Thereafter, soil-fitting operations are designed to provide desired moisture, density, and strength profiles throughout the test section.

A useful knowledge of soil conditions and soil behaviors relative to the soil-fitting process can only be achieved by trial and error. This knowledge and practical experience in handling different soils can be used to fit soils, within limits, into desired conditions. Experience gained with one soil might not be applicable to another soil, even though it appears to be quite similar in nature. Small soil bins have an advantage in that the moisture of high-clay-content soils may be removed from the bin and adjusted before executing the next soil-fitting operations. Deere, WES, and NSDL experiences bear this out.

Fitting large volumes of soil has been mechanized in different manners. Preparation of large soil volumes of clay soil required by the WES Small Scale Test Facility were handled with heavy duty bulk handling equipment (McRae et al, 1965). The Deere soil-processing equipment includes a rotary tiller, two types of rollers, a vibrating blade, a leveling blade, and a water spray to adjust soil moisture. Batchelder et al (1970) developed a soil bin that operated on a linear path with a continuous cycle of soil preparation and tillage tool to force measurements. Soil moving on the feed belt was compacted by a roller and leveled with a blade prior to reaching the position of the test tool. The NSDL circular soil bin similarly continuously fitted soil with a rotary tiller, leveling blades, and a packing roller.

The nature of the soil-fitting process might lend itself to the development and use of knowledge-based systems to provide inexperienced soil fitters with the benefit of historical experiences (Smith et al, 1993).

The Soil Fitting Process

There is no standard soil-fitting process. However, the following can be considered as a general procedure that could be followed, depending on the circumstances.

The soil-fitting sequence usually begins with the leveling of the soil surface with a blade to refill irregularities, pits, and furrows, and to make sure there is an even distribution of soil side-to-side and end-to-end of the bin. Wet soils may need to be dried before rotary tilling to avoid puddling (the destruction of soil structure). Dry soils, after loosening, mixing and crushing, may be moistened with natural rainfall or with a sprinkler system as necessary. Following moisture adjustment, the soils should be rotary mixed to create a uniform moisture distribution to the desired depth. Typical sequences of the final soil fitting operations include releveling and packing with various types of rollers to attain the desired density and releveling with a scraper blade to create a final grade level.

Other soil-fitting procedures have also been used. WES researchers (McRae et al, 1965) used a large pug mill to prepare cohesive soil for use in the segmented soil bin of its Small Scale Test Facility.

Tajima et al (1992) devised a novel horizontal tubular soil bin that employed tumbling the soil as the soil-fitting operation. Since a loose soil condition was desired, the tumbling action sufficed. A curved section, comprising about one-third of the tube circumference, was cut longitudinally from the tube to create the open top of the soil bin. When the top was replaced for the tumbling action, the soil was completely contained in the tube. The bin was rotated around its longitudinal axis to prepare the soil. Removing the top of the soil bin exposed the interior of the soil bin for test measurements.

Soil preparation operations that run the length of the bins may leave invisible bands of soil with low penetration resistance. The consequence of bands of unintended soil weakness depends on the nature of the test device and the relative position of its path to the low-strength band. A narrow coulter, operating in a band of low-strength soil, may require an unusually low draft, while the draft of a wide device, such as a moldboard plow, may exhibit a negligible effect. Cross-tilling devices, equipped with stirring chisels that move laterally as well as longitudinally along the soil bin improve the longitudinal soil bin uniformity.

Compaction increases soil strength to levels found in natural field conditions. Two types of packing rollers are useful in soil-packing operations. One is a wide, smooth roller which packs the surface layer of soil. Ideally, the roller should be as wide as the soil bin. The other is a subsurface packing roller made up of narrow packing wheels spaced about 500 mm apart. Being narrow, the packing wheels sink into the soil and pack the subsoil beneath. The combined use of both rollers produces a thicker compacted layer.

Soil-fitting operations can be devised to create the test conditions needed. Fo
example, deep compacted layers simulating traffic pans can be created with soil-fitting
equipment. The NSDL creates this condition by opening a deep clean furrow with a
special moldboard plow and compacting the soil at the bottom of the furrow by running a
heavily loaded wheel directly behind the plow. Each facility must develop its own
soil-fitting procedures to provide test sections with desired soil conditions.

Soil Uniformity

Soil-fitting operations usually create strength differences in the vertical profile. A
important quality control in soil bin studies is soil uniformly prepared across the wid
and along the length of a soil bin. Soil uniformity should be quantified with data on she
strength, cone index, texture, structure, soil moisture, and soil density.

Testing programs themselves can be responsible for creating nonuniform
conditions. What may appear to be spotty soil preparation may be due to a previou.
program. For example, a randomized block experiment to determine the effect o.
dynamic tire loads of 4, 12, and 20 kN on drawbar pull can create different levels of soil
compaction in the tire ruts. These differential conditions may be so difficult to remove
that it is almost justifiable to assume that that soil "has a long memory." NSDL
personnel have experienced this phenomenon on many occasions. When soil fitting does
not produce uniformity, fitting operations should be repeated or modified as needed to
improve uniformity.

When soil measurements indicate that a soil pretest condition is considere
acceptable, the soil measurements define the initial soil conditions for only that one soi
fitting. Following each test, a new set of measurements will define the new soil condition
that has been produced; it is impossible to reproduce exact soil conditions on a regular
basis. After fitting operations are completed, judicious use must be made of some type of
movable bin covers to moderate the influence of possible moisture gains or losses,
particularly in the case of outdoor bins.

Manual post-soil-fitting operations are often used when burying sensors in the soil.
Some sensors may be pressed into the soil, from a remote location, with a "push rod"
that is extracted, leaving the sensor at a specified location. When the soil is in a
compatible condition, the burying of sensors must be done from suspended platforms or
large sheets of load-spreading decking placed directly on the soil. Small pits, excavated
to bury measuring devices, must be refilled in a manner that duplicates the initial soil
uniformity.

In the NSDL soil bins, as well as in most other deep soil bin facilities, routine
soil-fitting operations are carried out from the bin surface downward into the soil. Major
soil renovation by the removal of bin soil in order to invert or deeply mix the soil in the
bin is not considered to be soil fitting, and it requires special equipment and methods.

Soil Protection

Soil protection should be a constant concern when dealing with bin soils to ensure
that the purity and friability of the soil will be maintained over long periods. Machine
components of soil preparation and measurement cars must be cleaned of soil before
being moved from one soil bin to another. The soil must be kept free of oil or other

contaminants that may change soil properties. Solid materials left in the soil may become missiles when struck by high-speed equipment. When there are industrial contaminants, dust storms, or other atmospheric pollutants, outdoor soils may need to be covered with plastic sheets.

Excessive soil working often destroys soil structure, which can be manifested as "balling." Procedures should be developed to prevent soil working at high soil moisture contents where structural damage usually occurs. Soil-reconditioning operations should be devised and executed to counter the structural degradation that accompanies severe soil handling operations. In outdoor bins, these operations may include the growing of crop plants and/or the incorporation of plant residues. The effect of weed control chemicals on soil is unknown, but extreme care should be exercised in their use until better information is available.

Measuring Equipment

Commercially available measuring and recording equipment should be used when possible. Some parameters required to characterize soil-machine relations have yet to be identified. New measuring devices should be developed as new measurable parameters are identified, so that their importance in soil-machine relations can be determined by physical measurements. Direct access to instrument makers, who share in the development of new measuring devices, provides an effective way of securing the best designs. An overall goal of soil dynamics research is to understand the soil-machine interaction and thus enable the design of systems that will permit manipulation of soil from an initial known condition into a new and specified condition; digging, cutting, loading, and transport of soil in effective and efficient ways; attainment of adequate tractive forces in effective and efficient manners; mobility across terrain with a variety of conditions; and prediction of soil behavior under the action of dynamic loads applied by machines and vehicles.

Soil Manipulation

Soil manipulation is the skilled application of mechanical forces to create a new and desired soil condition. The primary function of tillage tools is to manipulate soils. The effect of materials, shapes, motions, and velocities on tool performance can be easily studied in small soil bins. Soil-manipulating actions when visualized in simple terms, such as cutting, displacing, throwing, inverting, segregating, crumbling, or mixing, define the characteristics that conceptually need to be measured to assess performance. Similarly, the manipulating actions of earth movers that excavate, move soil, or create new surface configurations; rock pickers, stump up-rooters, and root crop harvesters that sort and segregate; and residue handling tools that execute cutting and mixing actions, can define conceptual characteristics that need to be measured to assess performance.

Soil Traction

Traction and transport devices are designed to develop traction forces or to provide vehicle flotation while optimizing rutting or soil disturbance and energy requirements. When the primary function of traction devices is to secure traction, the coincidental action of soil deformation becomes part of the soil-traction device relation and must be considered. Soil bin equipment and properly devised procedures provide the means for systematically studying soil-wheel relations (Reidy and Reed, 1966).

Soil Measurements

Conventional soil sampling devices used to collect soil for gravimetric analysis for soil moisture and bulk density are destructive (Revut and Rode, 1969); measurements of soil moisture and density with radiation are nondestructive (Black, 1965). A search is on for nondestructive measurements of stress, displacement, shear, penetration resistance, internal and external friction, soil-metal friction, and other dynamic soil properties. Since few nondestructive methods of measurement are available, the development of nondestructive methods can be viewed as an integral component of soil bin research (Hadas and Shmulevich, 1990; Raper et al, 1990; Morgan et al, 1991; Wang et al, 1991; Tollner, 1993).

Measuring devices intended to measure the soil's reaction to mechanical forces must be placed in the soil with great care. Electrical leads connecting sensors to data acquisition systems must be thin and flexible so as not to disturb the soil excessively. Leads must be placed so that test devices will not disturb them; transducer movements should be due only to soil deformation (Bailey and Burt, 1988). When inert markers are used, they may be traced for relocation and recovery by attaching lengths of thread, the ends of which remain exposed on the soil surface. Inert markers may also be colored or numbered; determination of pre- and postlocations indicate their total movement (Gill, 1969).

In the absence of nondestructive measuring methods, standard devices such as the soil penetrometer are being used increasingly in soil bin research (ASAE, 1992). The NSDL developed a special self-propelled penetrometer car to operate in the soil bins. It has a self-contained microcomputer-based data acquisition and analysis system to provide an instantaneous display of a map of the soil bin and the measured results (Sun et al, 1986).

Machine Measurements

A variety of measuring devices can be mounted on soil bin measuring cars. The type and number depend on the nature of the measurements that are required for the specific problems under study. Dynamometers (transducers) are designed to simultaneously isolate and acquire data of specific forces, deformations, strains, or machine parameters. Quantities frequently measured include: three-dimensional forces on test devices and the moments about the principal axes; torque on wheels or rotating components; displacements or positions of test devices and/or the soil; stress distributions on specific soil or machine surfaces; test condition indices such as tire inflation pressure; angular displacement and velocity of test devices; forward speed and distance of travel of the soil bin cars; performance parameters such as drawbar pull or tool draft; and other parameters or quantities that are important to the evaluation of the soil-machine relation.

All measuring devices must have the sensitivity, accuracy, capacity, and electronic characteristics necessary for the acquisition, analysis, and evaluation of data and to provide feedback to auxiliary controls, programmed to modify the parameters of test devices.

Soil-Machine Geometry

The forces exerted on a test device are an important component of the soil-machine mechanics. Changes in the soil-test device geometry significantly affects machine operation. The most obvious example is the mutual deflection of tires and soil when flexible tires operate on deformable soils. Strain transducers buried in the soil and inside the tire can measure the changing geometry of the dynamic soil-tire interface (Burt et al, 1987; Yu and Xu, 1990). The soil-test device geometry of rotating tillage tools is defined by the kinematics of the system. Programmed changes of the depth or width of cut and the direction of rotating tillage tools simultaneously change soil-machine geometry.

Changes of tillage tool orientation may cause new surfaces to engage the soil: a decrease in disk approach angle of a disk harrow blade may cause the back side of the disk to engage the soil; changes in rate and direction of movement of curved rotary tiller tines, under some circumstances, may cause the back side of the tines to begin to engage the soil. Changes of soil-machine geometry may alter the soil forces or behaviors. Thus, the simultaneous measurement of the change of soil-machine geometry and the soil forces on a test device is required to properly analyze soil-machine interactions.

New measurement techniques, aided by video and machine vision systems, are increasingly being used to provide electronic feedback from soil engaging elements. These data may be used to optimize machine performance by assisting in the assessment of the physical details of soil-machine interactions.

Instrumented Soil Bin Cars

The NSDL instrument car includes a mobile data acquisition and processing system, that is included in trains of bin cars assembled for intensive soil bin research measurements. During tests instrument cars permit the on-site assembly of test equipment; setup of instrumentation; computer programming of test sequences and variables; operator control and supervision of the data acquisition, recording, and play-back system; on-line calculations and analysis, including comparisons with previously collected data; operation in conjunction with a central computer or as independent "work station"; and transmission of data from sensors to the data acquisition system by wired connections or telemetry.

Special-purpose independent instrument packages can be tailored for special purpose activities. The NSDL penetrometer car was completely instrumented to operate independently of the central data acquisition system (Sun et al, 1986).

Field Measurement Systems

Soil bin expertise is being used in making field measurements. Mobile telemetry-computer systems are used in the field to evaluate the performance of a wide range of vehicles and machines (Wismer et al, 1978; Dwyer et al, 1990; Armbruster and Kutzbach, 1989; Schafer et al, 1979; Upadhyaya et al, 1986). Soil conditions in the field are not necessarily uniform, yet field measurement systems provide a practical means of overcoming the limited range of soil conditions found in soil bins including the availability in the field tests of naturally bonded, in-situ conditions not generally possible in soil bins.

Test Procedures

Measurement programs may be carried out in a number of ways, depending upon the degree of sophistication of the equipment. A wide range of control possibilities increase the research capability of soil bin research. Thus, an increase in the volume and quality of research results can be secured from each soil bin fitting.

The traditional procedure was to determine the effect of a single constant parameter on performance, eg, fixed levels of velocity or fixed levels of depth, during a single test run. Computer based systems permit the simultaneous variation of several parameters. An example of this procedure would be the simultaneous control of the angular and forward velocities of a tire in order to measure drawbar pull as a function of travel reduction (Wells and Buckles, 1987; Burt et al, 1980). Another example would be the varying of dynamic load on a tire, simultaneously with the varying of inflation pressure to control tire deflection and varying of travel reduction for optimization of tire performance (Lyne et al, 1983).

Models

Considerable savings of time and materials are possible with the use of models (Freitag et al, 1970; ASAE, 1977). Physical models require less space for testing, and because soil bin space is always at a premium they present a practical replacement for expensive, full-scale testing in large soil bins (Jin et al, 1986). In other cases, such as occur with lunar mobility studies, there are no alternatives to models (Freitag, 1971).

Modeling and similitude studies must be considered when interpreting soil bin test data. Nartov and Shapiro (1971) have suggested that soil bins should be considered as models of field testing and that the scale models of soil bins must meet similar criteria. During the selection of the type of soil bin and the type of tests, it is necessary to make an analysis of the conditions of similitude (Balovnev, 1969).

Soil Bin Facility Management

Management goals must ensure the planning, operation, and control necessary for the conduct of timely and effective soil bin research. An experienced manager responsible for a facility should identify research projects that are feasible based on the capabilities of the facility, the equipment, and the available personnel. An experienced manager can promote maximum use of the facility by ensuring timely scheduling of work to be done with the facility; accepting only projects that fall within the capabilities of the facility and personnel; ensuring a knowledgeable input by facility personnel into each planned research project; ensuring that only qualified personnel operate the facility; and by maintaining the facility in a reasonable state of repair and readiness.

The success of the facility depends upon establishment of research objectives and procedures that will provide reliable research results. Therefore, direct participation of the facility manager is important in planning and selection of the projects to be undertaken; selection of soils and soil conditions to be used; selection of test devices or models to be used; selection of the soil-machine parameters to be studied and their range; selection of statistically reliable methods of measurement; supervision of the research;

verification of research results and analyses and interpretation of the research data; and presentation and release of the results.

Upgrading should be a natural process that incorporates new developments in controls, data acquisition and analysis, measurements, and computers into the soil bin program. Upgrading activities should include updating the professional proficiency of the facility staff as well as the physical equipment.

Chapter 2

Soil Physical Properties

William J. Chancellor

The purpose of this chapter is to report on advances since the mid-1960s in information on soil physical properties that affect both soil-machine interactions in tillage and traction and physical soil-root interactions. The list of properties selected for review was formulated with the concurrence of the editorial board for *Advances in Soil Dynamics*. For each property it was intended that a description or definition of the property be provided along with information on advances in both measurement techniques and the values of pertinent parameters. When available, information will be presented on quantitative or qualitative relationships between a given physical property and other physical properties. In a number of cases, background information generally available prior to the mid 1960s will also be presented if this was regarded as necessary for the understanding of the more recent material included.

Nearly all of the illustrations and much of the quantitative data presented have been taken from reference sources. The illustrations have been copied (with permission) directly. Consequently, the units used in the illustrations are those selected by the original authors and may not be in the accepted, SI, format. In a number of cases data presented in text or tabular form are also not in the SI format because their units could not be converted while retaining the quantitative significance of the data.

A portion of the information presented here was developed at the University of California, Davis. The author is indebted to Professor Emeritus Walter H. Soehne, Agricultural Machinery Department, Technical University, Munich, and Professor Emeritus James A. Vomocil, Department of Soils, Oregon State University, Corvallis, for their leadership, guidance, and advice relative to the work at Davis.

The outcome of a dynamic process in which there is an interaction between elements of tillage or traction systems and an agricultural soil depends in part on the physical characteristics of the soil itself.

Elemental Soil Material Composition

Any discussion of the dynamic properties of soil must be based on information about elemental soil properties.

Grain Size Distribution

The determination of the grain size distribution for agricultural soils is usually made with a two-stage process (Archer and Marks, 1985). The first stage is a sieve analysis of finely crushed, oven-dried material (ASTM D421-85). The second stage is either a hydrometer or a pipette analysis of a suspension of deflocculated soil materials that pass through a sieve with 75 μm openings (ASTM D422-63).

There are several systems of classifying soil grains according to their size into categories designated as clay, silt, sand, and gravel, however, most systems designate as clay, those particles having a diameter less than 2 μm and as gravel those having a diameter greater than 2 mm (Baver and Gardner, 1972; Kézdi, 1974). The grain-size distributions shown in figure 2.1 (Chancellor, 1977) are categorized according to the system used by the International Society of Soil Science. The textural designation of the soil composite is made using a triangular diagram such as the one in figure 2.2 (Meyer and Knight, 1961)

The advances since the early 1960s with regard to grain size distribution have not been so much in methods for determination, but rather in information linking grain size distribution to other aspects of soil dynamics. One parameter of the grain size distribution frequently linked to other dynamic properties of soil is the uniformity coefficient, U.

$$U = \frac{d_{60}}{d_{10}} \tag{2.1}$$

in which U = uniformity coefficient (Kézdi, 1974), d_{60} = grain size of which 60% of the sample is smaller, d_{10} = grain size of which 10% of the sample is smaller. Higher uniformity coefficients (a more broad range of particle sizes) are associated with higher soil densities or lower porosities achieved by a given level of compaction effort (figs 2.3 and 2.4) (Kézdi, 1979; Chancellor, 1977).

The intrinsic permeability of porous media, k, has been found to be linearly correlated with the product of the soil porosity and the square of the product of the uniformity

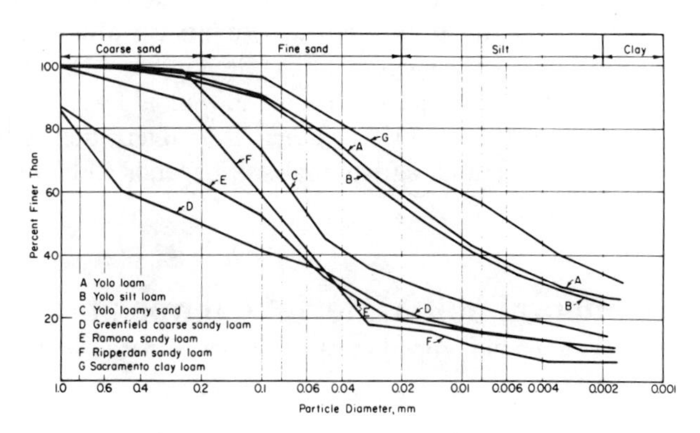

Figure 2.1–Grain-size distribution graphs for a range of soil textures. Sand, silt, and clay designations are those of the International Soil Science Society (from Chancellor, 1977).

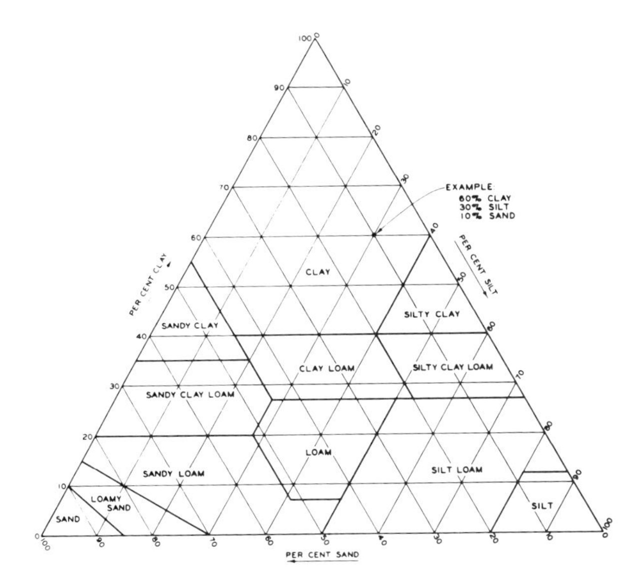

Figure 2.2–Textural triangle based on the USDA classification system for sand, silt, and clay (from Meyer and Knight, 1961, US Army Corps of Engineers).

coefficient, U, and the volume-surface mean diameter of the particles in the media (Pall and Mohsenin, 1980b).

Weight fractions of silt and clay have been linked by empirical equations (Hirschi and Moore, 1980) to estimates of soil moisture suction along a soil-water wetting front. This suction is a key element in the Green-Ampt equation, which relates this suction and other, more easily measured soil and water parameters to the infiltration rate of water into the soil surface (Mein and Larson, 1973).

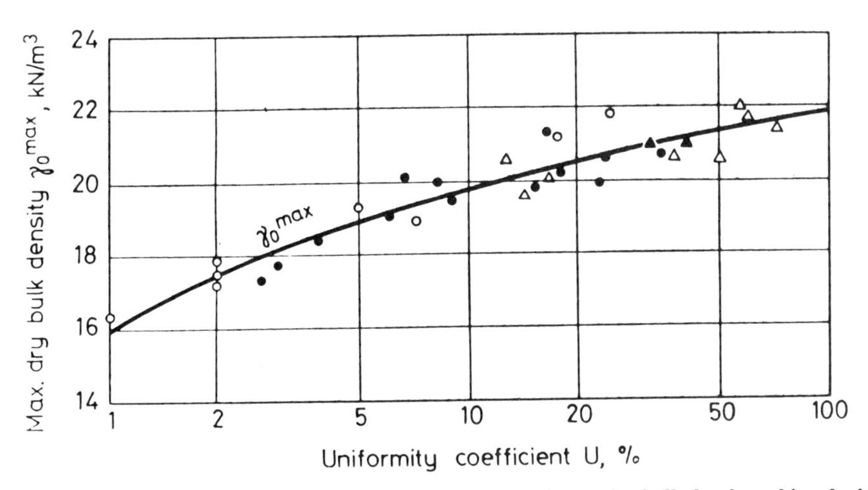

Figure 2.3–Effect of grain-size uniformity coefficient on the maximum dry bulk density achieved with a given compaction effort applied to three different soils groups (from Kézdi, 1979).

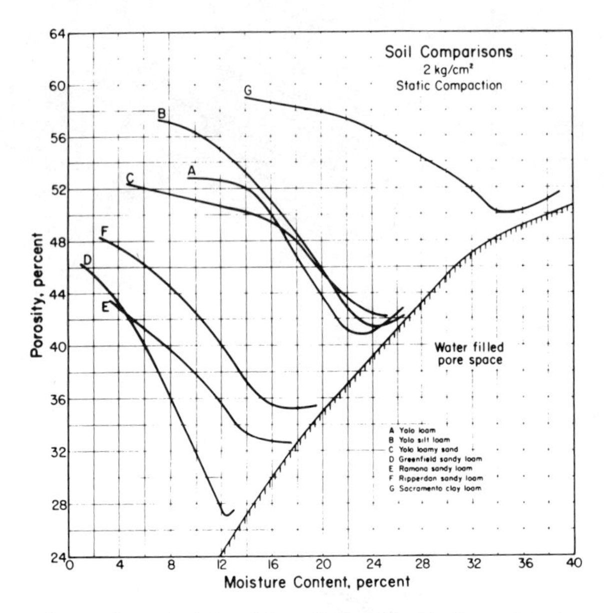

Figure 2.4–Porosity-moisture characteristics of the soils described in figure 2.1 when subjected to a given stress level in uniaxial compression (from Chancellor, 1977).

The distinction between soils classified texturally as either sands or clays has been the basis for the formulation of two completely different sets of relationships (one for sand and the other for clay) describing the performance of pneumatic tires on one or the other type of soil surface (Freitag, 1965).

Clay Mineralogy and Adsorbed Cation Species

There are two basic microcrystal forms of alumnosilicate clay minerals—a tetrahedron of four oxygen atoms surrounding a central cation, usually Si^{4+}, and an octahedron of six oxygen atoms or hydroxyls surrounding a larger cation of lesser valency, usually Al^{3+} or Mg^{2+} (Hillel, 1980) (see fig 2.5). Like types of microcrystals formed together into flat sheets are then arranged in one of two main sequences of tetrahedral and octahedral sheets. These are designated as 1:1 or 2:1 types of clays. The 2:1-type minerals are further classified into expanding and nonexpanding types.

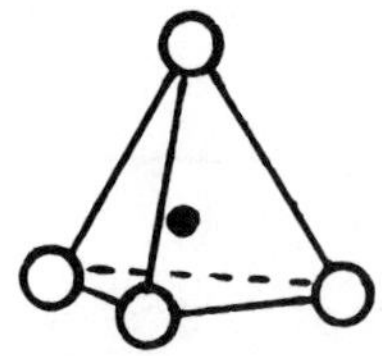

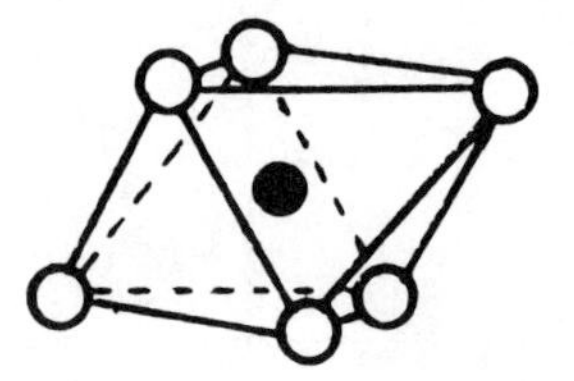

Figure 2.5–Basic structural units of aluminosilicate clay minerals: a tetrahedron of oxygen atoms surrounding a silicon ion (left), and an octahedron of oxygens or hydroxyls enclosing an aluminum ion (right). Reprinted with permission of Academic Press, Inc. from Hillel, 1980. Copyright © 1980, Academic Press, Inc.

A typical 1:1 material (all types of which are nonexpanding) is kaolinite, while a typical expanding 2:1 material is montmorillonite. A common nonexpanding 2:1 material is illite. Depending on the cations in the crystal and on the arrangement of the microcrystals in the sheet, there are varying levels of negative charge associated with the surfaces of clay particles.

Typical properties of the three above-mentioned clay materials are presented in table 2.1 (Hillel, 1980). Kaolinite usually has 90% of its particle sizes approximately uniformly distributed over diameters between 2.0 and 0.2 μm, while montmorillonite may have have 80% of its weight in particles less than 0.2 μm in diameter (Kézdi, 1974).

Clay minerals are identified by three common methods, electron microscope examination (Kittrick, 1965); x-ray diffraction, in which the spacing of the sheets or layers can be detected (Whittig, 1965); and differential thermal analysis, in which each type of material exhibits a distinctive pattern of temperatures at which endothermic reactions take place upon heating (Barshad, 1965 and Kézdi, 1974).

Because of the electrical charges on the clay particles, positively charged cations are attracted to the surface of these particles by both electrostatic forces, the strength of which are inversely proportional to the square of the distance from the cation to the particle surface, and by van der Waals forces, which are inversely proportional to the seventh power of the distance. In disperse suspensions the cations that are very close to the particle surface are very tightly bonded, while those some distance away are much less so, even to the extent of cations in the solutions being repelled by the charges of those bonded to the clay surface. This "double layer" effect, which interacts with clay types, pH, cation type, moisture content, and clay particle arrangement is a major determinant of the physical properties of the clay material and thus of the soil in which the clay is one of the most active elements. The extent to which cations can be absorbed on a clay is measured by the property termed cation exchange capacity which is determined by methods described by Chapman (1965).

High valence cations such as Al^{3+}, CA^{2+}, and Mg^{2+} tend to be attracted to clay particle surfaces more strongly than do monovalent cations such as $NH_4{}^+$, K^+, H^+, and Na+. When a confined body of clay is allowed to absorb water, swelling pressures develop, which may be as high as several bars. The swelling pressures are related to the osmotic pressure difference between the double layer and the solution. The osmotic attraction for external water is generally twice as high with monovalent cations as with

Table 2.1. Typical properties of selected clay minerals (from Hillel, 1980)*

Properties	Clay Mineral		
	Kaolinite	Illite	Montmorillonite
Planar diameter (μm)	0.1-4	0.1-2.0	0.01-1.00
Basic layer thickness (Å)	7.2	10	10
Particle thickness (Å)	500	50-300	10-100
Specific Surface (m^2/g)	5-20	80-120	700-800
Cation exchange capacity (meq/100 g)	3-15	15-40	80-100
Area per charge ($Å^2$)	25	50	100

*With permission of the publisher and copyright holder, Academic Press Inc.

divalent cations because there are normally twice as many of the former. Hence, swelling and repulsion will be greatest with monovalent cations such as sodium, and with distilled water as the external solution. With calcium as the predominant cation in the exchange complex, swelling is greatly reduced, as is the case at low pH when Al^{3+} is present (Hillel, 1980). The species of cation absorbed on the clay is frequently determined using flame photometry (Rich, 1965).

When wet clay is dried, shrinkage occurs. This is particularly noticeable in expanding-type clays such as montmorillonite. Cracking often occurs and the materials become very hard, particularly if the soil was puddled prior to drying. Sodium clays tend to soften and swell upon being rewetted although the puddled condition remains. In some cases with calcium clays that become very dry, the clay particle surfaces may come so close together that a single layer of calcium ions becomes bonded to clay plates on either side (fig 2.6). This condition is called *plate condensation*, and the bonding may be so strong as to prevent reabsorption of water.

The edges of clay platelets have a positive electrical charge at pH values below 7. If the extent of the double layer is small, the positively charged edges may approach the negatively charged surfaces of the platelets sufficiently to form weakly bonded floccules. A floccule structure is shown in figure 2.7. Since the stability of the floccules decreases with increasing extent of the double layer, a calcium clay will flocculate at a lower salt concentration of the soil solution than a sodium clay (Koorevaar et al, 1983).

In some cases these floccules will coalesce into a gel-like structure and exhibit considerable strength. However, when this structure is mechanically agitated the clay particles will become dispersed, and the material will become a liquid again (Kézdi, 1974). This property, known as *thixotropy,* is sometimes measured using the remolding index. When a dispersed or puddled clay is dried, it forms a hard crust. If the wetting and drying cycle is repeated, hard, large clods are formed. When the clods are rewetted, they form a sticky mud without structure. This is typical of sodium clays but not of calcium clays (Koorevaar et al, 1983).

Both clay type and cation species have major effects on many measures of dynamic properties of clay soils. However, even in soils with small proportions of clay, the clay plays such an important role in the soil matrix that the properties of sandy loams may also be strongly influenced by clay type and cation species.

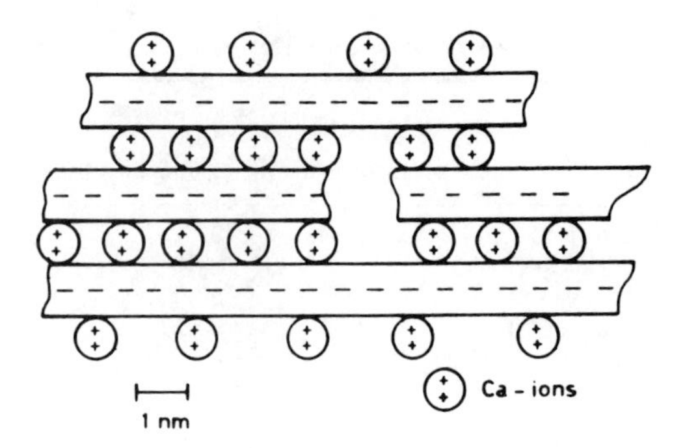

Figure 2.6–Plate condensation of a calcium clay (from Koorevaar et al, 1983).

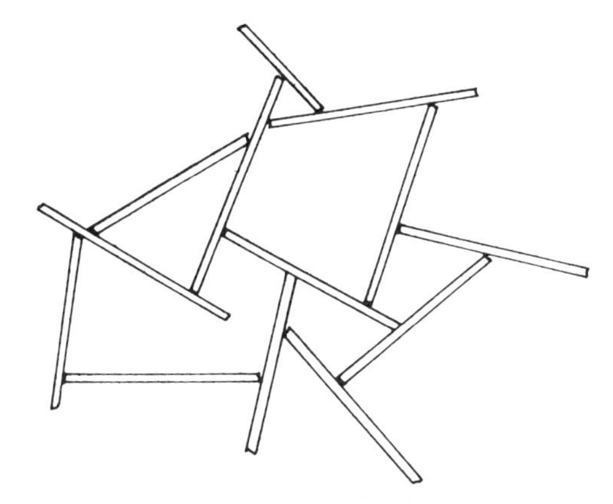

Figure 2.7–"Card-house" structure of a floccule (from Koorevaar et al, 1983).

The cation-exchange capacity has been found to be closely correlated with the modulus of rupture, plastic index, specific area, and the sticky point for a number of soils (Gill and Reaves, 1957). When normal cations in various silt-clay mixtures were replaced with sodium ions and those mixtures were compared with equivalent mixtures without the sodium ions, it was found that the modulus of rupture of the high sodium soil briquettes was somewhat higher than than of the normal briquettes—particularly when rapid drying of the briquettes took place (Gerard, 1965).

The inverse of this was found when high-sodium soils were treated with gypsum ($CaSO_4{\cdot}2H_2O$) or calcium chloride ($CaCl_2$) to replace some of the sodium ions in the soil with calcium ions, there occurred a 30% reduction in the energy required for the operation of tillage and cultivation implements (Schaefer et al, 1989).

Specific Gravity of Soil Solids

Soil solids may include gravel, sand, silt, clay, and organic-matter-based humus. There are two main methods used for measuring the composite value of the specific gravity of soil solids—the water pycnometer (Bowles, 1978; Blake, 1965a; Wray 1986; ASTM D854-91) and the air pycnometer (Pall and Mohsenin, 1980a). The air pycnometer method can also be used for measuring the porosity or dry bulk density of soil bodies (Pall and Mohsenin, 1980b). In both cases the principle of measurement is to place the oven-dried soil in a container of known volume and to fill the container to its calibrated volume with a measured volume of fluid under conditions in which the test fluid fills all pores.

For soils with small amounts of organic matter, a generally accepted specific gravity value of 2.65 is commonly used for estimating other parameter values (Baver and Gardner, 1972). Most clay materials found in soil have a specific gravity in the range of 2.2 to 2.6 while quartz (sand) tends to have values in the 2.5 to 2.8 range. Humus has a typical value of 1.37 (Ghildyal and Tripathi, 1987; Kézdi, 1974). Thus, the general value of 2.65 is not appropriate for soils with high percentages of organic matter.

The specific gravity of soil solids is an essential parameter in linking the measure of dry bulk density to values of soil porosity (see section on dry bulk density).

Mineral Grain Hardness and Angularity

The crushing strength of stone materials (Withey and Aston, 1950) is quite variable ranging from 34 to 310 MPa with typical values in the 138 to 172 MPa range as determined in conventional compression tests. Values for the modulus of elasticity in compression range from 12 to 90 GPa with typical values on the order of 41 GPa (Withey and Aston, 1950). Hardness of stone materials as measured with a Shore scleroscope (which measures hardness and resiliency simultaneously) ranges from 50 to 90 in Shore hardness numbers. Plate glass has a rating of 116 and mild steel a rating of 12 in these same units (Withey and Aston, 1950). The hardness of quartz grains was given as 10.3 GPa (Lambe and Whitman, 1969).

The angularity of soil grains is observed using microscopic techniques (Cady, 1965). Upon observation soil grain shape is frequently categorized as either bulky or flaky, or needlelike (Sowers and Sowers, 1961). Bulky grains are approximately spherical in shape, while flaky or needle-like grains may have ratios of length to thickness ranging up to 100:1. Bulky grains are frequently further subclassified as angular, subangular, subrounded, rounded, or well-rounded (see fig 2.8).

A quantitative procedure for characterizing sand grain shape (Kédzi, 1974) involves using sieves with circular openings and sieves with rectangular openings. A particle size distribution is plotted for the same soil when passed through the two series of sieves. The sphericity (ζ) of the particles is then designated as:

$$\zeta = 100\left(1 - \frac{\log u}{\log \sqrt{2}}\right) \tag{2.2}$$

in which

$$u = \frac{l_{50}}{m_{50}} ,$$

l_{50} being the spherical opening sieve opening diameter which will just pass 50% of the soil sample, and m_{50} being the rectangular opening sieve opening width that will just pass 50% of the soil sample. The sphericity (ζ) may be interpreted as:

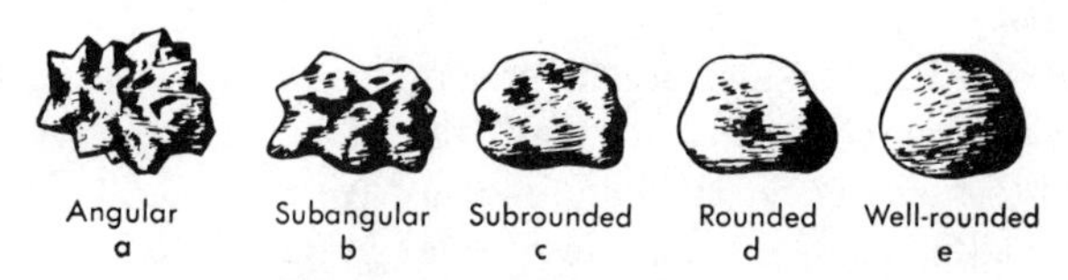

Figure 2.8–Typical appearances of bulky grains with varying degrees of angularity. Reprinted with permission of Macmillan Publishing Company, Inc. from *Introductory Soil Mechanics and Foundations*, 2nd Edition, Sowers, G. F. and G. B. Sowers. Copyright © 1961 Macmillan Publishing Company.

$$\zeta = \frac{dc}{di} \qquad (2.3)$$

in which dc is the diameter of the circle with an area equivalent to the area covered by the grain when laid on its long side, and di is the diameter of the circle inscribed into the same area (Kézdi, 1974).

> Soils composed of bulky grains behave like loose brick or broken stone. They are capable of supporting heavy, static load with little deformation, especially if the grains are angular. Vibration and shock, however, cause them to be displaced easily. Soils composed of flaky grains tend to be compressible and deform easily under static loads, like dried leaves or loose paper in a basket. They are relatively stable when subjected to shock or vibration. The presence of only a small percentage of flaky particles is required to change the character of the soil and produce the typical behavior of a flaky material.*

It may be shown that coarse-grained soils with angular particles have greater strength and bearing capacity than those with rounded particles, while in clays the inclusion of interlocking needlelike particles is a source of added strength (Hough, 1957). The shape of bulky grains has an important bearing upon the part of the shear strength of a soil which is influenced by internal friction. The angular grains offer much greater resistance to sliding over each other than do the more rounded grains. Soils containing appreciable quantities of flaky grains have relatively low internal friction because of the relative ease with which the flat grains slide over each other. Flaky grains do not tend to interlock as do bulky grains (Spangler, 1960).

Organic Matter

The percentage of organic matter in a soil is frequently estimated from measurements of the organic carbon content of the soil (Nelson and Sommers, 1982) using a formula such as:

$$\%\ \text{organic matter} = 0.35 + \left(a_i \times \%\ \text{organic C}\right) \qquad (2.4)$$

Various researchers have recommended that the coefficient, a_i, have values ranging from 1.724 to 2.5 depending on various factors. The value $a_i = 2.0$ is generally accepted, however, when a basis for using some other coefficient is not present (Nelson and Sommers, 1982).

Three different laboratory procedures are commonly used for determining the organic carbon content of soils:

---

* Reprinted with the permission of Macmillan Publishing Company, Inc. from *Introductory Soil Mechanics and Foundations*, 2nd Edition, Sowers, G. F. and G. B. Sowers. Copyright © 1961, Macmillian Publishing Company.

1. Analysis of a soil for total carbon and inorganic carbon, followed by subtraction of the inorganic C concentration from the total C present.
2. Determination of total C of the sample after destruction of inorganic C.
3. Reduction of $Cr_2O_7^{2-}$ by organic C compounds and subsequent determination of unreduced $Cr_2O_7^{2-}$ by oxidation-reduction titration with Fe^{2+} or by colorimetric techniques (Nelson and Sommers, 1982).

In mineral soils the organic matter content typically ranges from 1 to 10% by weight although the most common range is from 1 to 3% (Hillel, 1980). Organic matter particles have a specific gravity of about 1.3, while that of mineral soil particles is about 2.65 (Hillel, 1980). The presence of increased amounts of organic matter results in the increase of the moisture contents (upper and lower plastic limits) at which the soil displays plastic characteristics (see section on Atterberg limits) (Baver and Gardner, 1972).

Under wet conditions, soils with high organic matter contents are more resistant to compaction than soils low in organic matter; under dry conditions, the reverse holds true. This means that during drying, the soil resistance to compaction increases most in soils low in organic matter (Koolen and Kuipers, 1983).

In quartz sand soils the presence of organic matter tends to disrupt the effectiveness of the intragrain water layer as a lubricant and thereby increases intergranular friction (Lambe and Whitman, 1969).

In sandy loam and clay loam soils with organic matter contents increased from 3 to 17% by the addition of peat moss (Ohu et al, 1985a) it was found that soil shear strength as measured with a shear vane device was slightly higher for the low organic matter soils at low moisture content than for the high organic matter soils, but dropped to levels well below those of the high organic matter soils at high moisture contents. Furthermore, a great resistance to compaction in terms of retention of macroporosity was exhibited by the high organic matter soils, while the soils low in organic matter underwent significant reductions in macroporosity as compacting energy was applied (Ohu et al, 1985b). For these same soils, cone penetrometer resistance was essentially unaffected as moisture content was increased over a broad range for those having high organic matter, while the soils at 3% organic matter sustained major decreases in resistance as moisture content increased (Ohu et al, 1985b).

Increased levels of organic matter create a more stable soil structure. In the beginning of organic matter fermentation, mycelial filaments encircle the soil particles, promote aggregation and give certain cohesion to the structural units thus formed. However, it is principally the polysaccharides and polyuronides, formed mainly by microbiologic synthesis, that are active in structural stabilization (Bonneau and Souchier, 1982). A significant correlation was noticed between the percentage of soil aggregates larger than 0.05 mm and carbon content of a large number of soils (Baver and Gardner, 1972). The stability of soil aggregates is usually measured using a wet sieving technique (Kemper and Chepil, 1965)

In-field determination of soil organic matter percentages has been investigated for the purpose of providing a quantitative basis for very localized adjustments in real time for the rates of application of fertilizers and herbicides as application equipment operates in the field. Among the methods used have been those which use reflectance of visible light (660 nm) (Shonk and Gaultney, 1989) and the reflectance of near-infrared radiation (1640 to 2640 nm) (Sudduth et al, 1989). The instrument response in either case was found to be affected by moisture content of the soil. The near-infrared system was able to detect and compensate for moisture content while the visible light system was capable of operating in a subsurface environment where moisture content variation was less than at the soil surface. In general, reflectance decreases as organic matter content of soils increases. Typical ranges of variation in reflectance due to differences among common field-level moisture contents (reflectance in the 5 to 40% range with commonly 10 to 15 percentage points difference due to changes in soil moisture content) are approximately comparable to the ranges of variation of reflectance due to organic matter content differences (5 to 15 percentage points difference due to an organic matter increase from 1 to 3%). The spectral reflectance, as measured from the air, from the ground, and in the laboratory of soil derived from five different parent materials was studied by Schreier (1977) using a multi-channel spectro-photometer. A comparison with chemical analyses revealed that the spectral reflectance correlated well with the carbon content of the soil (the main distinguishing element in organic matter) as well as with the soil content of Fe and exchangable Mg.

Soil Mass Static Physical Properties

Dry Bulk Density, Porosity, and Void Ratio

The total volume of soil is made up of the volume of soil mineral grains and organic matter particles plus the volume of the pores between the grains and particles. The pore volume is usually partially filled with water, with air occupying the balance. Larger pores are usually occupied by air, smaller pores by water; the determining pore size decreases as moisture content also decreases. The most direct measurement of the status of solids vs pore-space distribution of a soil is the dry bulk density. Dry bulk density is the mass (or the weight) of dry soil materials per unit of total soil volume, reported in grams per cubic centimeter, tonnes per cubic meter, or pounds per cubic foot. Dry bulk density measured in metric units can be conveniently compared with the density of water, 1 g/cm^3 or 1 $tonne/m^3$.

The porosity of soil is the ratio of total (air plus water) pore volume to total soil volume and is usually expressed in percent. Porosity can be determined once dry bulk density and the density of individual soils grains are known (fig 2.9) by using the following formula:

$$\text{porosity} = 1 - \frac{\text{dry bulk density}}{\text{density of soil grains}} \tag{2.5}$$

For most mineral soils the density of individual soil grains ranges from 2.55 to 2.70 g/cm^3 (159 to 168 lb/ft^3) and when more specific information is not available it is assumed to be 2.65 g/cm^3 (165 lb/ft^3).

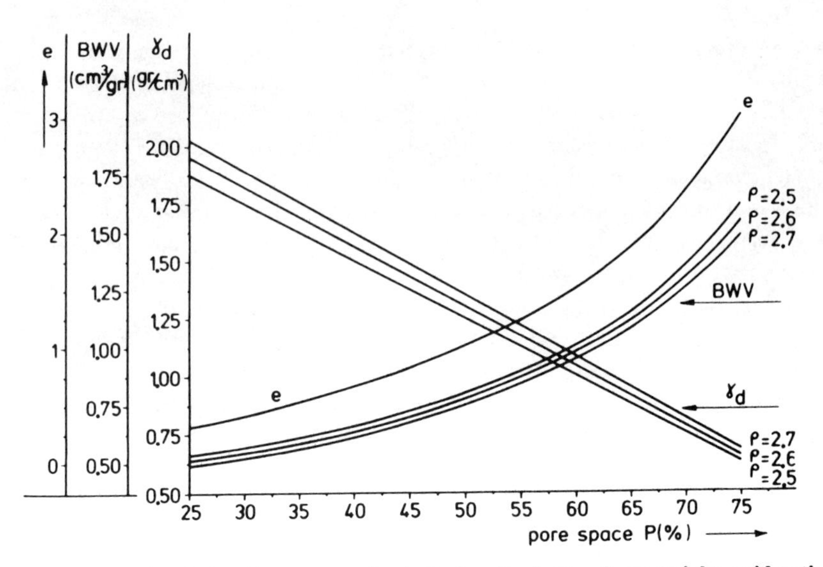

Figure 2.9–Computed relationships among dry bulk density, bulk-volume-weight, void ratio, and porosity, depending on the specific gravity of soil grain solids (from Koolen and Kuipers, 1983).

γ_d	=	**dry bulk density**
P	=	**porosity**
e	=	**void ratio**
BWV	=	$1/\gamma_d$

A third parameter of solid vs pore-space distribution is the void ratio which is commonly used in engineering of earth structures. Void ratio is the ratio of the volume of the total pore space to the composite volume of the individual solid particles. Void ratio is related to porosity by the following formula:

$$\text{void ratio} = \frac{\text{porosity}}{(1 - \text{porosity})} \tag{2.6}$$

The proportion of total soil volume occupied by air-filled pores can be found using the following formula:

$$\frac{\text{volume of air-filled pores}}{\text{total soil volume}} = \text{porosity} - \frac{(\text{dry bulk density}) \times (\text{moisture content})}{\text{density of water}} \tag{2.7}$$

in which the moisture content is determined as follows:

$$\text{moisture content} = \frac{\text{weight of water in a soil sample}}{\text{dry weight of soil in sample}} \tag{2.8}$$

and is usually expressed in percent. The product of the dry bulk density and moisture content, when divided by water density, represents the proportion of total soil volume which is composed of water-filled pores.

Thus, when determining the dry bulk density it is frequently convenient to find moisture content. Four of the most common means (Blake, 1965b) of determining dry bulk density and sampling for moisture content are, core sampler, excavating tube, volume excavation, and gamma-ray attenuation.

A core sampler cuts a cylindrical sample from the soil, disturbing the material inside the sample as little as possible. The core should be short in comparison with its diameter (to avoid side friction effects), and the tool edge should be as thin as possible and sharpened so the cutting edge is on the inside (fig 2.10). The edge bevel of such a tool does not laterally compress the sample and displaces only a small amount of soil, which is pushed away from the sample.

In some cases, the core may be removed from the soil by merely withdrawing the sampler. In other cases (such as with the tool shown in fig 2.10), the core must be dug out and carefully trimmed so that the end surfaces are parallel to those of the ends of the cylinder encircling the core sample. If the core sample is immediately weighed, it may then be dried at any time—the dry weight of the soil is used in computing the dry bulk density and the moisture content.

The core sample method is considered to be one of the most fundamental and reliable of the field methods for determining dry bulk density. Other methods are usually validated by being checked against the core sample method (Gameda et al, 1987; Ayers and Bowen, 1988; NeSmith et al, 1986). Nevertheless, some degree of variation exists among core sample results. Erbach (1987) indicated that between 10 and 30 replications of core sample tests were required to detect, with 5% probability of error, a difference of 0.1 g/cm^3 in a specific set of tillage experiments. To detect a 0.2 g/cm^3 difference, 3 to 8 replications were required. In another set of forest soil experiments cited by Erbach (1987), from 5 to 26 core samples were needed to estimate bulk density to within 0.1 g/cm^3 at a 0.05 probability of error.

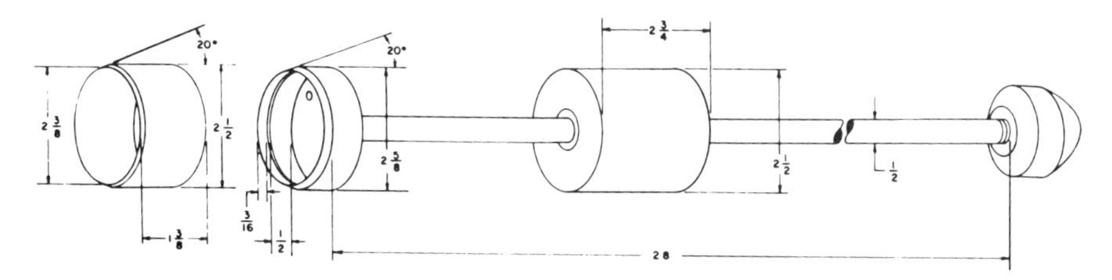

Figure 2.10–Core sampler (from Chancellor, 1977). All dimensions are in inches. All materials are steel except for the handle knob and the core cylinder which are brass. The core cylinder has a volume of 100 cm^3.

An excavation tube (fig 2.11) can be used to measure solid dry bulk density when holes must be dug to take samples at considerable depth (Abernathy et al, 1975). The 88.9 mm (3.5 in.) core tube is first driven into the soil beyond the depth to be sampled.

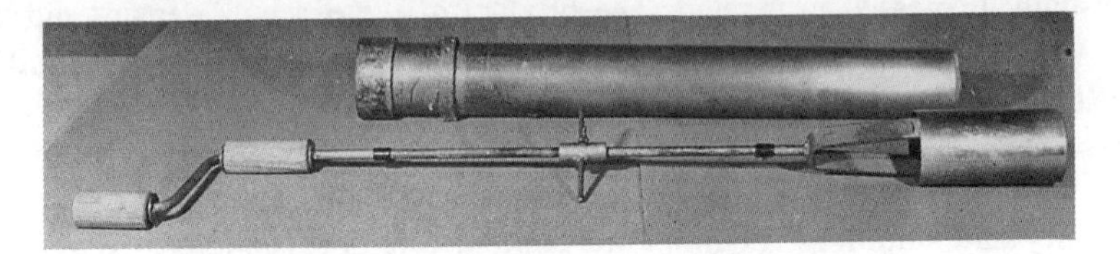

Instrument for sampling soil bulk density. Above, 3 1/2 in. diameter core tube, and below, close-fitting auger bucket. Adjustable cross-stop on auger handle is used to gage depth to which overburden is removed, and then depth to which sample is taken.

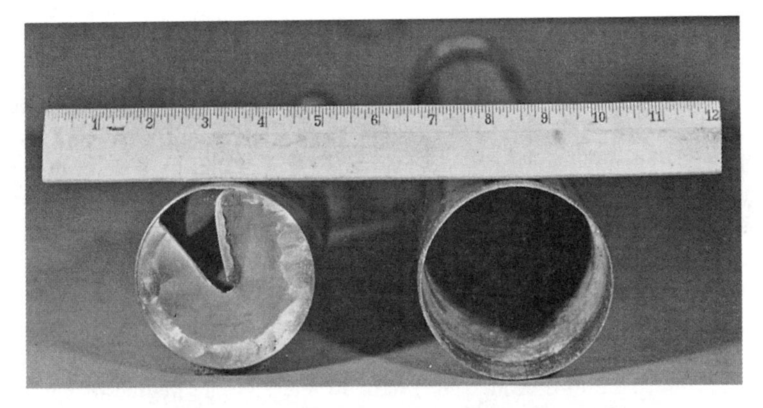

End view of core tube and auger bucket of soil bulk density sampling tool.

Detail of cutting lip on auger bucket of soil bulk density sampling tool.

Figure 2.11–Auger bucket for sampling soil density (from Abernathy et al, 1975). Dimensions shown are inches.

The close-fitting auger bucket is then used to remove overburden to the desired sampling depth. The auger bucket is used again to remove the soil to the maximum sampling depth. This soil is dried and weighed.

Volume excavation can be used if there is a flat horizontal surface at the soil depth to be tested. A small excavation is made beneath the horizontal surface, and all the material removed during the excavation is carefully saved, weighed, dried, and reweighed. The volume of the excavation is obtained by placing a thin, flexible membrane in the excavation and measuring the volume of a fluid required to fill the membrane to the original soil surface level. An apparatus for carrying out this type of measurement is illustrated in figure 2.12 (Bowles, 1978).

Two basic approaches have been used in employing gamma-rays to measure soil bulk density, gamma-ray back scattering (fig 2.13) and gamma-ray attenuation (fig 2.14). Numerous tests and comparisons (Ayers and Bowen, 1988; Greene and Stuart, 1984; Gameda et al, 1987) have shown the gamma-ray attenuation method to provide values with the least variation from core-sample values of wet bulk density. The gamma-ray attenuation method uses two probes: one containing the radioactive isotope (usually CiCs 137) and the other contains the detector. In some cases a Geiger-Müller tube was used (Vomocil, 1954), and in other cases a NaI(Tl) crystal was used, the latter of which tended to give a higher degree of resolution (Soane, 1968). The readings with the gamma-ray attenuation method seemed to be as accurate as with core sampling (Erbach, 1987), but the time required per value obtained was reduced by a factor of 3. In some cases with gamma-ray attenuation, the unit could be used directly with the calibration curve supplied by the manufacturer (Ayers and Bowen, 1988), while in other cases, the unit required calibration with samples of the soils on which it would be used—

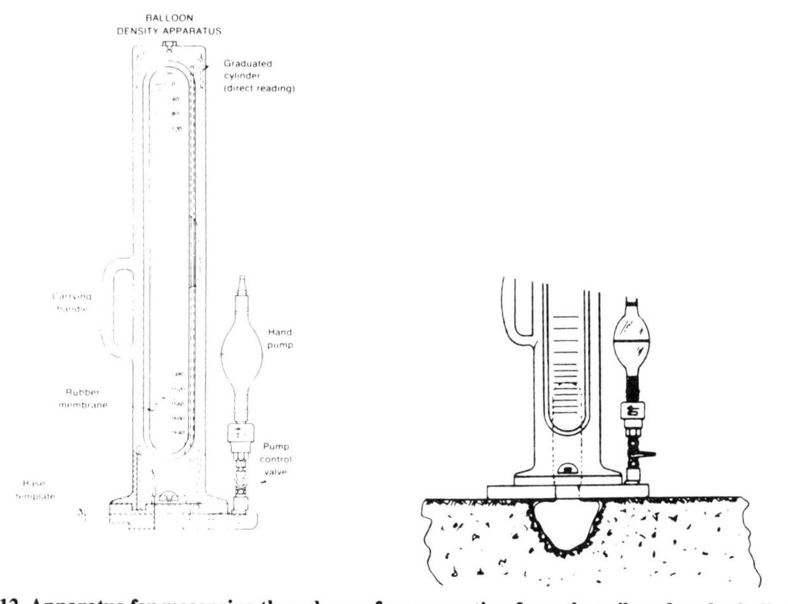

Figure 2.12–Apparatus for measuring the volume of an excavation from the soil surface for bulk density determination. Reproduced with permission of McGraw-Hill, Inc. from *Engineering Properties of Soils and their Measurement*, 2nd Edition, J. E. Bowles. Copyright © 1978, McGraw-Hill, Inc.

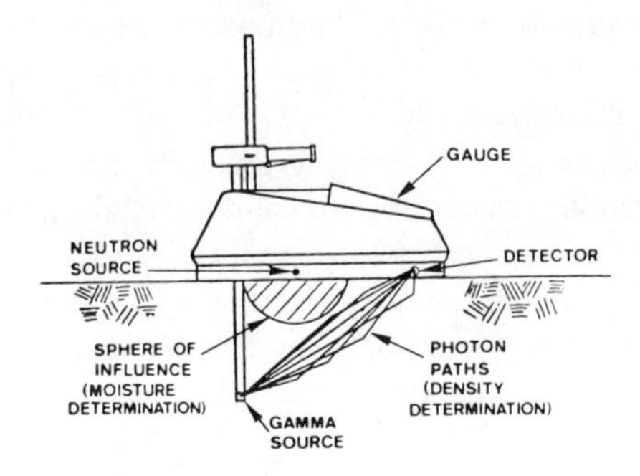

Figure 2.13–Schematic diagram of a single-probe neutron meter for soil density determination (from Gameda et al, 1987).

compacted to represent a range of typical density values. Because the unit measures wet bulk density, it is necessary to determine the moisture content of the material being tested if dry bulk density values are required.

The porosity or dry bulk density that soils will assume under a given set of conditions depends, among other things, on the grain size distribution. Minimum porosities for several agricultural soils with varying grain size distributions loaded to 200 kPa (29 $lb/in.^2$) are shown (in the section on grain size distribution) to range from 27 to 50%.

Moisture Content

It is most common to express the moisture content of agricultural soils as the ratio of the mass of water contained to the mass of dry material (dry-basis moisture content). The most usual way to determine moisture content of a soil sample is to place a weighed sample in a ventilated oven held at a temperature between 100 and 110°C (212 and

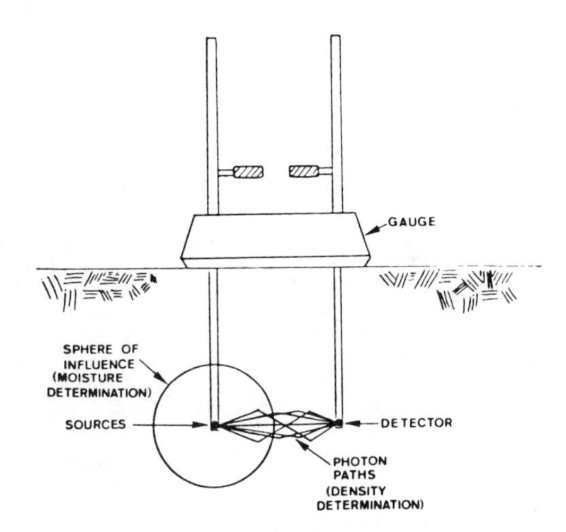

Figure 2.14–Schematic diagram of a dual-probe neutron meter for soil density determination (from Gameda et al, 1987).

230° F) until the sample weight becomes constant (Gardner, 1986). For some soils this constant level is never achieved exactly, however, for most low- to medium-clay-content soils, 24 h is a commonly used oven time.

The moisture content as a proportion of total wet soil weight (wet-basis) moisture content is related to the dry-basis moisture content by:

$$\frac{\text{mass of moisture}}{(\text{mass of dry soil} + \text{mass of moisture})} =$$

$$\frac{\text{mass of moisture}}{\text{mass of dry soil}} \times \frac{1}{[1 + (\text{mass of moisture / mass of dry soil})]} \tag{2.9}$$

The moisture content of the soil on a proportion-of-total-volume basis is obtained as the ratio of the product of dry bulk density and dry-basis moisture content to the density of water:

$$\frac{\text{vol of moisture}}{\text{bulk vol of soil}} =$$

$$\frac{\text{mass of moisture}}{\text{mass of dry soil}} \times \frac{\text{mass of dry soil}}{\text{bulk vol of soil}} \times \frac{\text{vol of moisture}}{\text{mass of moisture}} \tag{2.10}$$

Other methods of direct determination of moisture content include microwave-oven drying, the procedure for which varies with soil type and the characteristics of the oven used (Gardner, 1986) and the reaction of the moisture with calcium carbide in a fixed-volume vessel; moisture content is related to the pressure of the acetylene formed (Soiltest, 1973).

Because the moisture content of soil is such an important factor in determining the physical characteristics of the soil, many indirect methods have been developed for assessing moisture content. A sizable number of these indirect methods involve *in situ* measurements (Erbach, 1987). Some of the principles involved in these are:

1. Soil moisture tension as determined by a tensiometer
2. Neutron scattering
3. Gamma-ray attenuation
4. Electrical resistance or capacitance of absorbent materials in moisture tension equilibrium with the soil
5. Thermoelectronic methods using a heat dissipation sensor
6. Nuclear magnetic resonance
7. Time-domain reflectometry
8. Microwave back-scattering and forward scattering (Wallender et al, 1985)
9. Thermocouple psychrometry (Riggle and Slack, 1980)
10. Infrared reflectance and photography (Kano et al, 1985)
11. Microwave emissivity (Bausch, 1983; Jackson et al, 1984)

12. X-ray tomography (Tollner and Verma, 1987a)
13. Gamma-ray emission (Gutwein et al, 1986)
14. Electromagnetic wave reflection (Parchomchuk and Wallender, 1986)
15. Soil electrical conductivity (Freeland, 1989)

Succeeding sections will discuss the soil properties on which signal response is based for many of the above methods.

Many of these methods require that some calibration be made between soil moisture content and signal response. This calibration may vary depending on soil type, porosity and/or other properties.

The soil moisture content at which an agricultural soil may be found under common crop-growing conditions in which there is a normal supply of water to meet crop needs, depends on the grain size distribution of the soil material, with higher moisture contents being associated with more-fine-grained soils.

Soil Moisture Tension

When water is enclosed in an open vertical glass tube, the lower end of which is below the surface of a water reservoir (fig 2.15), the water wets the inside of the tube, and the resulting air-water interface surface tension causes the water to rise in the tube according to the equation (Kirkham and Powers, 1972):

$$h_t = 2\,\sigma \cos\theta \,/\, r \cdot \rho \cdot g \tag{2.11}$$

where

- σ = the surface tension in an air-water interface (about 73 dynes/cm under standard atmospheric conditions)
- θ = the wetting angle (approximately zero for most soil materials with water)
- r = radius of the tube in cm
- ρ = the density of water (about 1 g/cm^3)
- g = acceleration of gravity (about 980 cm/s^2)

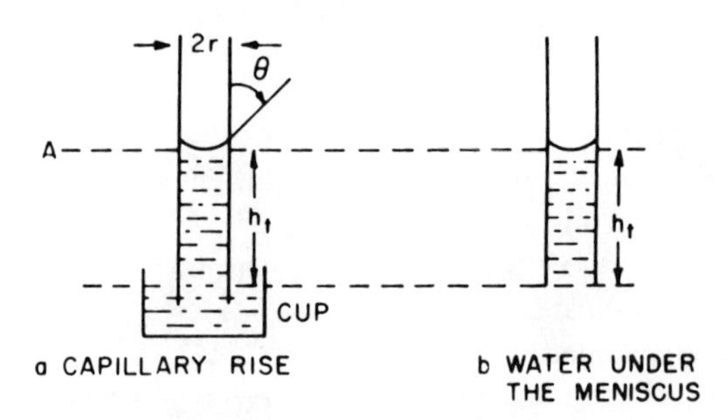

Figure 2.15–Illustration of the wetting angle, θ, and how surface tension of the water-air interface causes capillary rise of water in a tube. Reprinted with permission of Wiley-Interscience Division, John Wiley &Sons, Inc. from *Advanced Soil Physics*, Kirkham, D. and W. L. Powers. Copyright © 1972, John Wiley & Sons, Inc.

Consequently, as soil begins to drain, the water tension required to drain each pore is a function of the pore radius–smaller pores requiring greater tensions for drainage. Thus, unsaturated soils tend to contain pore water under tension. As soils are drained or dried to lower moisture contents, the soil moisture tension (sometimes called soil moisture suction) becomes greater (fig 2.16). However, if the soil were allowed to reabsorb water, the moisture content at any given tension level will be less than during drainage because of irregularities in the diameters of soil pores (fig 2.17).

When moisture is drained or dried from soil until most of the larger pores become air-filled, the remaining moisture becomes isolated in the smaller pores as might exist between adjacent soil grains at their point of physical contact (fig 2.18). Under these circumstances in which the particle radius is much larger than the water body radius the negative pressure or tension in the water can be represented by (Kirkham and Powers, 1972):

$$P = \sigma\left(3r_1 - 2R\right) / r_1{}^2 \tag{2.12}$$

in which P = pressure in dynes/cm^2, r_1 = the minimum radius of the water body (cm), and R = the radius of particle (cm).

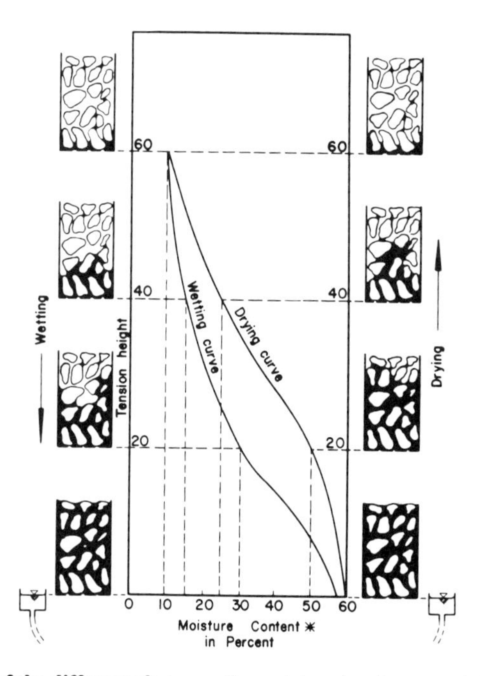

Figure 2.16–Illustration of the difference between the moisture tension vs moisture content relationships for wetting as opposed to that for drying of a soil (see fig 2.17). Reprinted with permission of Wiley-Interscience Division, John Wiley & Sons, Inc. from *Advanced Soil Physics*, Kirkham, D. and W. L. Powers. Copyright © 1972, John Wiley & Sons, Inc.

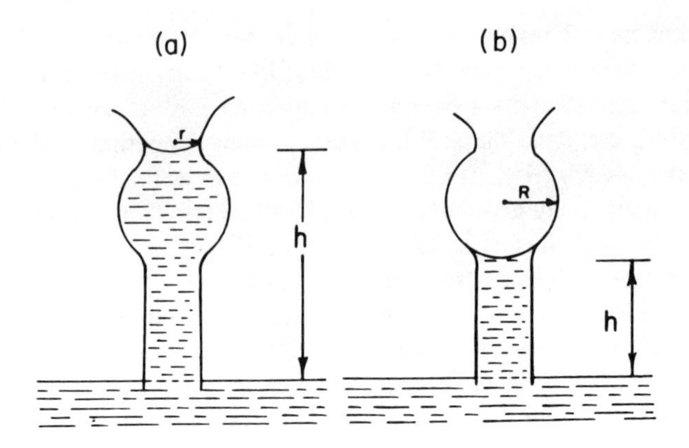

Figure 2.17–Schematic diagram of the different water contents of an irregular tube when at a given moisture tension for drying (left) as opposed to that for wetting (right). Reprinted with permission of Academic Press, Inc. from Hillel, 1980. Copyright © 1980, Academic Press, Inc.

The resulting force bonding the two particles together, as caused by surface tension in the skin of the water body, and the effect of the negative hydrostatic pressure (or moisture tension) inside the water body, amounts to (Kirkham and Powers, 1972):

$$F = \pi\sigma\left(2R - r_1\right) \tag{2.13}$$

in which $\pi = 3.1416$.

Note that the radius of curvature of the water body, r_2, can be calculated from:

$$r_2 = r_1{}^2 / \left[2\left(R - r_1\right)\right] \tag{2.14}$$

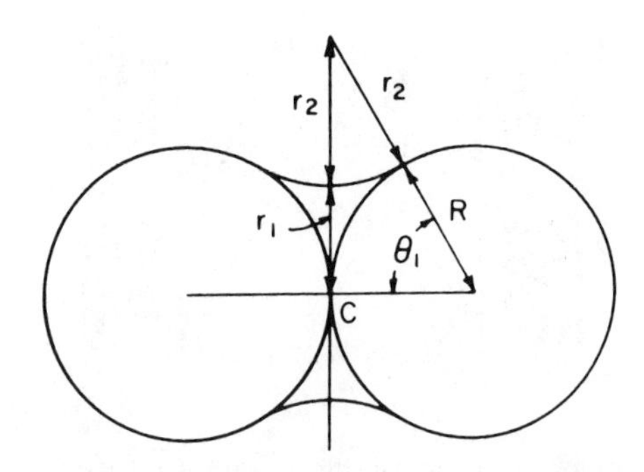

Figure 2.18–Illustration of the shape and dimensions of a water body that would bond two spherical particles together. Reprinted with permission of Wiley-Interscience Division, John Wiley & Sons, Inc. from *Advanced Soil Physics*, Kirkham, D. and W. L. Powers. Copyright © 1972, John Wiley & Sons, Inc.

As a soil becomes progressively drier, these forces tend to increase. However, a moisture content may be reached at which the water body becomes so small that the forces decrease (fig 2.19). This point is more commonly reached in the range of normal agricultural soil moisture contents for coarse-grained soils than it is for fine-grained soils.

These relationships among pore size, soil grain size, and soil moisture tension are responsible for the high moisture content of fine-grained soils held at higher tensions than is the case for coarse-grained soils (fig 2.20).

The tension with which moisture is held in the soil also represents one component of the potential difference, across which plant roots draw water and dissolved nutrients from the soil. It is believed that plants in general can extract moisture from the soil at soil moisture tensions up to 15 bars (217 lb/in.^{2} or 1.5 MPa). The moisture content at which soil moisture tension reaches 15 bars is called the *permanent wilting point.*

When a saturated soil is allowed to drain naturally without evaporation from the surface or transpiration from the root zone, the soil moisture tension tends to assume a stable value of approximately ⅓ bar (4.9 lb/in.^{2}, or 33.3 kPa) (Bruce and Luxmore, 1986). The soil moisture content at which a soil moisture tension of ⅓ bar is reached is frequently used as an estimate of the "field capacity" of the soil or the equilibrium level of soil moisture that would prevail after saturation but without extraction. The difference in soil moisture contents between the field capacity and permanent wilting point levels is frequently referred to as the *water holding capacity* or the *available water content* of the soil, because it represents the amount of water potentially available to plant roots.

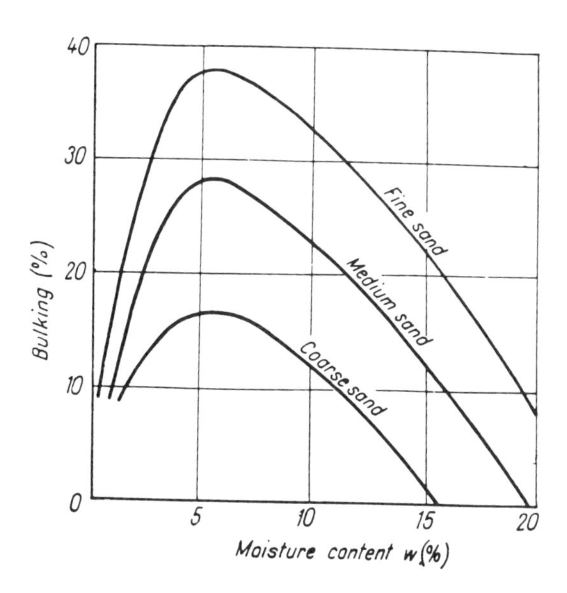

Figure 2.19–The effect of moisture content on the strength of sand. Bulking is the increase in settled volume of the material after it has been agitated. At very low moisture content, water bonds are so small as to not have much strength, while at high moisture contents, the moisture tensions are so low that there is little strength. At intermediate moisture contents, the maximum bonding effect is noticed (from Kézdi, 1974).

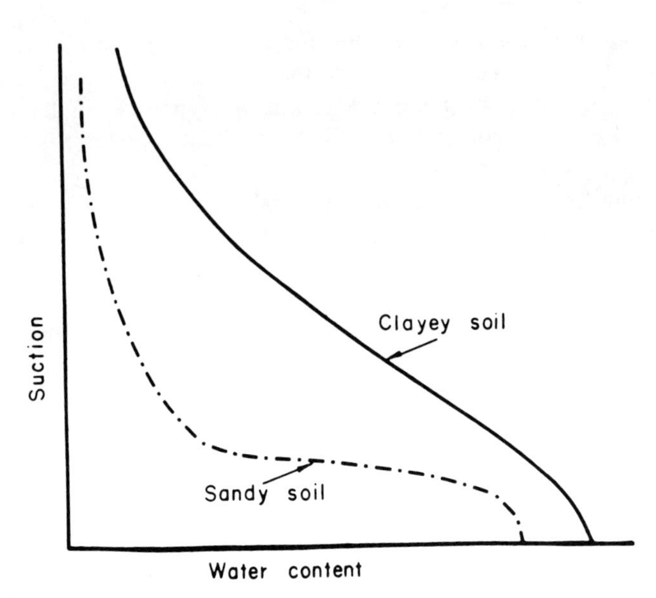

Figure 2.20–Soil moisture suction (tension) vs. moisture content for a fine grained soil (clayey) and a coarse grained soil (sandy). Reprinted with permission of Academic Press, Inc. from Hillel, 1980. Copyright © 1980, Academic Press, Inc.

In the laboratory, soil moisture tension at low levels (0.0 to a maximum of 0.8 bars) can be measured using a porous plate apparatus (fig 2.21). For higher levels of tension (up to 15 bars) a pressure plate apparatus (fig 2.22) is used (Klute, 1986). For *in situ*

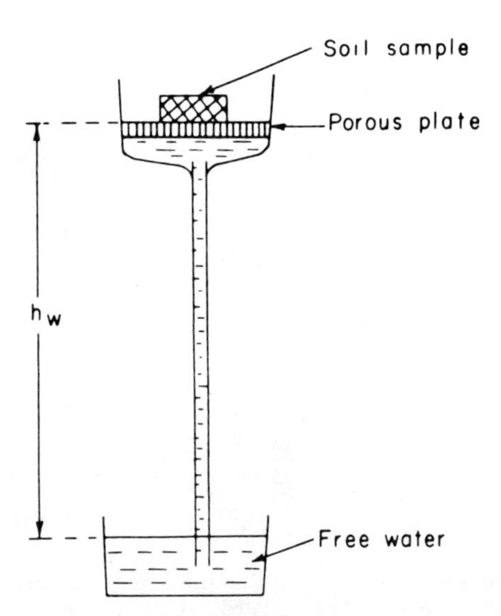

Figure 2.21–Schematic diagram of an apparatus to place the moisture in a soil sample at a given level of suction (after which moisture content is measured). Reprinted with permission of Academic Press, Inc. from Hillel, 1980. Copyright © 1980, Academic Press, Inc.

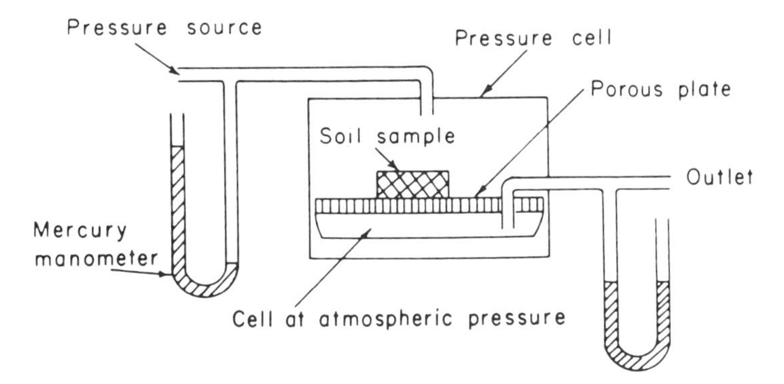

Figure 2.22–Pressure plate apparatus to subject the soil moisture content to high tension levels. Reprinted with permission of Academic Press, Inc. from Hillel, 1980. Copyright © 1980, Academic Press, Inc.

measurements of soil moisture tension, several methods are used. One of these methods is the tensiometer (fig 2.23) which has a porous ceramic cup through which moisture continuity is established between the water inside the tensiometer tube and the soil moisture matrix. Water is extracted from the tube until the suction inside the tube equals

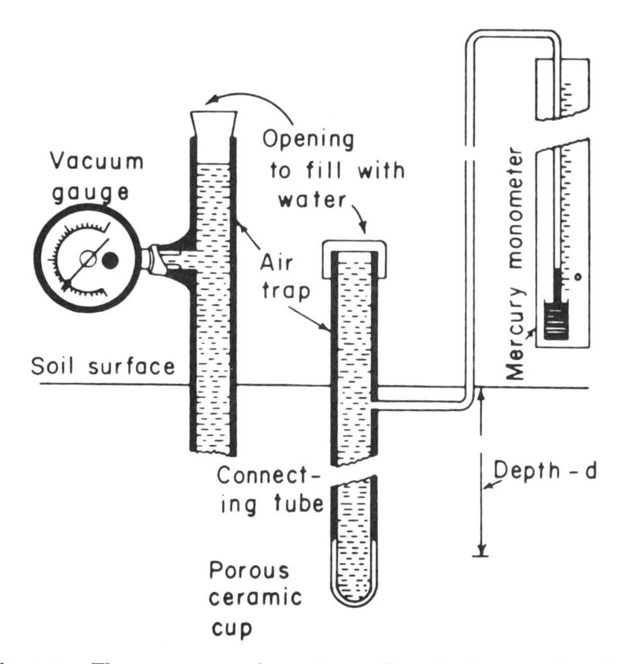

Figure 2.23–Tensionmeter. The porous ceramic cup transmits water between the soil moisture matrix and the inside of the tube until the suction in both are in equilibrium. Suction is then measured with the vacuum gage or the manometer. Reprinted with permission of Academic Press, Inc. from Hillel, 1980 as derived from Vomocil, 1965. Copyright © 1980, Academic Press, Inc.

the tension in the soil moisture. The vacuum in the tube is then measured with a gage or a pressure transducer. This apparatus is suitable for tensions from 0.00 to 0.85 bars (Klute and Cassell, 1986).

The resistance between two electrical conducting material grids embedded in a porous block tends to change in a more or less linear manner with the tension of the moisture within the block when represented on a log-log plot. Installation of these blocks with their moisture content in continuity with the soil moisture matrix then allows the determination of electrical resistance parameters, which can be calibrated against soil moisture tension. Gypsum blocks tend to be more responsive for tension in the 0- to 1-bar range, while fiberglass or nylon blocks have a particularly regular response at tensions greater than 1 bar (Campbell and Gee, 1986).

Heat dissipation sensors have an electric heater circuit for furnishing a heat pulse to a porous reference matrix, and have a temperature sensor for monitoring the temperature at the center of the ceramic before and after the pulse (fig 2.24). The temperature response is a function of the thermal conductivity and thermal diffusivity of the porous reference matrix. Conductivity and diffusivity are in turn affected by the moisture content of the matrix which is in moisture tension equilibrium with the soil-water matrix. Calibrated relationships between temperature response and soil moisture tension tend to be linear within the range of 0.2 to 1.0 bars tension. This sensor is not affected by the salt content of soil moisture (Campbell and Gee, 1986).

Thermocouple psychrometry (Erbach, 1987; Rawlins and Campbell, 1986) uses a fine-wire thermocouple placed inside an air-filled ceramic cup. Air can be exchanged through the pores of the cup with the surrounding soil atmosphere (fig 2.25). By applying a current across the thermocouple junction in the proper direction, the junction is cooled until water condenses on the junction. Once the current is stopped, the junction tends to remain at the wet bulb temperature (sensed by noting the electrical potential across the junction). This temperature is then compared with the dry bulb temperature

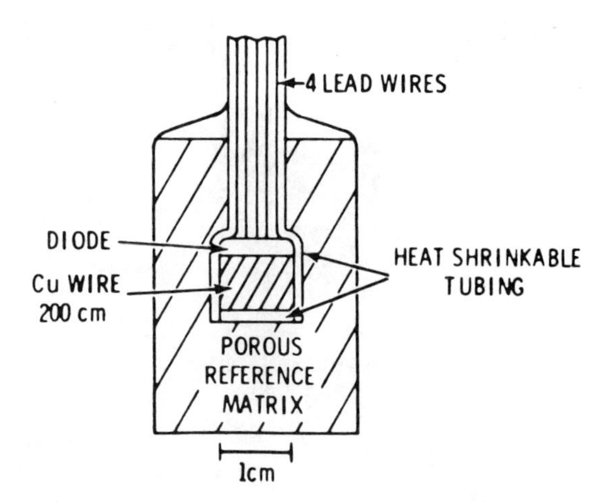

Figure 2.24–Cross-sectional sketch of a heat dissipation moisture sensing unit (from Campbell and Gee, 1986).

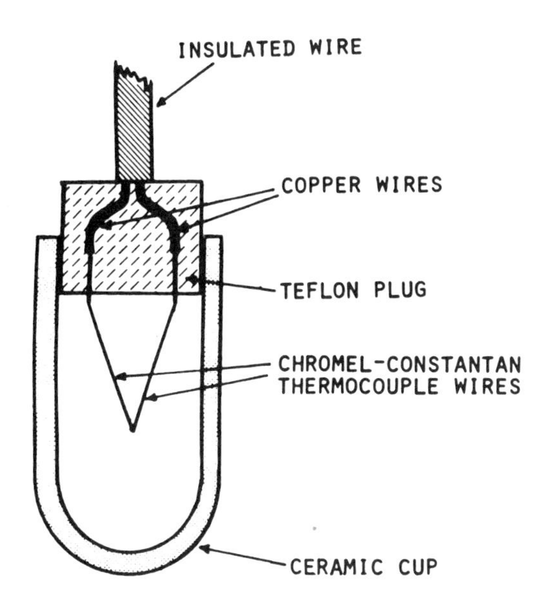

Figure 2.25–Cross-section of a thermocouple psychrometer contained in an air-filled ceramic cup. Reprinted with permission of Academic Press, Inc. from Hillel, 1980. Copyright © 1980, Academic Press, Inc. (Trade names are used solely to provide specific information. Mention of a trade name does not constitute an endorsement of the product to the exclusion of other products not mentioned.)

measured at a similar thermocouple (insulated from moisture transfer) in the adjacent soil. The relative humidity of the soil atmosphere (P/Po) is determined from:

$$P / Po = 1 - \left[\left(s + \gamma *\right) / Po\right] \Delta T \qquad (2.15)$$

and the soil water potential (ϕ) is:

$$\phi = \overline{R}T \ln \left(P / Po\right) \quad \left(\text{Hillel, 1980}\right) \qquad (2.16)$$

where

- P = vapor pressure of soil water and soil atmosphere water
- Po = vapor pressure of pure free water at the same temperature and air pressure as for P above
- $\overline{R}$ = specific gas constant for water vapor
- T = absolute temperature
- s = slope of the saturation vapor pressure curve
- $\gamma*$ = the apparent psychrometer constant, the product of the thermodynamic psychrometer constant and the ratio of vapor to heat transfer resistance
- ΔT = wet bulb temperature depression

Because most agricultural soils that support plant growth have a relative humidity of the soil atmosphere in the range 99 to 100%, extreme precision is needed in taking the temperature measurements required since the differences between wet- and dry-bulb temperatures will be very small.

Temperature has been found to have an effect on soil moisture tension (fig 2.26). At a given soil moisture tension, less water is held by a soil at high temperatures than at low temperatures (Taylor and Ashcroft, 1972).

The effect of various tillage operations on *in situ* soil moisture tensions at various depths throughout the growing season for a corn crop were measured using tensiometers (Kanwar, 1986). Moisture tension changes during one three-week interval are shown in figure 2.27.

Pore Size Distribution

One of the objectives of tillage is to loosen soil so that water may flow more rapidly through it and so that growing root tips may find pores large enough to enter. On the other hand, one of the main concerns about soil compaction is that many of the larger pores will be eliminated. Thus, pore size distribution is an important parameter of soil physical condition relative to tillage and traction actions on the soil.

There exist a number of bases for categorization of pore sizes. One of these (Koorevaar et al, 1983) categorizes pores by their diameter as follows: macropores > 100 μm; mesopores 30 – 100 μm; micropores < 30 μm.

A generalized characterization of these pore size groups (Koorevaar et al, 1983) describes macropores as conducting water only during conditions in which there is flooding, ponding, rainfall, etc. They soon drain after such water supply ceases. Thus, they affect aeration and drainage. Mesopores are effective in conducting water also, after

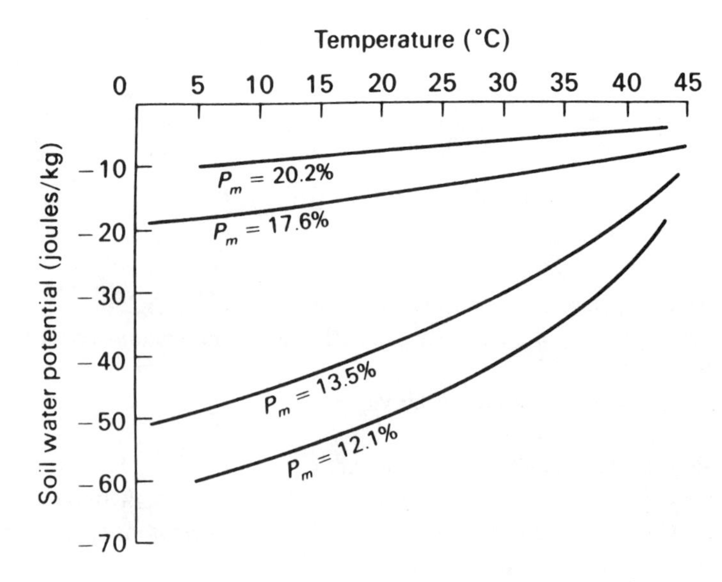

Figure 2.26–The influence of temperature on the soil water potential at consistent pressure, and at several different constant water percentages (P_m) in a silty clay loam (from Taylor and Ashcroft, 1972).

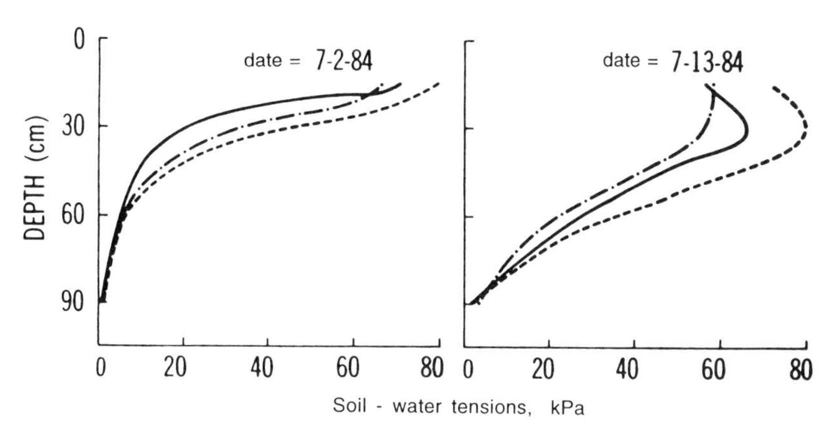

Figure 2.27–Soil water tensions as a function of depth on selected days, showing the increases in tension due to moisture removal by evapotranspiration (from Kanwar, 1986).

the macropores have become empty. The remaining soil solution is retained or moves very slowly within the micropores. Part of this water can be taken up by plant roots.

The hydraulic conductivity of a soil may be calculated on a theoretical basis from information about soil porosity and pore size distribution. Such values depend on a summation of terms, one for each pore size fraction, in which each term has the pore radius squared as a component (Marshall and Holmes, 1988; Koorevaar et al, 1983). Thus, the largest, the macropores, play a disproportionately large role in determining hydraulic conductivity. Hydraulic conductivity values which approximate the actual initial conductivity of field soils may be determined from such information (Schmidt, 1963).

The pore size distribution is linked to the grain size distribution of the soil materials (fig 2.28) and thus can be expected to be different for sands than for loams (fig 2.29).

The application of compressive loads to soils can also result in changes of the pore size distribution for soils with a given grain size distribution (figs 2.30 and 2.31). Compaction is responsible for removal of many of the larger diameter pores (Sommer, 1976) and thus has a major impact on the hydraulic conductivity of the soil.

Tillage can have the effect of increasing macroporosity (Pikul et al, 1988) as is shown in table 2.2.

The measurement of the pore size distribution has in the past been carried out by either the water desorption method or the mercury intrusion method (Danielson and Sutherland, 1986). Both methods are based on the relation between surface tension for an air-water or air-mercury interface. The basic relationship is:

$$\Delta P = 2\sigma \, r_p^{-1} \qquad (2.17)$$

in which ΔP = the pressure difference across the air-water or air-mercury meniscus (N/m^2), σ = the surface tension of the fluid (N/m), and r_p = the radius of the pore (m).

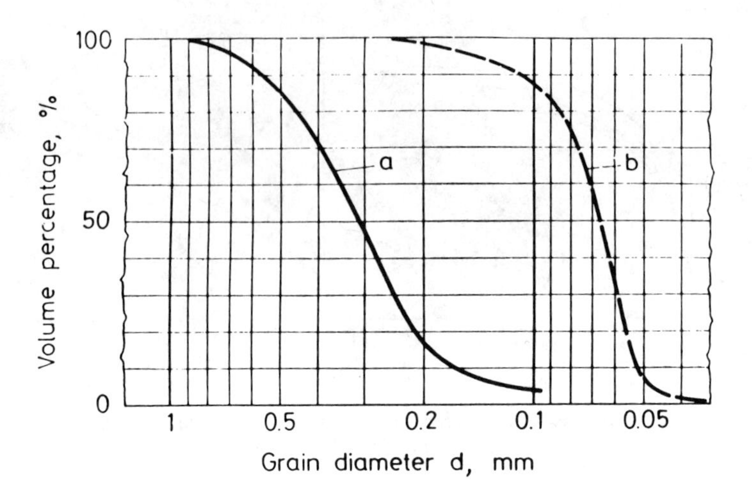

Distribution curves
a grain size
b pore size

Figure 2.28–Illustration of the similarity between grain size distribution and pore size distribution in a sand (from Kézdi, 1979).

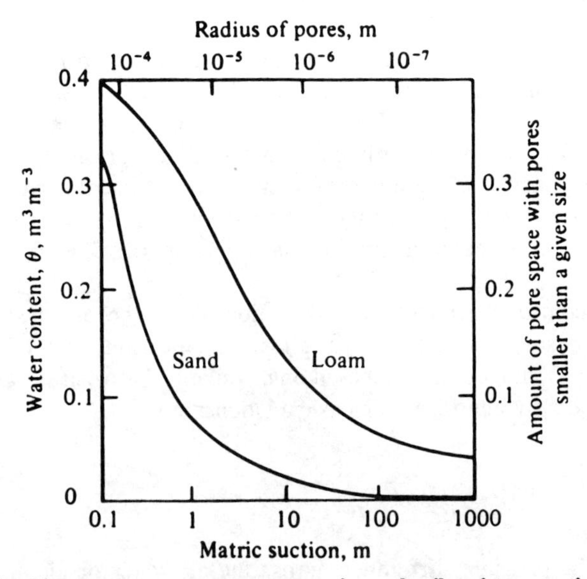

Figure 2.29–Relationships among water content, pore size and soil moisture suction for a sand and a loam soil. The relationship between pore radius and soil moisture suction is obtained by linking the intercepts of any vertical line with the top and bottom horizontal scales (from Marshall and Holmes, 1988).

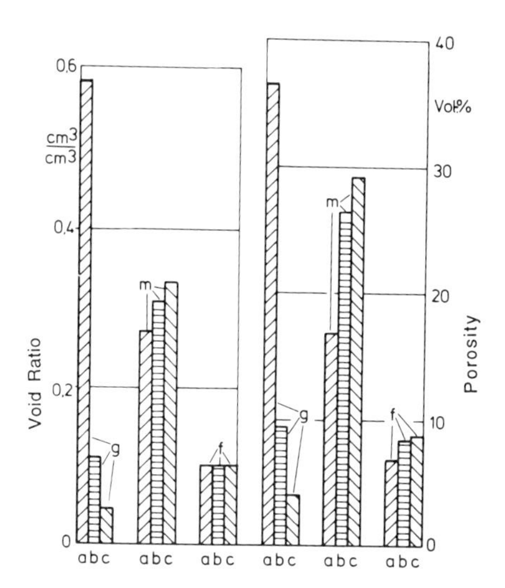

Figure 2.30–Void ratio (left) and porosity (right) changes upon loading for soil aggregates smaller than 1 mm for applied pressures of a = 0 MPa, b = 0.1 MPa and c = 0.4 MPa. Effects on large (g) pores (>10 mm), medium-sized (m) pores (10 to 0.2 mm), and small (f) pores (<0.2 mm) are shown (from Sommer, 1976).

With water desorption the soil is fully wetted and the amount of volume extracted at each increment of applied moisture tension is recorded using an apparatus which resembles, in principle, that in figure 2.32.

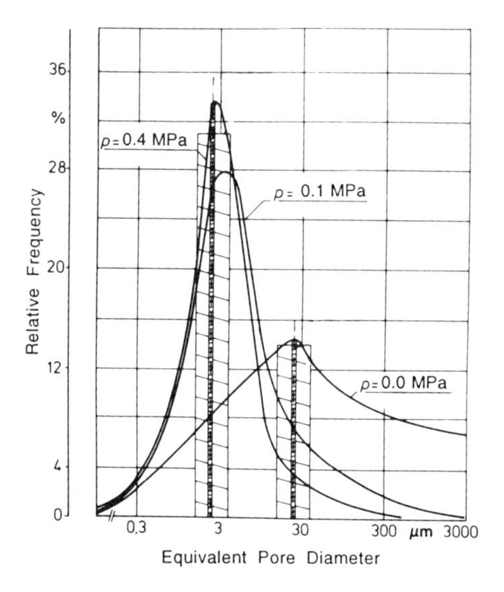

Figure 2.31–Relative frequency of pore size categories for a soil of aggregates smaller than 1 mm when subjected to three different compressive loads (0, 0.1, and 0.4 MPa) (from Sommer, 1976).

Table 2.2. Percentage of soil cross-sectional area consisting of pores with an area of 0.25 mm^2 or more*

	Treatment		
Soil Depth (cm)	No-Till	Paraplow	Chisel
7.6	<1	14.5	20.7
10.2	<1	12.4	13.7
12.7	<1	6.9	14.8
15.2	<1	9.7	10.8
17.8	<1	7.9	7.1
20.3	<1	11.0	5.1
25.4	<1	17.2	0.5
Average	<1	11.4	10.4

* From Pikul et al, 1988, equivalent pore diameter = 0.564 mm.

For the mercury intrusion method, the mercury does not wet the soil materials, so pressure must be used to force the mercury into the pores of an evacuated soil sample. The pressure required is then related to the pore size as shown above, while the volume of mercury absorbed by the soil is measured for each increment of applied pressure.

With the advent of electronic vision systems, it has become possible to detect the fraction of the total cross-sectional area of a soil material impregnated with paraffin or a

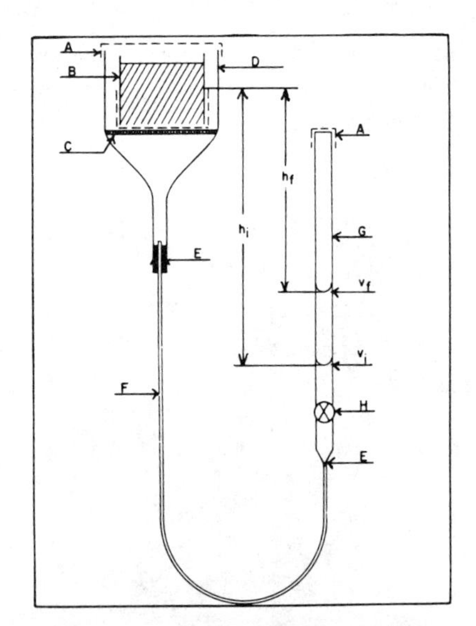

A = Aluminum foil covers.
B = Sample in cylinder with cheesecloth bottom. Volume, V_b. (Omit cheesecloth for suctions >150 cm. of water.)
C = Fritted glass porous plate (part of D).
D = Büchner funnel with porous plate.
E = Joints must be secure.
F = Flexible tubing.
G = Burette, least division not more 0.1% sample volume.
H = Stopcock of burette.
h_i = cm. of water suction, initial.
h_f = cm. of water suction, final.
v_i = Burette reading, initial.
v_f = Burette reading, final.

Figure 2.32–Schematic diagram for pore-size distribution determination (from Vomocil, 1965).

plastic polymer, which consists of macropore area (pore areas greater than 0.25 mm^2) (Pikul et al, 1988). Computerized tomography x-ray scanners have also been used to detect macroporosity (pores larger than 1 mm in diameter) in core samples of soil (Warner and Nieber, 1988). This technique permits scans to be made at several depth levels below the field surface of the otherwise undisturbed core.

The scanning electron microscope may also be used to obtain data on not only the quantitative distribution of pore sizes, but also on the location of pores of various sizes relative to the soil surface (Chen et al, 1980 and Upadhyaya et al, 1988).

In situ measurement of pore size distribution – particularly of macropores with diameters greater than 0.25 mm – can be accomplished with a tension infiltrometer (Everts and Kanwar, 1989).

Air Permeability

It has been shown (Vomocil and Flocker, 1961; Bowen, 1966) that oxygen availability and the associated necessity of gaseous exchange between the atmosphere and the soil pore space is a critical factor governing both seed germination and plant performance. Air permeability of soil material has been one of the main measurements of this property of soil. Both constant air volume (Kirkham, 1946) and constant air pressure (Grover, 1955) methods have been proposed. The constant air pressure method has found application in both devices for *in situ* measurements (figs 2.33 and 2.34) (Bowen and Liang, 1988) and apparatus for laboratory measurement of core samples (fig 2.35) (Morgan et al, 1988).

For *in situ* measurements the intrinsic air permeability, k_a, has been computed (Brooks and Reeve, 1959) from:

$$k_a = V\eta\ L / t\ AP \qquad (2.18)$$

where

k_a = intrinsic permeability determined by air flow (cm^2)
V = volume of air (cm^3)
η = viscosity of air (dyne s cm^{-2})
L = length of sample, cm
t = time for volume (V), to flow through the sample (s)
A = cross-sectional area of the sample cm^2
P = gage pressure of air on the high-pressure side of the sample (dynes cm^{-2})

For the case of an apparatus for which pressures at both entrance and exit of the sample are known (Morgan et al, 1988), P can be determined from:

$$P = \frac{P^2_{entrance} - P^2_{exit}}{2 \cdot P_{exit}} \qquad (2.19)$$

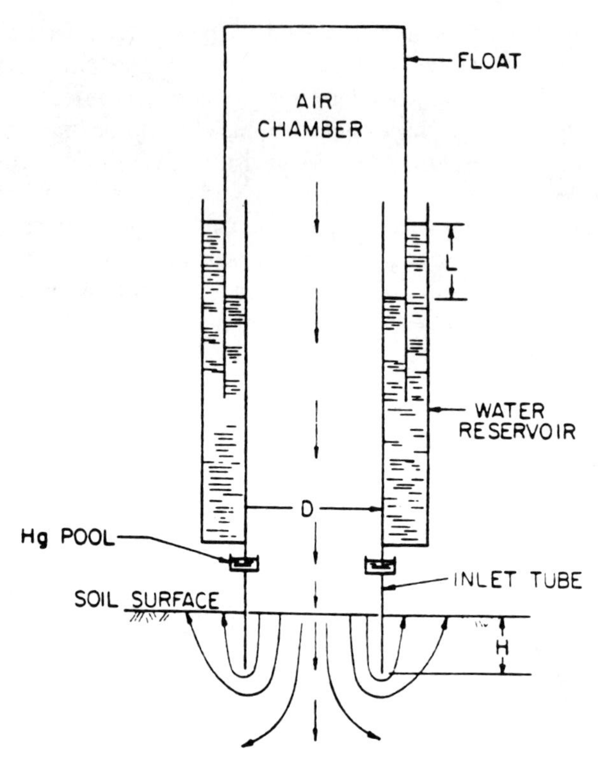

Figure 2.33–Schematic diagram of an apparatus to measure *in-situ* air permieability of soil using the constant pressure method (from Grover, 1955, as reproduced by Bowen and Liang, 1988).

Attempts to devise a system using a probe to test air permeability of soil from a moving vehicle (Boedicker and Bowen, 1976) were not fully successful because of changes in the soil immediately adjacent to the moving probe.

In general, it has been found that air permeability of agricultural soil decreases as soil moisture content increases (fig 2.36) (also see fig 2.37 for the response to wetting between days 0 and 1). However, for soils at moisture contents below the field capacity moisture content, agitation of soil causes higher moisture soils to assume lower dry bulk densities (sometimes called *bulking*, which is particularly noticeable in sands) and thus have higher air permeability values (figs 2.37 and 2.38). Even the application of compacting stresses to previously agitated soil cannot prevent this interaction among moisture content, bulking, and air permeability (fig 2.38). Nevertheless, it can be seen from figure 2.39 that compacting stresses do cause reduction in air permeability as a result of increased soil dry bulk densities (and thus reduced total porosities). An example of the relationships among air permeability, porosity, and dry bulk density, as determined from data on core samples from various depths at a specific soil site, are given in figure 2.39 (Morgan et al, 1988) in which it can be seen that air permeability parallels porosity but has an inverse relationship with dry bulk density (as does porosity).

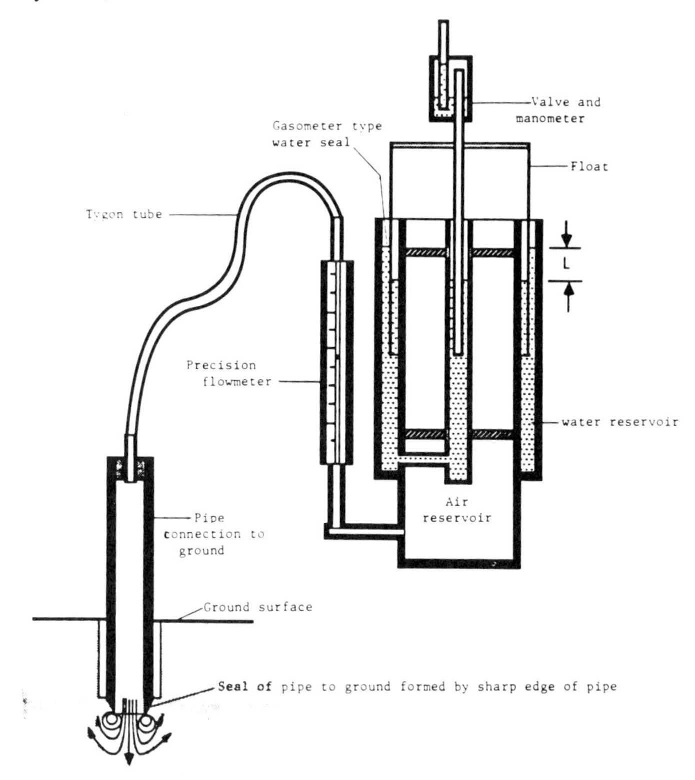

Figure 2.34–Modified constant-pressure air permeability instrument. The flowmeter allows rapid readings of relative, rather than absolute, air permeability (from Bowen and Liang, 1988).

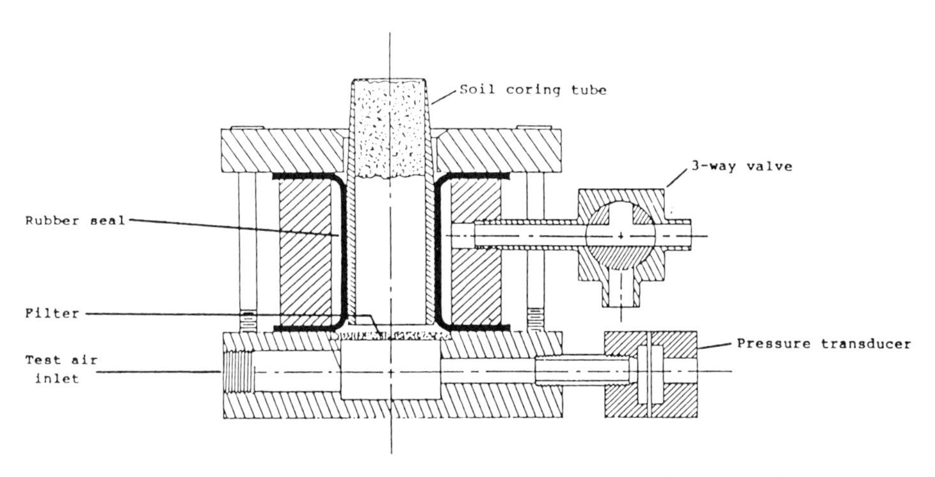

Figure 2.35–Cross-section of soil core test cell for constant-pressure determinations of air permeability (from Nau, 1987, as reproduced in Morgan et al, 1988).

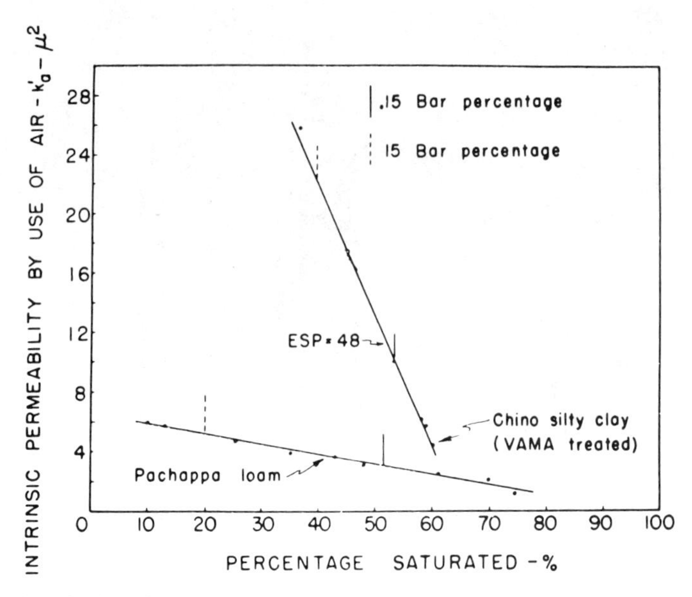

Figure 2.36–Effect of soil moisture content on air permeability as measured on soil cores. Air passing through initially very moist soil material was allowed to gradually dry the soil permitting permeability observations at reduced moisture contents. Saturation moisture contents were 35% for the Pachappa soil and 94% for the Chino soil (from Brooks and Reeve, 1959).

When the air permeability of soil falls to zero, it is likely that plant growth will be adversely affected, irrespective of whether air permeability has been caused to reach this zero level by high moisture levels in the soil or by compaction. However, it has been generally found that when air permeability is zero, there still remains about 10% of the soil volume constituted by air-filled pores. It is believed (Bowen and Liang, 1988) that these remaining air-filled pores are discontinuous to the extent that there is no communication between them and the atmosphere. Thus they are not available to serve in the capacity of routes of gas exchange, which are essential to the growth of most plants.

Pressure Wave Propagation Velocity

The velocity of propagation of a pressure wave (C) through a solid, elastic medium can be described by (Eshbach, 1952):

$$C = \sqrt{\frac{B + 4G/3}{\rho}} \tag{2.20}$$

in which C = propagation velocity (m/s), B = bulk modulus of the medium (Pa), ρ = density of the medium (kg/m^3), and G = shear modulus of the medium (Pa). In a long, narrow rod wave propagation velocity is given by:

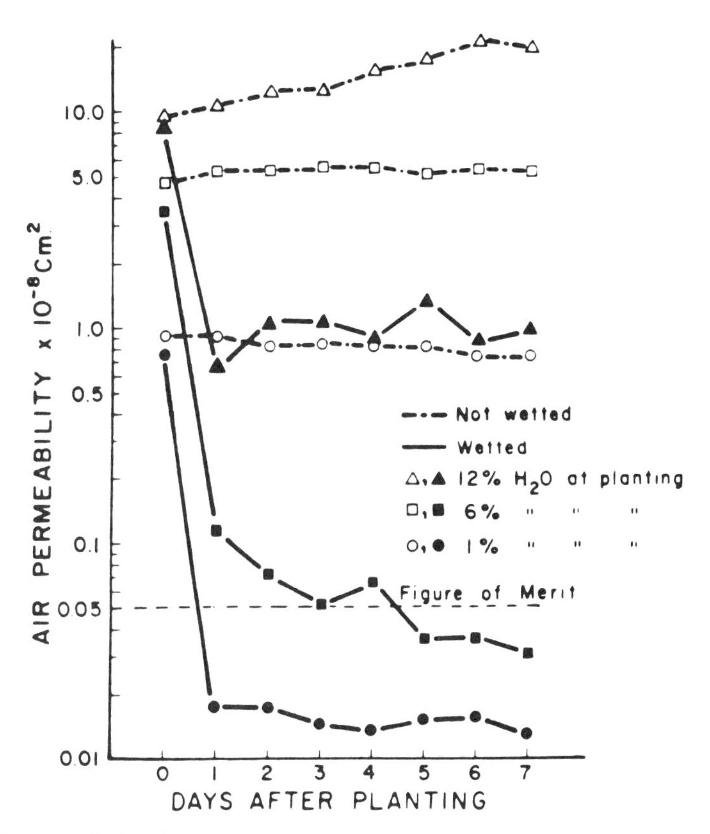

Figure 2.37–Decreases in the air permeability (between day 0 and day 1) as the soil was wetted after planting. Note: At planting high moisture content, soil had higher air permeability than low moisture content soil. It is believed that this is due to greater bulking or bulk density reductions of agitated soil of higher moisture content than for soil of lower moisture content (from Bowen and Liang, 1988).

$$C = \sqrt{\frac{E}{\rho}} \tag{2.21}$$

in which E is Young's modulus.

Thus, the measurement of the propagation velocity provides some information on the physical properties of the soil material. Agricultural soils are somewhat different from typical elastic media in that they contain both liquid and gas fractions and that these fractions are free, to a certain extent, to move relative to the solid fraction and to each other. These complicating factors can be addressed analytically (Brutsaert, 1964); however, it is possible to establish some direct links between propagation velocity and physical soil characteristics.

The general method used to establish the values for propagation velocity is to place two accelerometers a measured distance from each other in the soil and then send either a singular pulse or a vibrating pulse through the soil. The difference in the time necessary

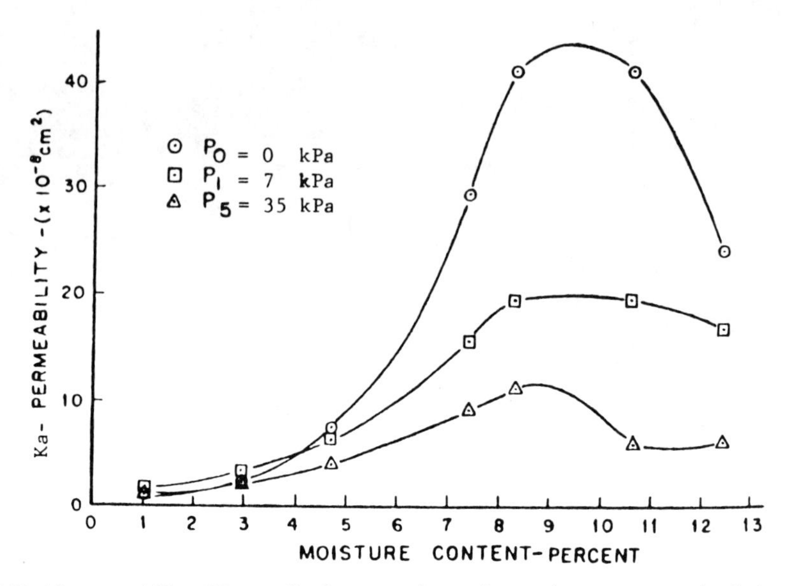

Figure 2.38–Air permeability of Ruston fine loamy sand at various moisture contents after being agitated and then compacted by pressures of 0, 7, and 35 kPa (from Bowen, 1985).

for the sound signal to arrive at the two accelerometers is related to the propagation velocity. Gupta and Pandya (1967) compared the stress applied to a moving flat plate, which generated the test pulses, to the average draft per unit furrow cross-sectional area of moldboard plow bottom, and found a good correlation.

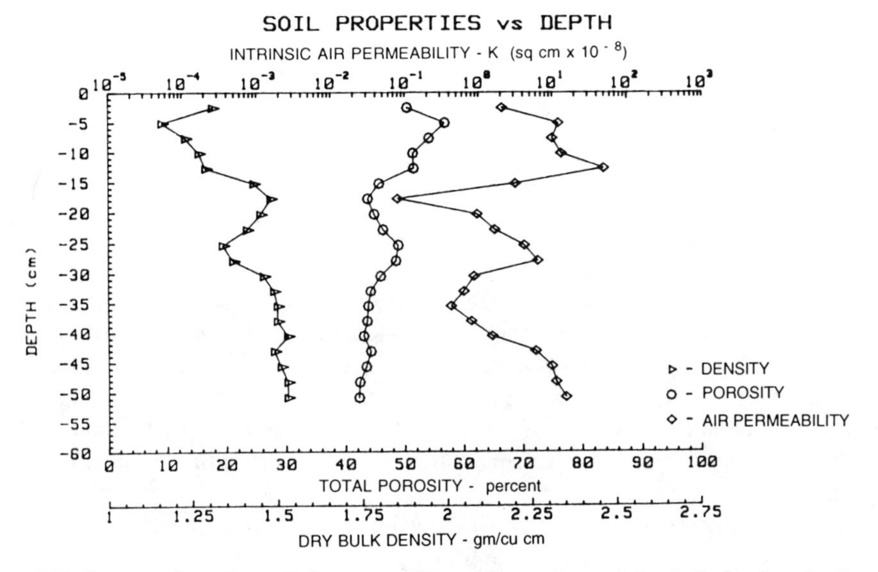

Figure 2.39–Comparative values of air permeability, soil porosity, and dry bulk density of soil core samples taken at varous depths on an uncompacted test plot (from Morgan et al, 1988).

$$\sigma_c = \rho CV \text{ and } \sigma_c = \frac{P}{wd} \tag{2.22}$$

where

σ_c = compressive stress on plate
ρ = wet soil density
C = propagation velocity of the pressure wave in the soil
V = velocity of the translating plate
P = draft force for plow bottom
w = width of furrow
d = depth of furrow

Compression wave velocities ranging from 97 to 250 m/s were measured with values decreasing with increasing moisture content and with increasing plate velocity. Later Krishna and Gupta (1972) developed equations to relate the compression wave propagation velocity to the tool velocity, the moisture content, and the wet density of a single soil. Their results covered velocities ranging from 40 to 400 m/s.

DeRoock and Cooper (1967) measured compression wave propagation velocities in the range from 200 to 700 m/s for a broad range of density and moisture content conditions of a single soil. They found a linear relationship between wave propagation velocity and penetrometer resistance that held true irrespective of moisture content or density, even though wave propagation velocity varied greatly with density at low moisture content and very little with density at high moisture content.

$$s = -3.94 + 0.0226 \times C \tag{2.23}$$

where

s = penetrometer resistance (MPa)
C = compressive wave propagation velocity (m/s)

Using monolithic field core samples of soil with accelerometers bonded to each end of each sample, Kocher and Summers (1988) passed cyclical pulses through the soil samples. From the phase angle shift between the top and bottom ends and the resulting instantaneous ratios of sample top acceleration values to those on the sample bottom, they were able to determine which of several stress-strain-time models best fitted the physical characteristics of the soil. All models involved the quasistatic determination of the equivalent of Young's modulus (E). They found that the most representative form was a second order viscoelastic one:

$$\sigma = E\varepsilon + \alpha \frac{d\varepsilon}{dt} + \xi \frac{d^2\varepsilon}{dt^2} \tag{2.24}$$

where

σ = magnitude of stress wave at any instant (Pa)
E = the elastic modulus in the axial direction (Pa)
ε = unit strain in the axial direction
α and ξ = viscoelastic constants (Pa • s and Pa • s^2)
t = time

It was found that the viscoelastic constants could be represented by:

$$\alpha = e^{a1} (\omega)^{b1}$$
$$\xi = e^{a2} (\omega)^{b2}$$

in which e = the natural logarithmic base (2.71828__), ω = frequency (radians/s), a_1, b_1 = numerical constants related to α, and a_2, b_2 = numerical constants related to ξ.

Soil core samples were taken in both the vertical and horizontal direction in the field. The mean values found in tests for the numerical constants that applied to one soil at one moisture and density condition were as shown in table 2.3. For all but a_1, the constants were significantly different for horizontal samples than for vertical ones.

Womac, et al. (1988) excited three different soils in torsion until resonance occurred for both remolded and minimally disturbed soil core samples 36 mm in diameter and 72 to 83 mm in length. From the physical parameters of the test apparatus, it was possible to compute the shear modulus of the soil. At resonance, shear strain amplitudes ranged from 5×10^{-5} to 12×10^{-5} cm/cm. Shear modulus values which decreased with increasing shear strain amplitudes, ranged from 13 to 38 MPa. Damping ratio values increased with strain amplitude and ranged from 0.9 to 4%. Vibration responses of minimally disturbed specimens indicated greater shear modulus values (particularly at high strain amplitudes) than was the case for the remolded samples. Resonant frequencies ranged from 181 to 238 Hz.

Dielectric Constant

The capacitance of a capacitor is a measure of the amount of charge difference that can be maintained between two parallel plates for a potential difference of 1 V. This charge difference is a function of the area of the plate (A), the thickness of the material separating the plates (d), and the properties of the material separating the plates. These properties are represented by the product of two values, a constant which represents the

Table 2.3. Mean values for numerical constants for soil core samples taken in horizontal and vertical directions

Numerical Constant	Vertical Samples	Horizontal Samples
a_1	10.88	11.43
b_1	–0.289	–0.322
a_2	15.05	16.06
b_2	–1.558	–1.638

properties of a vacuum or of free space (ε_v) and a constant which accounts for the differences in properties from those of a vacuum, for any given material (ε_r).

$$C = \varepsilon_r \varepsilon_v \frac{A}{d} \tag{2.25}$$

where

C = capacitance of the capacitor

$$\left(\text{farads or } \frac{\text{coulombs}}{\text{volt}}\right)$$

A = area of capacitor plate (m^2)
d = thickness of material between the plates (m)
ε_v = eielectric constant for free space (farads/m)
ε_r = relative dielectric constant

Thus, the dielectric constant for some material other than free space can be measured, in an ideal sense, by determining the capacitance of a capacitor in a vacuum (C_v) and then determining the capacitance with the other material between the plates of the same capacitor (C_m). The dielectric constant (ε_r) is then the ratio:

$$\varepsilon_r = \frac{C_v}{C_m} \tag{2.26}$$

The product $\varepsilon_r \times \varepsilon_v$ is also referred to as the permittivity of the material. For alternating potentials, and particularly at very high frequencies, the relative complex permittivity, ε_r, is considered to consist of two components (Nelson, 1983):

$$\varepsilon_r = \varepsilon_r' - j\,\varepsilon_r'' \tag{2.27}$$

where

ε_r' = the term frequently called the dielectric constant
ε_r'' = the dielectric loss factor
j = $\sqrt{-1}$ (representing complex notation)

Because water molecules are dipoles, water exhibits considerably higher permittivity than free space. For pure water the dielectric constant, ε_r, is approximately 80 (Taylor and Ashcroft, 1972; Truman et al, 1988). When water freezes, its dielectric constant drops from 80 to about 3.0 as a result of the immobilization of the molecules (Hayhoe et al, 1986). Dielectric constants, ε_r', for dry sand, loam and clay are 5.0, 2.5, and 2.6 respectively (Truman et al, 1988). In addition to texture, the dielectric constants for soils are affected by dry bulk density, temperature, frequency, and by the moisture content

(Topp et al, 1980; Hallikainen et al, 1985; Wong and Schmugge, 1980). The square root of the dielectric constant, ε_r', (and the cube root, as well) tends to be linearly correlated with the density of air-particulate mixtures (Nelson, 1983). However, it is the major effect of moisture content on the dielectric constant of soils which has resulted in many investigators making efforts to link microwave phenomena associated with soils—to the remote or nonintrusive determination of soil moisture content, both at the surface and at some depth within the soil (Wallender et al, 1985; Jackson et al, 1984; Parchomchuk and Wallender, 1986; Truman et al, 1988; Wheeler and Duncan, 1984). A generalized relationship between the dielectric constant for mineral soils, ε_r', and the volumetric water content (in decimal form), θ_v, in the frequency range from 1 MHz to 1 GHz (Krishna and Gupta, 1972) is:

$$\varepsilon_r' = 3.03 + 9.3\left(\theta_v\right) + 146\left(\theta_v\right)^2 - 76.7\left(\theta_v\right)^3 \qquad (2.28)$$

Several methods of measuring the dielectric constant of soils have been used, and measurements made with a given method do not always agree with those made by other methods (Topp et al, 1980). Time-domain-reflectometry was used by Topp et al (1980) with frequencies ranging from 1 MHz to 1 GHz. They used an instrumentation system (fig 2.40) that incorporated a cylindrical sample holder (fig 2.41) with a sample diameter of 5 cm. Hallikainen et al (1985) used both a waveguide transmission technique and a free-space transmission technique over frequencies ranging from 3 GHz to 18 GHz and found good agreement between the two methods at 6 GHz. The free-space transmission technique (fig 2.42) was better adapted to the higher frequency levels. This method used as a sample a slab of soil 30 cm in diameter and 3.7 cm in thickness.

Electrical Conductivity or Resistivity

There are a number of reasons for which electrical currents are caused to flow through agricultural soil. Among these are:

1. Detection of the depth of various soil strata (Bethlahmy and Zwerman, 1959; Carter, 1970)
2. Heating of soil for sterilization (Tavernetti, 1935)
3. Recirculation of current imbalances from electrical power branch circuits (Gustafson et al, 1987; Surbrook et al, 1982)
4. Electroreclamation of saline soils (Gibbs, 1966; Zareian, 1989)
5. Electro-osmosis to reduce implement draft (Weber, 1932; Gerlach, 1953)
6. Conductance current flowing in response to streaming potentials caused by water-flow in soil (Edwards and Monke, 1968)
7. Sensing of soil moisture content (Carter, 1970; Freeland, 1989)

A property of the soil that describes its current-carrying characteristic is its resistivity (ρ) or electrical conductivity ($1/\rho$). The resistance that a material will display to the passage of electrical current may be measured from:

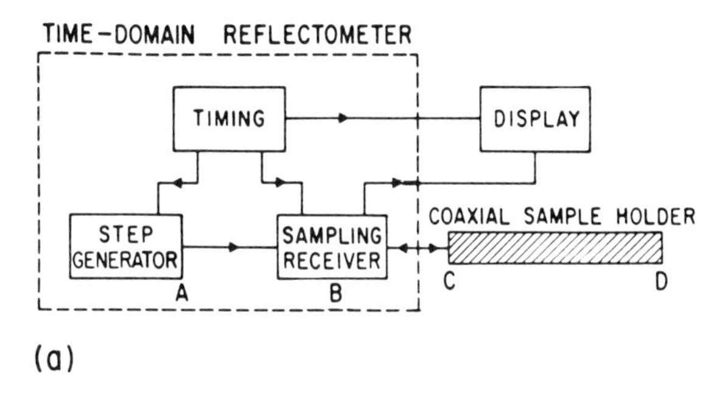

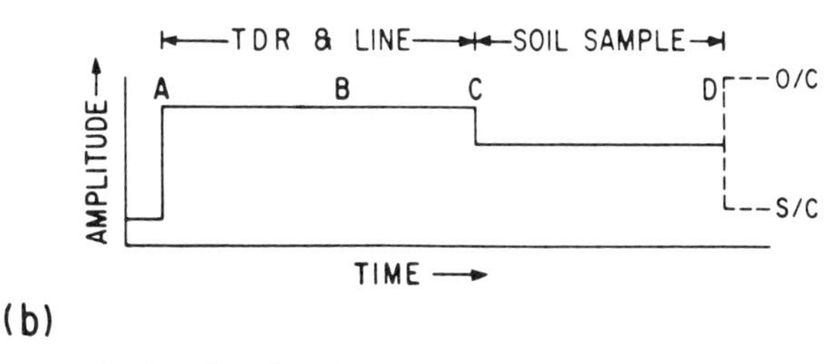

Figure 2.40–Block diagram of a time-domain reflectometer connected to a soil sample holder (a) and an idealized representation of the time-domain reflectometer output from measurement on a soil sample (b); time interval C-D represents the travel time in the soil sample; O/C, S/C indicate open circuit and short circuit, respectively (from Topp et al, 1980).

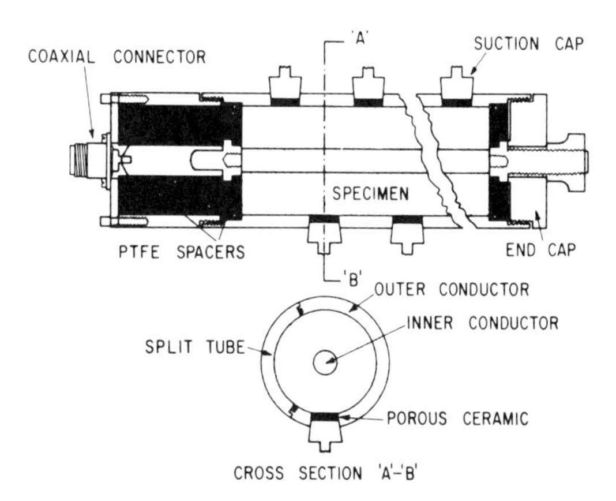

Figure 2.41–Diagram of a coaxial transmission line soil column showing the position relationship of the ceramic-capped cups (from Topp et al, 1980).

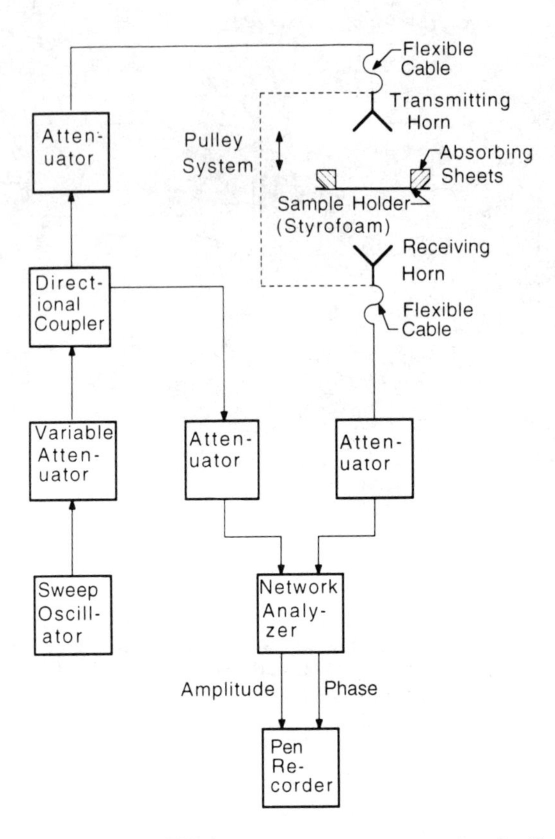

Figure 2.42–Block diagram of a 4- to 18- GHz free-space measurement system for the dielectric constant of a soil sample (from Hallikainen, © 1985, IEEE).

$$\text{resistance (ohms)} = \frac{\text{E (volts)}}{\text{I (amperes)}} \tag{2.29}$$

Because resistance is usually in proportion to the length of the electrical path through the material and inversely proportional to the cross-sectional area of that path, a characteristic parameter of the material, resistivity (ρ), may be defined by:

$$\rho\ \text{(ohm-cm)} = \text{resistance (ohms)} \times \frac{\text{area}\left(\text{cm}^2\right)}{\text{length (cm)}} \tag{2.30}$$

There have been some general studies on earth resistance (Tagg, 1964; Freeland, 1989), while others (Tavernetti, 1935) have examined agricultural soils in particular. It has been found (Tavernetti, 1935) that the resistivity of agricultural soils increases with: (1) decreases in moisture content (fig 2.43); (2) decreases in soil density (fig 2.44); (3) decreases in soil temperature (figs 2.43 and 2.44); and (4) decreases in the salt content of the soils (fig 2.45). Furthermore, different soil textures exhibit different levels of resistivity even when at the same moisture content, temperature, and density levels (Tavernetti, 1935). Frozen soil exhibits different resistivity values than does the same soil in a nonfrozen condition (Bethlahmy and Zwerman, 1959).

The general method for determining resistivity of soil is by placing electrodes of a given surface area (or connecting with a soil body of a given cross-sectional area) in the soil a fixed distance apart. Then a voltage is applied to the electrodes, and the resultant current is measured. Both alternating and direct currents have been used, but the use of direct current gives rise to both moisture and cation movement in the soil carrying the current (Gibbs, 1966; Zareian, 1989), both of which cause localized changes in resistivity. If large currents are used there is a heating of the soil and a subsequent evaporation of moisture. It has been found (Tavernetti, 1935) that there is also a certain amount of resistance associated with the interface between the electrode and the soil. In one case this resistance—normalized according to the reciprocal of the soil-electrode interface area (Tavernetti, 1935)—amounted to 8648 ohm-cm^2, 2279 ohm-cm^2 and 1462 ohm-cm^2 at temperatures of 10, 60, and 100° C, respectively. In other cases interface resistances considerably lower than these were observed. In a number of cases, minor changes in the moisture content or salt content of the soil immediately adjacent to the electrode surface caused major changes in the total resistance measured.

The usual approach to minimizing the effect of variations of interface resistance between electrode and soil during measurements of soil resistivity is to use a four-electrode method (Freeland, 1989). With the four-electrode method four equally spaced electrodes are arranged in a line in the soil. The current (usually alternating current) is passed between the two electodes at the ends of the line. The potential difference is then measured between the two inner electrodes using a high-impedance voltmeter. However, when alternating current at high frequency was used, the apparent conductivity increased

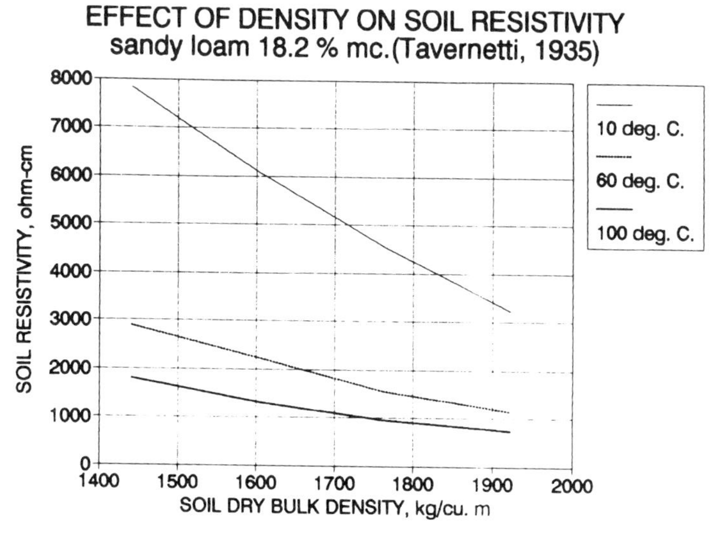

Figure 2.43–Effect of soil density on soil resistivity at three different temperatures (from Tavernetti, 1935).

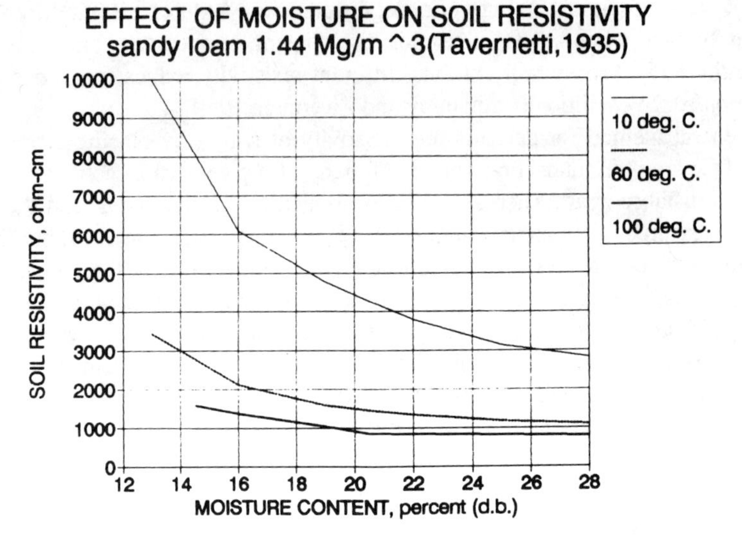

Figure 2.44–Effect of soil moisture content on soil resistivity at three different temperatures (data from Tavernetti, 1935).

rapidly with frequency once the frequency of 800 MHz was reached (Parchomchuk and Wallender, 1986).

In one nonsaline soil, resistivities ranged from 1923 ohm-cm at 22.3% moisture to 1324 ohm-cm at 30% moisture (Shonk and Gaultney, 1989). Resistivities measured in another nonsaline surface soil ranged from 487 to 1134 ohm-cm while those for dense fragipan materials underlying these soils ranged from 1460 to 1562 ohm-cm (Bethlahmy

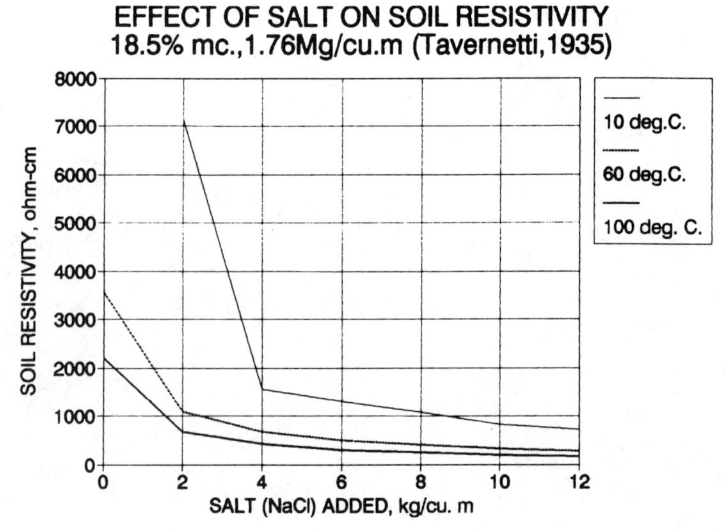

Figure 2.45–Effect of salt concentration on soil resistivity at three different temperatures (data from Tavernetti, 1935).

and Zwerman, 1959). On the other hand, values for a saline-alkalai soil ranged from 152 to 228 ohm-cm (Gibbs, 1966), with values tending to increase as the long-term application of direct currents tended to leach the sodium ions from the soil, allowing them to be replaced by less mobile calcium ions. This type of ion replacement has numerous effects on the physical properties of the soil; one effect is to reduce the force and energy values associated with soil tillage (Schaefer et al, 1989).

If water is allowed to move toward a drain at the cathode while a direct current is passed through the soil and soil moisture is not replaced, soil resistivity increases greatly as the moisture content decreases. In one case with saline soils the electro-osmosis process resulted in an increase in resistivity from 810 ohm-cm at a field-capacity moisture content to a range between 6670 and 12,000 ohm-cm as the soil dried (Zareian, 1989). This same sort of electro-osmosis-driven soil moisture movement resulted in a reduced friction coefficient between soil and steel (due to lubrication by moisture) and resulted in a 19% reduction in plow draft when a direct current level of 1 A was used for a single plow bottom moving at 1 m/s (Weber, 1932; Gerlach, 1953).

Electromagnetic Wave Transmission, Adsorption

The following is an excerpt from "Measuring Soil Moisture Electromagnetically" by Paul A. Wheeler and Graeme L. Duncan (1984).

> **Electromagnetic Waves**
>
> Electromagnetic waves consist of electric and magnetic fields oscillating in mutually perpendicular planes. In a vacuum, the wave travels at the same velocity for all regions of the electromagnetic spectrum: the speed of light. Typical of all waves, electromagnetic waves have four basic characteristics, amplitude, frequency, phase, and polarization. To measure soil moisture using electromagnetic waves, one or more of these basic characteristics must change between dry and wet soil.
>
> The amplitude of an electromagnetic wave is measured as the field strength of the wave. Field strength measurements can be passive or active. All objects emit electromagnetic energy. A passive system simply measures the field strength of the emitted energy at different frequencies. An active system uses a separate electromagnetic source or transmitter aimed at the soil. The field strength of reflected, transmitted, or absorbed energy is measured.
>
> The electromagnetic spectrum extends from frequencies close to zero (direct current) to those emanating from cosmic ray interactions with atmospheric nuclei at 10^{23} Hz. Table 2.4 shows frequency allocations and wavelengths for the electromagnetic spectrum. The spectrum is divided into many bands, each having unique characteristics. The bands are combined into seven groups that show similarities in interactions.
>
> The phase characteristic of electromagnetic waves is not readily used for detecting soil properties.

Polarization varies with reflection and transmission. A wave can be linearly polarized (horizontal or vertical), elliptically polarized, circularly polarized, or not polarized.

Electromagnetic Properties of Soil

Three electromagnetic properties are important in describing the interactions of electromagnetic waves with conductors and dielectrics: conductivity, permittivity, and permeability.

Conductivity in the soil depends on its ion content. Conduction currents in the soil result from motion of free electrons. Conductivity can be defined for fields using Ohm's law as the ratio of the conduction current to the electric field. Conductivity is important in soils for frequencies below about 1 MHz. In this region, soils act as conductors.

Permittivity is the ratio between the electric displacement density and the electric field strength. The relative permittivity is known as the dielectric

Table 2.4. Frequency allocations for electromagnetic spectrum*

Band	Frequency			Wavelength			Designation
1	3	–	30 Hz	100	–	10 Mm	AF
2	30	–	300 Hz	10	–	1 Mm	ELF
3	300	–	3000 Hz	1000	–	100 km	VF
4	3	–	30 kHz	100	–	10 km	VLF
5	30	–	300 kHz	10	–	1 km	LF
6	300	–	3000 kHz	1000	–	100 m	MF
7	3	–	30 MHz	100	–	10 m	HF
8	30	–	300 MHz	10	–	1 m	VHF
9	300	–	3000 MHz	1000	–	100 mm	UHF
10	3	–	30 GHz	100	–	10 mm	SHF
11	30	–	300 GHz	10	–	1 mm	EHF
12	300	–	3000 GHz	1000	–	100 μm	EEHF/IR
IR	30	–	50 THz	100	–	6 μm	Far IR
	50	–	200 THz	6	–	1.5 μm	Middle IR
	200	–	400 THz	1500	–	750 nm	Near IR
VIS	400	–	482 THz	750	–	623 nm	Red
	482	–	502 THz	623	–	598 nm	Orange
	502	–	519 THz	598	–	578 nm	Yellow
	519	–	609 THz	578	–	493 nm	Green
	609	–	658 THz	493	–	456 nm	Blue
	658	–	811 THz	456	–	370nm	Violet
UV	811	–	1000 THz	370	–	300 nm	Near UV
	1	–	1.5 PHz	300	–	200 nm	Far UV
	1.5	–	3 PHz	200	–	100 nm	Extreme UV
X-ray	3 PHz	–	30 EHz	100		0.01 nm	–
Gamma	3	–	3000 EHz	0.1		0.0001 nm	–

* From Wheeler and Duncan, 1984.

constant. The dielectric constant depends on soil texture as well as soil moisture. As water is first added to dry soil, it is tightly bound to the soil, not substantially increasing the dielectric constant. Further addition of water produces free water that increases the dielectric constant of the soil by an order of magnitude. The change in dielectric constant is the main discriminator for soil moisture. It is predominant for frequencies above 1 MHz.

Permeability is the ratio of magnetic flux density to magnetic field strength. The permeability values for soil are close to the values of free space except when traces of ferrous materials are found in the soil. It is not a strong discriminator for detecting changes in soil properties.

Electromagnetic Interactions with Soil

When an electromagnetic receiver is aimed at a sample of soil, it receives electromagnetic energy coming from the soil. The receiver is usually limited to a small band of frequencies in the electromagnetic spectrum. The energy coming from the soil is from emission of the soil itself, reflection from an external source, transmission through the soil, and reradiation of energy earlier absorbed. Transmission is negligible due to the size of the earth. At thermal equilibrium, absorption and re-radiation can be neglected.

The emissivity of soil is defined as the rate of radiant energy emitted from the soil divided by the rate of radiant energy from a black body at the same temperature. The emissivity depends on roughness and mineral composition of the soil. In the infrared region, the emissivity is primarily a surface effect. In the microwave band, the body as well as the surface emits radiant energy. This complex emissivity depends on many soil properties, including permittivity.

A wave traveling in air and striking a perfect conductor is completely reflected. The angle of reflection is equal to the angle of incidence. Because there is always some impedance in the soil, some of the energy will be absorbed, reducing the signal strength of the reflected signal.

A wave impinging on a pure dielectric is partially reflected and partially transmitted. The reflected wave obeys the law of reflection (as does a conductive surface), whereas the transmitted wave is refracted, obeying Snell's law. The reflected wave changes in amplitude, phase, and polarization. The reflection coefficient is a function of permittivity and permeability, so it changes as water is added to dry soil. If the surface of the soil is smooth compared to the wavelength of the incident wave, specular reflection occurs. If the surface is rough, diffuse reflection occurs, varying the expected angle of reflection. In addition to reflection from the surface, subsurface reflections can occur from layers within the soil.

Frequency Dependence of Electromagnetic Waves

Each of the seven frequency regions is discussed separately.

Very Long Wavelengths

Included here are all frequencies below 30 MHz. The physical size of an antenna used can be very large, ranging from tens of meters to many kilometers. The earth acts as a perfect reflector at these frequencies, with a depth of penetration ranging from about 2 m (6.6 ft) to thousands of meters, varying as a function of frequency. Higher frequencies penetrate less. This frequency region is not practical for on-farm use because of the size of antennas needed. It is well suited for total soil moisture detection over very large regions.

VHF

This band is also useful for total soil moisture measurements. In addition, the antenna size is such that it is practical for on-farm measurements. The depth of penetration is not very dependent on frequency and ranges from 2 m (6.6 ft) for good (high conductivity and high permittivity) earth to about 6 m (19.7 ft) for poor earth. This is the range desired for total soil moisture in the root zone.

Microwaves

In the microwave region, the dielectric constant describes the propagation characteristics of an electromagnetic wave. The dielectric constant of soils is determined by water in the soil; by soil characteristics such as porosity, texture, and composition; and by surface roughness. The signal is further modified by the effects of surface geometry, slope, vegetation cover, and sky and surface temperatures.

Passive Microwave Detection

A microwave radiometer is used as a passive microwave receiver. The radiometer detects both radiation from the earth's surface layer and reflection from sky radiation. The emissivity is determined by the complex dielectric constant, hence, the soil moisture. It is also affected by surface roughness, making the dielectric constant very difficult to calculate. Therefore, a homogeneous surface is assumed. The sky component is influenced by soil emissivity, surface roughness, sky conditions, and vegetation. The radiometer signal itself depends on the sensor frequency and polarization.

Active Microwave Detection

A microwave radar that emits its own microwave signal and then receives the reflected signal is used for active microwave detection of soil moisture. Only the moisture in the top few centimeters of the soil is measured. This method is more applicable to measuring the wetting front than total soil moisture.

Infrared

Two effects are of interest in the infrared band. In the near IR, solar reflectivity measurements are possible. In the middle and far IR soils act like black bodies, and the radiated energy is a measurable quantity. Both reflectivity and radiated energy are greatly affected by meteorological and canopy conditions. Because of heat transfer conditions between water and the soil, it is easier to detect water-soil boundaries during the night than during the day.

Visible

Detecting soil moisture in the visible band is a surface effect only. As an electromagnetic wave in the visible band interacts with soil, some of the wavelengths are reflected while other wavelengths are absorbed. The reflected wavelengths are considered the color of the soil.

An increase in water content of the soil will change the color of the soil. Soils generally darken on wetting. However, some soils become lighter upon wetting, whereas others show little change in color. Dark colored soils show a greater change in color than light colored soils.

In addition to color itself, the total reflectance can be used to detect soil properties. Dry surfaces have greater reflectance values than do wet surfaces. There are many other effects in the visible band. Weathered surfaces exhibit less reflectance with increasing wavelength than do fresh surfaces. Vegetative cover gives a composite reflectance. It also provides shadows. The atmosphere acts as a filter that affects the wavelengths of sunlight reaching the soil.

Polarization can be used to detect soil moisture in the visible band, because materials of low reflectance polarize strongly. Polarization also depends on wavelength and soil particle size. In addition, depolarization can be used for soil moisture discrimination.

Ultraviolet

Soil is not a strong reflector of ultraviolet radiation. There is an extreme scattering of UV in the atmosphere so that much of the solar UV is filtered out before it reaches the earth. The ultraviolet band offers no advantage over the infrared and visible bands for soil moisture detection. It may have some application in a multi-spectral approach.

Ionizing Radiation

Except for laboratory research, X-rays offer no practical means of remote detection of soil moisture. Gamma attenuation techniques have been successfully used for in-situ soil moisture monitoring. Remote monitoring using gamma rays depends on a gamma flux emitted from radio isotopes in the soil. Because soil density is increased by addition of water, the gamma rays are attenuated in wet soil.

When an electric current is passed through a conductor, a magnetic field is formed with lines of flux encircling the conductor. In one case where such a magnetic field was to be used for tractor guidance in an agricultural field (Pichon and Steinbruegge, 1965), the magnetic field surrounding a conductor carrying 241 mA at 20 kHz was measured in air and with the conductor buried at various depths. It was found that when the magnetic field strength was measured at a distance of 91.5 cm above the conductor, the field strength decreased very little with conductor burial (about 0.16% reduction in field strength with each centimeter of burial depth).

Images of signals obtained using ground-penetrating radar operating at 120 MHz (Truman et al, 1988) permitted the discerning of the depth of clay layers and of the water table when these depths were as much as 10 m below the soil surface. This apparatus sensed abrupt changes with depth in the dielectric constant of the soil materials.

At microwave frequencies the dielectric constant is usually calculated from the impedance of the material to an electromagnetic wave passing through it. For a plane electromagnetic wave propagating at normal incidence through a material (Parchomchuk and Wallender, 1986), the relationship between impedance, Z, and dielectric constant is:

$$Z = \eta_o / \varepsilon^{1/2} = \eta_o / \left[\varepsilon' \left(1\text{-tan } \delta\right)\right]^{1/2} \tag{2.31}$$

in which η_o = the impedance of free space (about 377 ohms), and $\varepsilon = \varepsilon' - j\,\varepsilon''$ in which j = $(-1)^{1/2}$, ε' = relative permittivity, frequently called the *dielectric constant*, and ε'' = $\sigma/2\,\pi\,f\,\varepsilon_o = \varepsilon' \tan\delta$; in which ε_o = permittivity of free space (about 8.854×10^{-2} F/m), σ = electrical conductivity, and f = frequency (Hz).

Thus, $\tan\delta = \varepsilon''/\varepsilon' = \sigma/(2\,\pi\,f\,\varepsilon_o\,\varepsilon')$ and is called the loss tangent. For a given soil the ratio $\varepsilon''/\varepsilon'$, or the value $\tan\delta$, does not vary appreciably with moisture content and may be considered to be a constant over a narrow frequency range. For soils, $\tan\delta$ ranges from 0.05 to 0.30 at 1 GHz, depending on conductivity and texture (Parchomchuk and Wallender, 1986). Individual determinations of ε' and ε'' on various soils at various frequencies and moisture contents are presented by Hallikainen et al (1985).

The impedance is most commonly measured from the reflection coefficient, R, which is the ratio of reflected to incident power of an electromagnetic wave at the interface between two dissimilar materials with impedances Z_1 and Z_2.

$$R = \left(Z_2 - Z_1\right) / \left(Z_2 + Z_1\right) \tag{2.32}$$

The value for the imaginary part of the complex relative dielectric constant ε'', and thus the determination of values for $\tan\delta$, can also be obtained during tests to measure the real part of the dielectric constant, ε'.

The depth at which the intensity of the electromagnetic field decreases to 1/e (about 37%) of its value at the soil surface is defined as the skin depth, d, and may be expressed by:

$$d = 1/\pi \, f \, \eta_o \left[2 \, \varepsilon' \left(1 + (\tan \delta)^2 \right)^{1/2} - 1 \right]^{1/2} \quad (2.33)$$

Thus, the skin depth is less for wet or dense soils as compared to dry or less dense soils and is decreased by increasing frequency (Parchomchuk and Wallender, 1986). For a dry soil at 0.1 GHz skin depth, d was approximately 2.4 m, while on a moist soil at 1.0 GHz skin depth, d was about 0.3 m.

The velocity of propagation of an electromagnetic wave in a transmission line, V, is (Marshall and Holmes, 1988):

$$V = C \left[1/2 \, \varepsilon' \left(1 + \left(1 + \tan^2 \delta \right)^{1/2} \right) \right]^{-1/2} \quad (2.34)$$

in which C = the velocity of light (about 3×10^8 m/s).

The emissivity of the soil surface for electromagnetic radiation in the 0.3- to 3-GHz (microwave) range tends to be negatively correlated with volumetric moisture content of the soil (Schmugge, 1983). Emissivities range from between 1.0 and 0.90 for very dry soils to between 0.6 and 0.7 for very moist soils.

The soil reflectivity of near-infrared radiation (1600 to 2700 nm or about 100 to 200 THz) has been used to assess both soil moisture content and soil carbon content (the latter as an indicator for soil organic matter content) (Sudduth et al, 1989). The reflected radiation was scanned using band-width segments of 40 and 60 nm to obtain the detail required for accurate prediction of both moisture and carbon contents. Reflectance values ranged from 0.1 to 0.5, with higher values being associated with lower levels of both moisture and organic matter. There was a strong absorption band centered at approximately 1920 nm.

Reflectance of visible light, 750 to 370 nm or 400 to 800 THZ, was also used to detect organic matter content of soil (Shonk and Gaultney, 1989). Correlations were influenced by the soil type and moisture content but were generally linear in the 0.5 to 5% organic matter range. Reflectance values ranged from 5 to 45% of that of a $BaSO_4$ standard.

Polarized light from a He-Ne laser (632.8 nm or 474 THz) was reflected from a soil surface (Zhang et al, 1989). It was found that soil moisture content could be predicted from the degree of depolarization of the light, while the degree of depolarization was not affected by the dry bulk density of the soil. However, an index, a, which correlated well with dry bulk density, was formulated as follows:

$$a = \frac{A_{F_d}}{1 - k} \quad (2.35)$$

in which A_{F_d} = range of variation of the luminous flux, and $k = R_{\parallel} / R\perp$; in which $R_{\parallel}$ = intensity reflection coefficient of the component parallel to the plane of incidence, and $R\perp$ = intensity reflection coefficient of the component perpendicular to the plane of incidence.

The index, a, was only slightly influenced by moisture content.

Computerized tomography (CT) scanners produce images by computing the x-ray (0.01 nm to 100 nm or 3 PHz to 30 E Hz) attenuation coefficient on a millimeter-by-millimeter basis over a preselected prism contained within the object under study (Tollner and Verma, 1987a). The customary units for x-ray CT (Hounsfield units) are computed by:

$$H_{(x,\,y)} = \left[\frac{\mu(x,\,y) - \mu_w}{\mu_w}\right] \times 1000 \qquad (2.36)$$

in which $H_{(x,\,y)}$ = computed Hounsfield units as a function of position, $\mu_{(x,\,y)}$ = x-ray attenuation coefficient as a function of position (L^{-1}), and μ_w = x-ray attenuation coefficient of water (L^{-1}).

For soil:

$$\mu_{(x,\,y)} = \mu_s \rho_b + \mu_w \, \theta_v \qquad (2.37)$$

in which ρ_b = dry bulk density of soil solids (ML^{-3}), μ_s = x-ray attenuation of soil solids (L^{-1}), θ_v = volumetric moisture content ($L^3\,L^{-3}$).

For a 7 × 9 × 22 cm soil test volume, x-ray unit settings used were 120 kV and 1881 MAS x-ray intensity (Tollner and Verma, 1987a). Hounsfield units and volumetric moisture content were positively correlated, and the moisture content profile could thus be tracked satisfactorily as drying occurred over an 8-day period to depths of approximately 0.2 m.

The attenuation of gamma radiation has been used to measure both soil moisture content (Smith et al, 1967; Ligon, 1969) and soil density (Gameda et al, 1987). The basic principles involved in the attenuation of monoenergetic gamma radiation passing through a soil can be expressed as follows (Gameda et al, 1987):

$$\ln I = \ln I_o - X\left(\mu_s\,\gamma + \mu_w\,\theta\right) \qquad (2.38)$$

in which

I = count rate through soil
I_o = unattenuated count rate
X = soil thickness (m)
μ_s = mass attenuation coefficient of soil (m^2/Mg)
μ_w = mass attenuation coefficient of water (m^2/Mg)
γ = soil dry bulk density (Mg/m^3)
θ = volumetric soil moisture content (m^3/m^3)

At the particle energy level of 0.661 MeV associated with gamma rays from Cs^{137}, the mass attenuation coefficients for hydrogen, oxygen, carbon, and water are

15.38 m^2/Mg, 7.75 m^2/Mg, 7.74 m^2/Mg, and 8.45 m^2/Mg, respectively. A typical mass attenuation coefficient for a moist soil would be on the order 7.75 m^2/Mg (Ligon, 1969).

One of the most common sources of gamma radiation is the radio isotope Cs^{137}. The particle energy of 0.661 MeV associated with gamma radiation from this source is linked to a wave length of 0.01 nm and a frequency of 30 EHz.

Airborne measurements of the natural gamma flux can be made to determine soil moisture. Two difficulties in using this method are caused by the nonuniformity of radioisotopes in the soil and interference from radon gas in the atmosphere (Wheeler and Duncan, 1984).

Because the nucleus of an atom behaves similarly to a spinning bar magnet, it is possible to add energy to certain atomic nuclei by exciting the atom with a sequenced electromagnetic pulse of exactly the correct frequency in combination with an external nonvarying magnetic field. This phenomenon is called nuclear magnetic resonance (NMR). For the hydrogen atom the appropriate frequency is 27 MHz (Tollner and Rollwitz, 1988). Once the pulsed field is turned off, the spinning nucleii return an electromagnetic signal, called the *free induction decay*. From analysis of the information from the free induction decay signal, it is possible to discern not only the quantity of hydrogen atoms present, but also the degree of molecular bonding of the hydrogen atoms. The volumetric moisture content of agricultural soils tends to correlate linearly to both the NMR free induction decay and NMR spin echo signal strengths (Matzkanin et al, 1984). However, a separate correlation is required for each soil texture if precise interpretation is required.

Temperature

Although it would seem that temperature is a physical property of soil that would be very easy to measure, and for variations in which there would not seem to be changes in soil characteristics, soil temperature and the processes by which temperature changes take place remains one of the areas in which research is still going on to define quantitatively the relationships involved.

Various methods of measuring soil temperature (Taylor and Jackson, 1986a) involve the placement of temperature measuring devices in the soil, while others (Allen and Sewell, 1973; Mitchell, 1988; Lambert and McFarland, 1987) make use of infrared and microwave radiation to estimate composite soil surface temperature. Because of dynamic heat-flux transitions in natural soil situations, temperatures change with time and position in the *in situ* soil matrix (fig 2.46) (Costello and Brand, 1989). These measurements are further complicated by the fact that if the measuring device is initially at a different temperature than the soil, moisture evaporation or condensation adjacent to the device may occur, allowing the latent heat of vaporization to influence the temperature reading for some time until vapor equilibrium is reestablished.

Generally high levels of soil temperature in unsaturated soils are associated with more rapid oxidation of organic matter and the attendant changes in soil properties related to the organic matter content (see section on organic matter). Each organism that decomposes organic materials has its own growth range and optimum. Mesophilic bacteria, actinomycetes, and fungi grow in the range of 0 to 40 or 45° C. Thermophilic organisms grow at temperatures of 45 to 60° C. The optimum growth temperature for

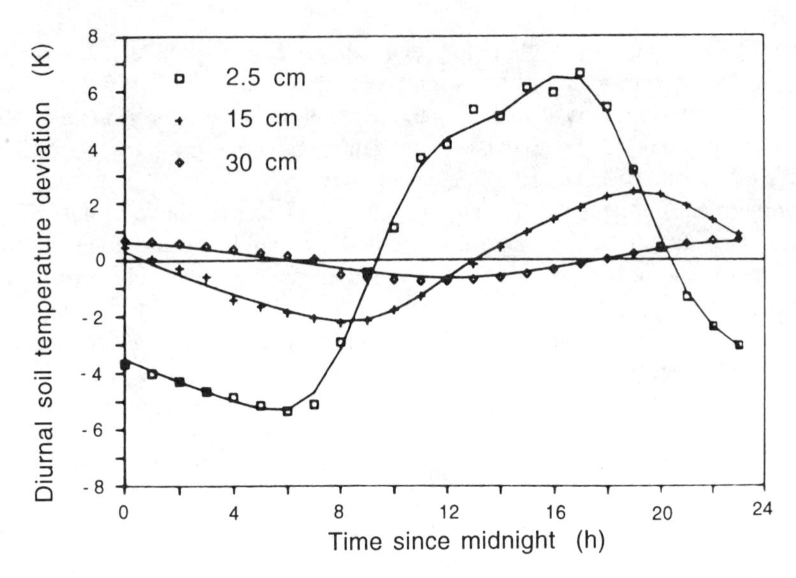

Figure 2.46–Diurnal soil temperature deviations at three depths (from Costello and Brand, 1989).

any organism is commonly within 3 to 8° C of its upper limit of growth. In general, organic decomposition increases two- or threefold for each rise of 10° C in temperature between minimum and optimum temperature (Larson and Gilley, 1976).

Temperature affects two physical characteristics of the water in soils: surface tension and viscosity. Both become lower as temperature increases as is shown in table 2.5.

Minerals in water cause surface tension to be lower than the values shown in table 2.5. As a result, shear strength of soils has been found to decrease as soil temperature increases. A temperature increase of 20° C may decrease strength as much as 10% (Koolen and Kuipers, 1983).

Thermal Conductivity

Temperatures in soil do change, primarily due to heat and vapor fluxes to and from the soil surface (figs 2.47 and 2.48). Three main thermal properties of soil are: (1) volumetric heat capacity, C, $J/m^3 \cdot K$; (2) thermal conductivity, λ, $W/m \cdot K$; (3)thermal diffusivity, α, $\alpha = \lambda/C$. Because the heat capacity of most materials is given on a per-

Table 2.5. Effect of temperature on surface tension and viscosity of water in soil*

Temp C°	Surface Tension (N/m)	Dynamic Viscosity (N-s/m^2)
10	0.0740	1353
30	0.0706	806
50	0.0673	555
70	0.0634	409
90	0.0605	315

* For pure or distilled water (Eshbach, 1952).

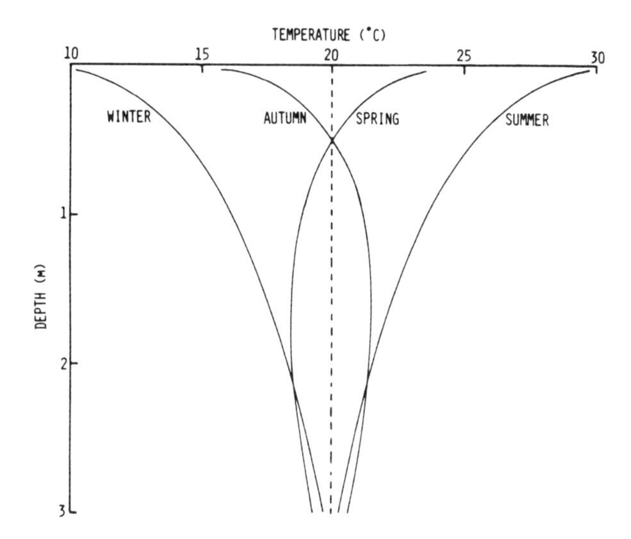

Figure 2.47–Seasonal soil temperature profiles. Reprinted with permission of Academic Press, Inc. from Hillel, 1980. Copyright © 1980, Academic Press, Inc.

unit-mass basis, the density of the material is needed to determine the volumetric heat capacity. Methods for determining the heat capacity of soils on a weight basis involve the use of a calorimeter, in which a known quantity of warm water of a known temperature is mixed with a known weight of cool soil at a known temperature, and the

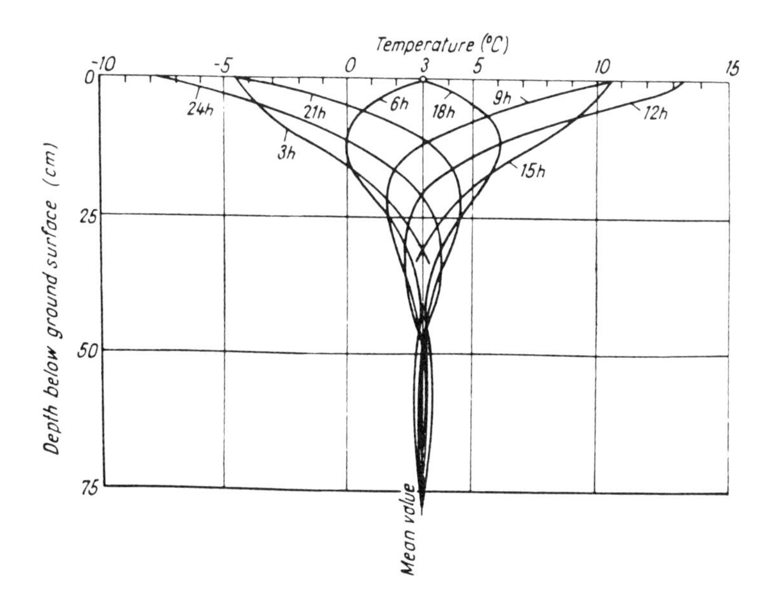

Figure 2.48–Diurnal soil temperature profiles (time given in hours (from Kezdi, 1974).

temperature change history with time is recorded before, during, and after the mixing process (Taylor and Jackson, 1986b).

Typical properties of soil constituent materials are given in table 2.6.

The unidirectional thermal conductivity of a homogeneous material can be derived from the Fourier law of conduction:

$$Q = -\lambda A \frac{\partial T}{\partial x} \tag{2.39}$$

or

$$\lambda = \frac{-Q}{A \dfrac{\partial T}{\partial x}}$$

in which

λ = thermal conductivity cal/cm•s•°K
Q = rate of heat transfer (cal/s)
$\partial T/\partial x$ = temperature gradient (°K/m)
A = cross-sectional area (m^2)

where the negative sign indicates that the heat moves from a zone of higher temperature to one of lower temperature.

With most materials, thermal conductivity can be measured by applying a known temperature difference across a fixed distance of the material and measuring the steady-state energy flow. This method may also be used for soil (Ghildyal and Tripathi, 1987; Tollner and Verma, 1987b), using a cylindrical apparatus (fig 2.49). For this case:

$$\lambda = \frac{Q \cdot \ln\left(r_2 / r_1\right)}{2\pi\, L \left(T_1 - T_2\right)} \tag{2.40}$$

Table 2.6. Thermal properties of soil constituents at 10° C (ice at 0° C)*

Constituent	Density (ρ) (kg/m^3)	Heat Capacity (C) (J/m^3•°K)	Thermal Conductivity (λ) (W/m•°K)
Quartz	2.66×10^3	2.0×10^6	8.8
Other minerals	2.65×10^3	2.0×10^6	2.9
Organic materials	1.3×10^3	2.5×10^6	0.25
Water	1×10^3	4.2×10^6	0.57
Ice	0.92×10^3	1.9×10^6	2.2
Air	1.25	1.25×10^3	0.025

* Data from DeVries and Afgan, 1975.

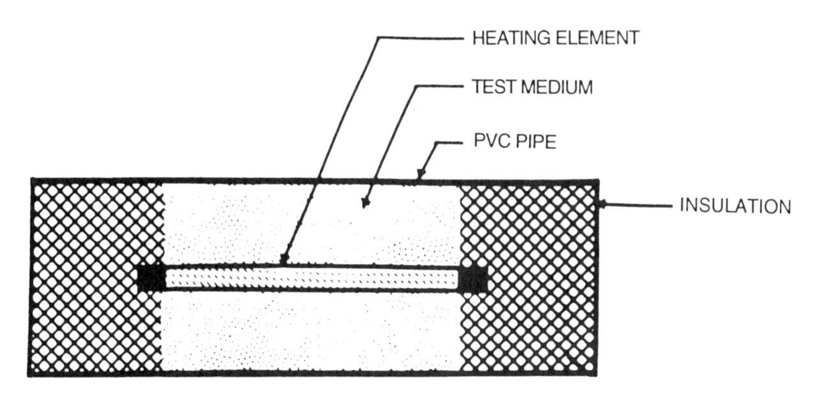

Figure 2.49–Schematic diagram of an apparatus for measuring the thermal conductivity of soil (from Tollner and Verma, 1987b).

in which

λ = thermal conductivity (cal/cm•s•°K)
Q = thermal energy input rate (cal/s)
L = length of the test chamber (cm)
r_1 = radius of the heating element (cm)
r_2 = outer radius of the test medium (cm)
T_1 = temperature of the test medium at the heater surface (°K)
T_2 = temperature of the test medium at its outer radius (°K)

However, the use of this steady-state method allows opportunity for vapor transfer from the zone of high temperature to the zone of low temperature and the resultant decrease of moisture content (due to evaporation) near the high-temperature surface and increase of moisture content (due to condensation) near the low-temperature surface. This vapor transfer not only causes the test medium to be nonhomogeneous but also causes heat transfer (in the form of the latent heat of moisture vaporization), which might not be applicable to cases in which temperature differentials might apply for only very short periods.

This vapor-medium heat transfer can be measured using a device of the type shown in figure 2.50 (Kézdi, 1974). Results (fig 2.51) show that the heat transfer rate by saturated water vapor increases rapidly with temperature, a phenomenon that occurs because the saturated vapor pressure of water (the relative humidity of the air-filled pore space is usually near 99%) also increases rapidly with temperature, as is shown in table 2.7 (Koorevaar et al, 1983):

To overcome some of those limitations the probe method of determining the thermal conductivity of soil (Jackson and Taylor, 1986; Ghildyal and Tripathi, 1987) is sometimes used. A probe, constructed as in figure 2.52 is usually inserted horizontally into the wall of a pit dug in the soil. The long, slender probe provides a heat source at one point and a temperature detector placed also on the probe surface but some distance away from the heat source. Thus, heat flow is assumed to be uniaxial.

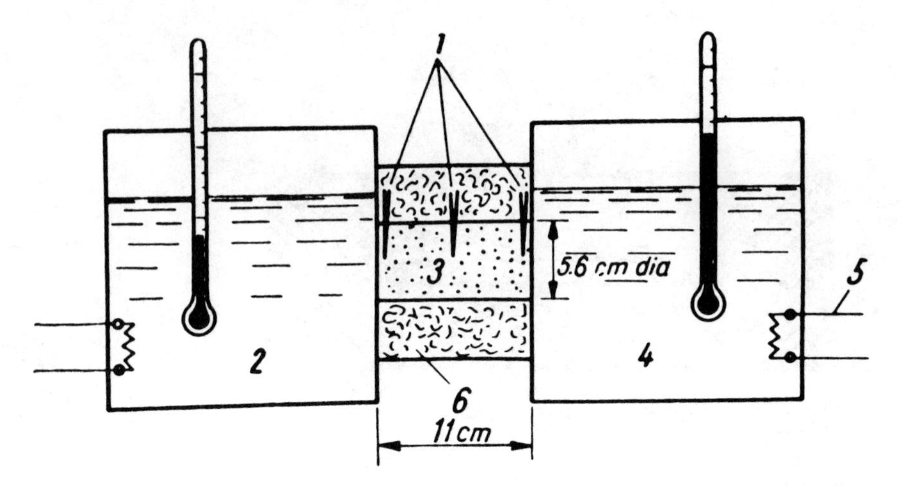

Figure 2.50–Schematic diagram for measuring the heat transfer in soils due to moisture vapor migration (1 - thermocouples, 2 - cold bath, 3 - soil sample, 4 - warm bath, 5 - heating coils, 6 - heat insulation) (from Kézdi, 1974).

After the probe is placed on the soil, the heater is turned on, and the current, voltage, and temperature sensor signal are recorded with time. The results are plotted as in figure 2.53. For two points (t_1, T_1) and (t_2, T_2) on the linear portion of the response curve, the following equation is applied to determine the thermal conductivity, λ:

$$T_2 - T_1 = \frac{2.303\, q_h}{4\,\pi\,\lambda} \sum \log_{10} \left[\frac{t_2 - t_c}{t_1 - t_c}\right] \tag{2.41}$$

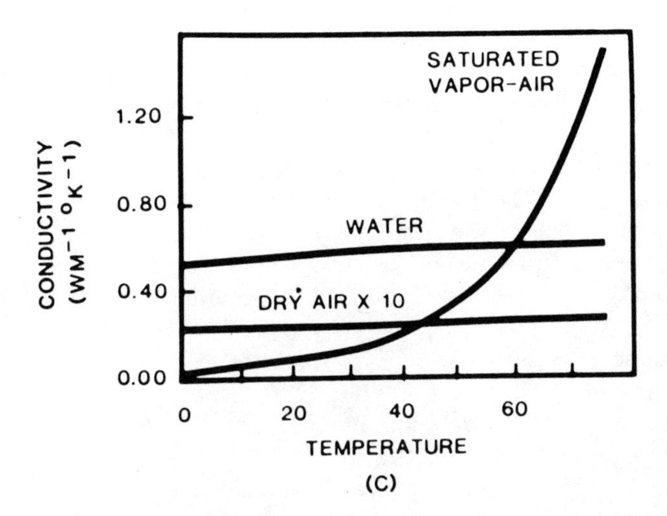

Figure 2.51–Comparison of thermal conductivity values for air, water, and saturated water vapor at various temperatures (from Tollner and Verma, 1987b).

Table 2.7. Effect of temperature on saturated vapor pressure of water

	Temperature (C°)									
	0	5	10	15	20	25	30	35	40	45
Vapor Pressure (kPa)	0.61	0.87	1.23	1.70	2.33	3.16	4.23	5.61	7.36	9.56

in which q_h = the rate of heat energy supplied by the heater (cal/s) and t_c = a probe time-constant factor.

The term t_c has to be determined experimentally for the material for which it is being used. For small diameter probes (<0.1 cm), t_c may be neglected. For larger probes, t_c must be evaluated graphically from the plot of change in temperature vs log t. The straight-line portion of the curve is extrapolated to smaller values of t, and a value of t_c is chosen such that the adjusted experimental points fall on the extrapolated line. Using the above equation, a linear relationship between temperature rise and logarithm of corrected time (t-t_c) is obtained for an even longer duration. The theoretical basis for this correction factor is that a finite time is required for the probe mass to be heated.

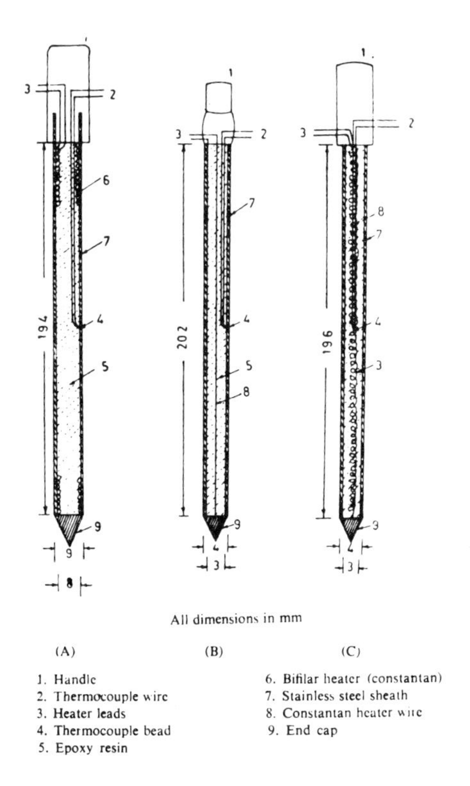

Figure 2.52–Typical designs of probes for measuring the thermal conductivity of soil. Reprinted with permission of Wiley Eastern, Ltd. from Ghildyal and Tripathy, 1987.

Researchers continue to try to devise a model that can predict the thermal conductivity of soil depending on various other measurable physical soil factors Some of these models attempt to examine the solid, liquid, and gaseous constituents of the soil and evaluate the heat transfer in each of these phases individually. One of these yields, for mineral soils, a three-phase diagram on which contours of equal thermal conductivity are drawn (fig 2.54) (Kézdi, 1974). Another such approach (DeVries and Afgan, 1975) examines the state of the constituents in some detail.

The table of thermal properties of soil constituents shows large differences in the thermal conductivity values of the various soil constituents. In general the thermal conductivity of a soil depends in a rather complicated way on its composition, in particular, its water content, x_w. Typical curves for a quartz sand, a loam, and a peat soil are given in figure 2.55.

The general trend of the curves for mineral soils can be understood qualitatively as follows. At complete dryness the heat flow passes mainly through the grains but has to bridge the air-filled gaps between the grains around their contact points (fig 2.56a). At very low water contents the soil particles are covered by thin adsorbed water layers (not shown in fig 2.56). The thickness of these layers increases with increasing water content. At a certain x_w liquid water rings start to form around the contact points between grains; they show a curved air-water interface (fig 2.56b). From this point on the thermal conductivity increases rapidly with increasing x_w until the rings almost completely fill the original gap (fig 2.56c). When x_w increases still further, complete pores are filled with water, up to saturation. This is reflected by the slower increase of λ with x_w.

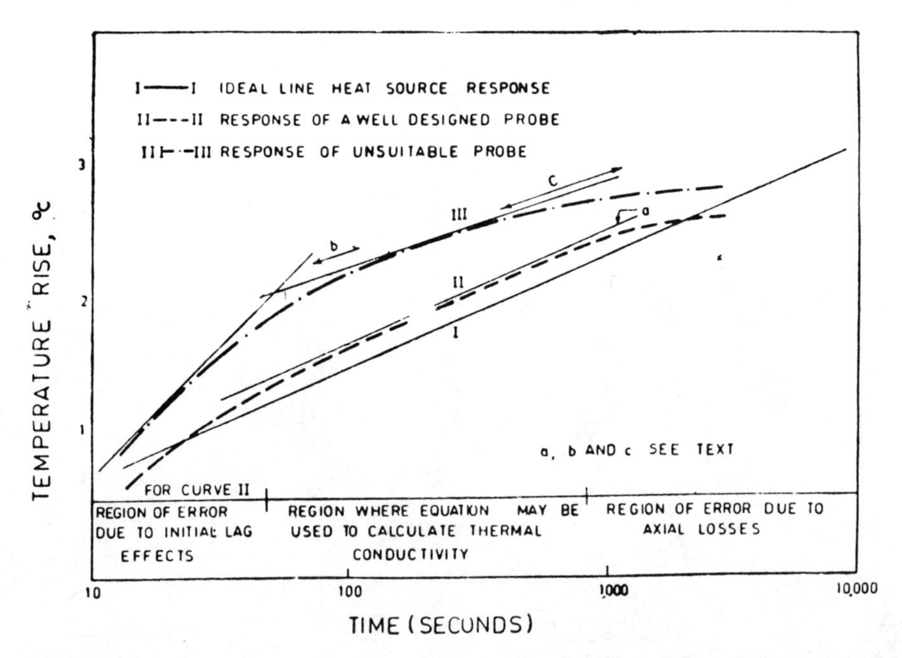

Figure 2.53–Temperature-time curves for a soil thermal conductivity probe and illustration of the method for determining t_c (probe time-constant factor) (from Wechsler, 1966, as cited and reproduced in Ghildyal and Tripathy, 1987).

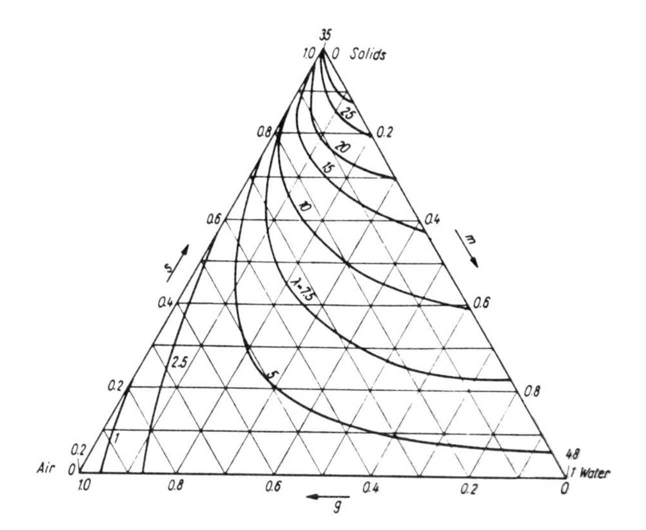

Figure 2.54–Effects on thermal conductivity (λ, cal/°C cm-h) of various soil porosity and moisture (volumetric) states: s = proportion of volume occupied by solids; g = proportion of volume occupied by gases; m = proportion of volume occupied by water (from Kézdi, 1974).

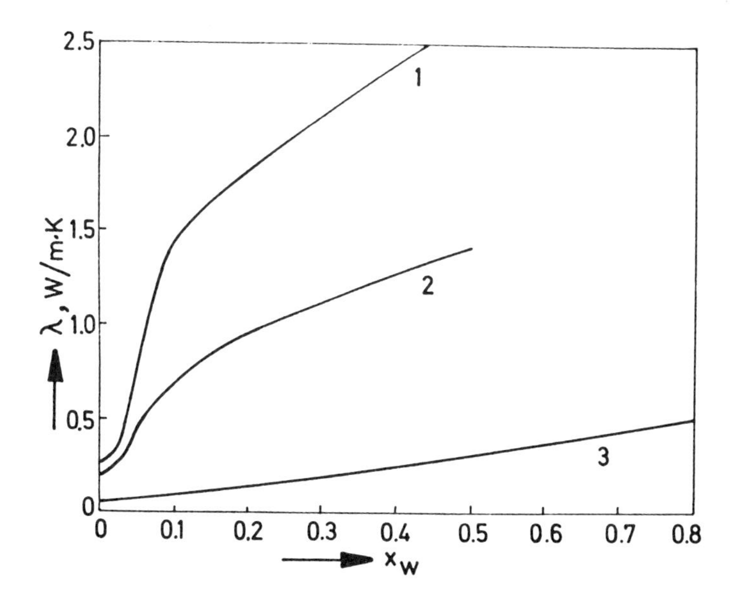

Figure 2.55–Thermal conductivity, λ, in relation to volumetric soil water content, x_w at 10° C. Curve 1: quartz sand with x_s = 0.55. Curve 2: loam with x_s = 0.50. Curve 3: peat soil with x_s = 0.20 (x_s = volume fraction of solids) (from DeVries and Afgan, 1975).

A physical model for calculating the thermal conductivity of soils as a function of their water content (DeVries and Afgan, 1975) is as follows. The soil is considered to consist of water as a continuous medium in which soil grains and small air pockets are distributed. The overall thermal conductivity of the soil can be expressed as:

$$\lambda = \frac{x_w \lambda_w + \Sigma_i k_i x_i \lambda_i + k_a x_a \lambda_a}{x_w + \Sigma_i k_i x_i + k_a x_a} \qquad (2.42)$$

in which λ = thermal conductivity (cal/cm•s•°K), x = volume fraction, and k = weighting factor; subscripts: w = water, a = air, and i = solid materials of various types.

Here the summation extends over the different solid soil constituents, which are now characterized by their thermal conductivity and shape. The multiplication factors k_i are easily seen to represent the ratio of the space average of the temperature gradient in the soil grains of kind i and the space average of the temperature gradient in the water. Similarly, k_a represents this ratio for the gradients in the air and the water in the soil. Of course, $k_i < 1$ if $\lambda_i/\lambda_w > 1$ and conversely. Hence, k is a weight factor depending on the thermal conductivity and the shape of the enclosure (DeVries and Afgan, 1975; Tollner and Verma, 1987b).

In most cases a good result is obtained by considering the soil grains to be spheroids with axes $a_1 = a_2 = na_3$. Then they are characterized by a single shape factor $g_1(n)$.

For the air enclosures this shape factor is deduced by interpolation between ⅓ (spherical enclosures) near water saturation to a value for an oblate spheroid with n of the order of 10 at low water contents. See DeVries and Afgan (1975) for elaboration on this point.

The influence of latent heat transfer in the air-filled pores is proportional to the temperature gradient in these pores and can therefore be taken into account by adding to the thermal conductivity of air an apparent conductivity resulting from the evaporation and condensation of water vapor. This apparent conductivity rises rapidly with increasing temperature.

The model breaks down at low moisture contents when water no longer can be considered as a continuous medium that envelops the other constituents. This usually occurs at x_w values somewhere in the lower half of the steepest part of the λ-versus-x_w curve.

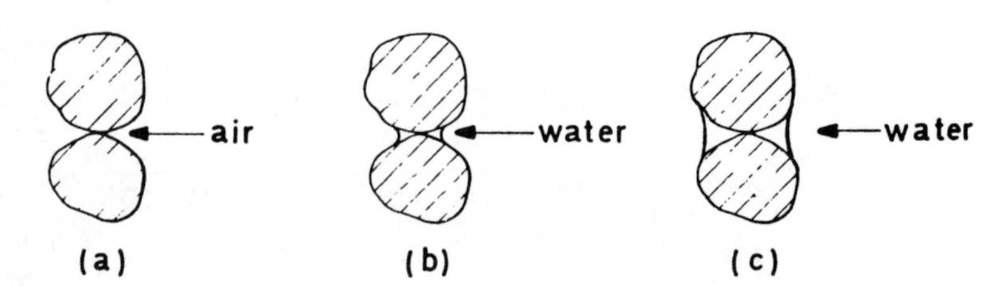

Figure 2.56–Sequence in the formation of water rings at increasing moisture contents (from DeVries and Afgan, 1975).

At complete dryness and at very low water contents, the same model applies with air as the continuous medium. However, the theory needs correction when the ratio of conductivity of the enclosures to that of the continuous medium becomes of the order of 10^2. This applies to $\lambda_{solids}/\lambda_a$ for mineral soils. The corrected values are found by introducing a semi-empirical multiplication factor of 1.25 on the right side of the above equation.

Another category of models is that of those referred to as *coupled models* (heat and mass transfer) because the heat flow and the vapor flow are linked by physical flow characteristics of the vapor passing through the medium coordinated with the heat transfer processes associated with both the applied heat flux and the evaporation-condensation cycle of the pore-space vapor (Matthes and Bowen, 1968; Butts et al, 1989).

Quantitative models of heat transfer processes from equipment to soil requires, in addition to the thermal conductivity of soil, information on the thermal contact conductance, H, in $W/m^2 \bullet K$, between equipment surfaces and the soil. It can be anticipated that such a conductance would tend to increase with increases of soil moisture content and dry bulk density and would change over time if the temperature gradient caused the soil next to the equipment surface to change in moisture content (and density if shrinkage is associated with moisture content change). Values found by researchers investigating this parameter (Redmund et al, 1988; Redmund and Schulte, 1989) ranged from 203 to 2411 $W/m^2 \bullet K$.

Actual values of the specific heat of solid soil materials ranged from 0.18 to 0.226 cal/g•°K, while values of heat capacity ranged from 0.175 to 0.720 $cal/cm^3 \bullet K$. Values of thermal conductivity ranged from 0.00045 to 0.0083 cal/s•cm•K (Merva, 1975).

Tillage practices were found to affect the thermal conductivity of soil. An equation which predicted thermal conductivity based on the volumetric water content, dry bulk density, and clay content of a soil was able to approximate moderately well the thermal conductivity of a soil subjected to various tillage operations. The main factor involved in changing the thermal conductivity was the change in dry bulk density produced by tillage. This change also affected the volumetric water content, although the gravimetric water content remained unchanged (Kenny and Saxton, 1988).

Soil Mass Dynamic Properties Definable within a System of Mechanics

Stress Levels for Yield or Failure

Soils, in general, exhibit great variation in physical properties, depending on numerous factors, the influence of each of which on soil physical properties, varies as well. Persons concerned with the physical outcome of soil-machine interactions are still engaged in research to establish the full range of quantitative understandings required to anticipate such outcomes accurately. In cases of tillage and traction operations, the most important aspects of soil-machine interactions are those associated with gross yield or failure of the soil material. Examples are the draft force required of a tillage implement or the maximum thrust that a traction device can develop at high levels of slip.

In cases such as these soil physical properties that are not necessarily conditional on the quantitative degree of soil deformation can suffice in predicting the outcome of the soil-machine interaction. Physical properties of this type can be measured in a process in which some sort of measured physical loading is applied to the soil until a given degree of failure or gross deformation is achieved. Results can be interpreted either (1) without a requirement to specify the mechanism (other than cause and effect correlation) to link the loading and deformation or (2) in a manner that uses stresses and supposed failure mechanisms to interpret the results so that they might be applied to physically varying practical situations. Measures involving the latter system of interpretation will be discussed first.

Soils are very weak in comparison with common materials of construction such as wood, steel, and concrete. In isotropic compression soils do exhibit yield and plasticity at low stress levels, but failure or loss of monolithic structure does not occur at low stresses. However, weaknesses that lead to soil failure under load are those associated with resistance to tensile loading, shear loading, and frictional loading of soil sliding on other materials. These characteristics in their fundamental forms have been discussed by Gill and VandenBerg (1968) and will be reviewed here in combination with recent related developments.

Shear Strength of Bulk Materials

When a cubic element of material with unit dimensions is subjected to a system of stresses, it is possible to find an orientation for this element such that none of the faces are subject to shear stress. In this orientation, one pair of faces of the cube will be subject to the highest possible stress, called the *major principal stress*, σ_1. Another pair of faces will be subject to the minimum possible stress called the *minor principal stress*, σ_3, while the third pair of faces will be subjected to an intermediate *principal stress level*, σ_2 (fig 2.57).

If attention is focused on a plane parallel to the direction of the intermediate principal stress, σ_2, but inclined to the face carrying the major principal stress, σ_1, by an angle α (fig 2.57) shear (τ_α) and normal (σ_α) stresses on this plane will be found to be (Sowers and Sowers, 1961):

$$\sigma_\alpha = \frac{\sigma_1 + \sigma_3}{2} + \frac{\sigma_1 - \sigma_3}{2} \cos 2\alpha \qquad (2.43)$$

and

$$\tau_\alpha = \frac{\sigma_1 - \sigma_3}{2} \sin 2\alpha \qquad (2.44)$$

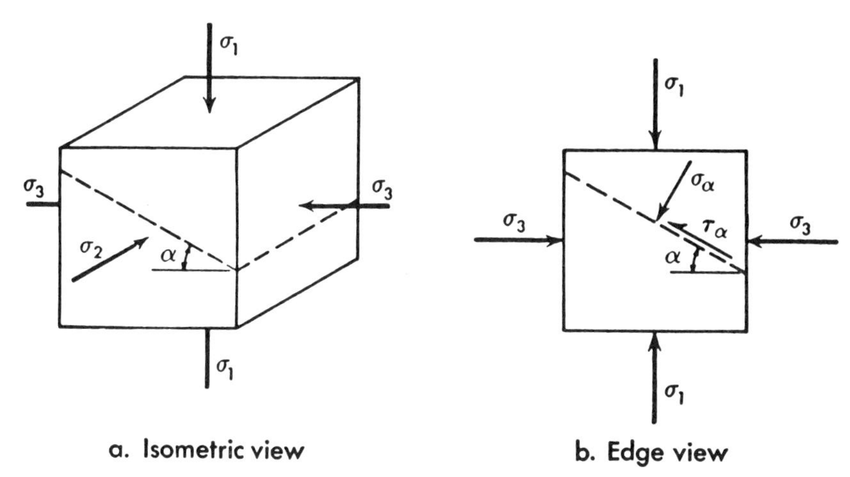

Figure 2.57–Stress in a cube of unit dimensions that is cut by a plane which is perpendicular to the plane on which σ_2 acts and which makes an angle of α with the plane on which σ_1 acts. Reprinted with permission of Macmillan Publishing Company, Inc. from *Introductory Soil Mechanics and Foundations*, 2nd Edition, Sowers, G. F. and G. B. Sowers. Copyright © 1961, Macmillan Publishing Company.

Physicist Otto Mohr suggested that these relationships could be represented by a circle on rectangular coordinates of σ (compressive-tensile stress) and τ (shear stress) (fig 2.58).

It is possible to load a sample of soil with varying levels of σ_1 and σ_3 using a triaxial soil test cell (fig 2.59). The stress state when shear failure takes place may be represented

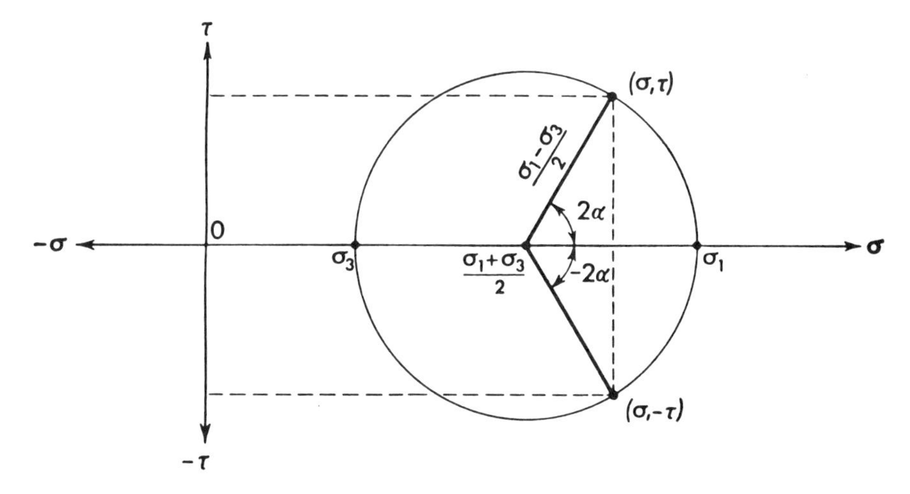

Figure 2.58–Mohr's coordinates and Mohr's circle of stresses. Reprinted with permission of Macmillan Publishing Company, Inc. from *Introductory Soil Mechanics and Foundations*, 2nd Edition, Sowers, G. F. and G. B. Sowers. Copyright © 1961, Macmillan Publishing Company.

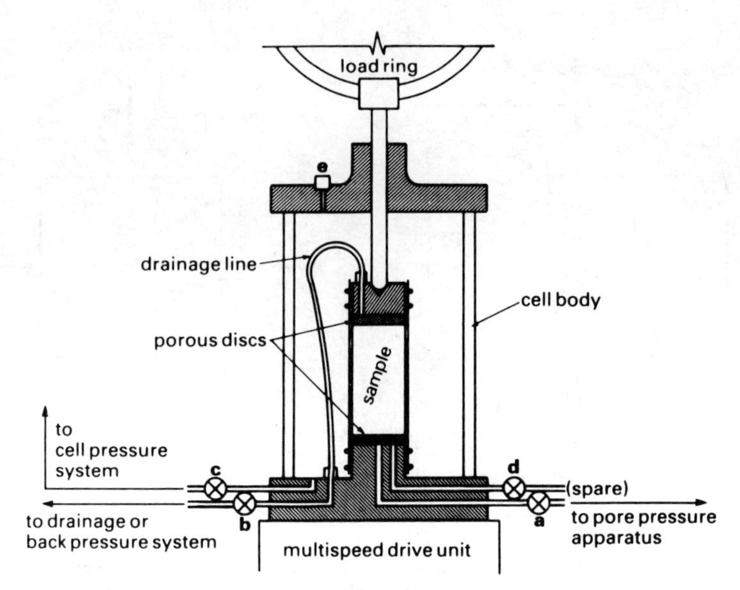

Figure 2.59–Triaxial soil test cell (from Head, 1986).

by a Mohr circle. When two or more triaxial tests are made on the same material, while using for each one a different level of the cell pressure (σ_3), the results may be plotted on a Mohr diagram and the Mohr failure envelope constructed tangent to the circles representing the states of stress at failure (fig 2.60). This envelope can be represented by the equation:

$$\tau_{max} = c + \sigma_n \tan \phi \qquad (2.45)$$

in which τ_{max} = the shear stress at failure, and c = the maximum shear stress on any plane when the normal stress is zero, σ_n = the summation of normal forces in a given

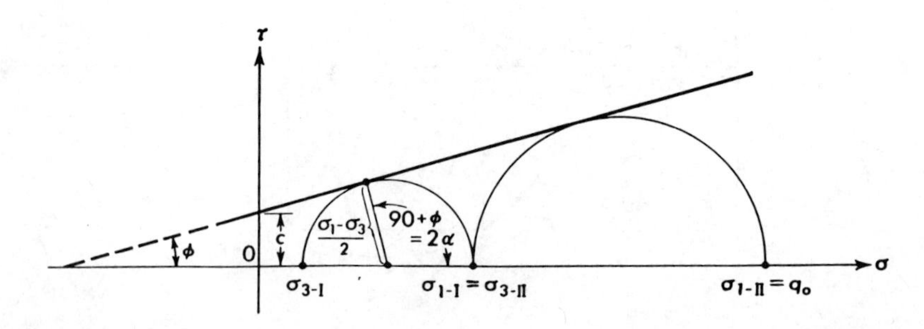

Figure 2.60–Mohr's circle for stresses at failure illustrating the realtions between σ_1 and σ_2 at failure Reprinted with permission of Macmillan Publishing Company, Inc. from *Introductory Soil Mechanics and Foundations*, 2nd Edition, Sowers, G. F. and G. B. Sowers. Copyright © 1961, Macmillan Publishing Company.

direction between soil grains divided by the cross-sectional area of the plane perpendicular to that direction, and $\tan \phi$ = the coefficient of friction of soil grains sliding on soil grains.

The point of tangency of a stress circle at failure with the Mohr envelope defines the state of stress (σ_n and τ) at which failure takes place and defines (from the geometry of the Mohr circle) the angle of the plane on which this state of failure stresses exists:

$$2\alpha = 90° + \phi \tag{2.46}$$

An independent evaluation of this result (Withey and Aston, 1950) may be obtained by analyzing the unconfined compression strength of a tall prism of material of unit cross-sectional area (fig 2.61). For any plane through the prism with an angle of inclination from the base of α:

$$\sigma_{1max} \sin \alpha = c \sec \alpha + \sigma_1 \cos \alpha \tan \phi \tag{2.47}$$

$$\frac{d\sigma_{1max}}{d\alpha} = c\left(\cos^2 \alpha - \sin^2 \alpha + 2 \tan \phi \sin \alpha \cos \alpha\right) = 0 \tag{2.48}$$

if failure is to take place at the lowest value of σ_1, which will produce failure. Thus,

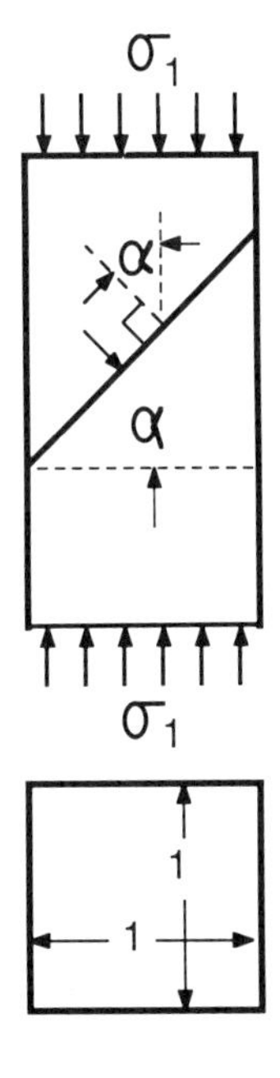

Figure 2.61–Rectangular prism with unit cross-sectional area illustrating a plane on which failure might occur (Withey and Aston, 1950).

$$\tan\phi = -\frac{\cos^2\alpha - \sin^2\alpha}{2\sin\alpha\cos\alpha} = -\frac{\cos 2\alpha}{\sin 2\alpha} = -\cot 2\alpha \tag{2.49}$$

$$= -\tan(90° - 2\alpha) = \tan(2\alpha - 90°)$$

thus

$$\phi = 2\alpha - 90°, \text{ or } 2\alpha = 90° + \phi, \text{ or } \alpha = 45° + \phi / 2 \tag{2.50}$$

When soil is in a state of stress representing conditions of failure (the Mohr's circle for the state of stress is tangent to the Mohr failure envelope) (fig 2.60) the relationship between σ_1 and σ_3 can be expressed (Sowers and Sowers, 1961) by:

$$\frac{\sigma_1 - \sigma_3}{2} = \left(\frac{c}{\tan\phi} + \frac{\sigma_1 + \sigma_3}{2}\right)\sin\phi \tag{2.51}$$

$$\sigma_1 = \sigma_3\left(\frac{1 + \sin\phi}{1 - \sin\phi}\right) + 2\,c\left(\frac{\cos\phi}{1 - \sin\phi}\right) \tag{2.52}$$

$$\sigma_1 = \sigma_3 \tan^2\alpha + 2c\tan\alpha$$

Some soil testing cells have been constructed in such a way as to control not only σ_1 and σ_3 but also to control σ_2 independently of either σ_1 or σ_3 (Ko and Scott, 1967; Dunlap and Weber, 1971; Sture and Desai, 1979; and Gibas et al, 1991). Two of these are illustrated in figures 2.62 and 2.63. Applications of these units in determining the shear strength characteristics of soils indicate results considerably different than those arrived at using a conventional triaxial cell. Dunlap and Weber (1971) found that when $\sigma_2 = \sigma_m$* or when $\sigma_2 = \sigma_1$ the shear stress at yield generally increased more rapidly with increases in the mean normal stress than was the case when $\sigma_2 = \sigma_3$ as is common to the conventional triaxial cell (fig 2.64). Ko and Scott (1967) found similar results (much higher values of σ_1/σ_3 were tolerated prior to failure) and that axial strain at these high levels of σ_1/σ_3 was much reduced in the $\sigma_2 = \sigma_m$ case as compared to the conventional $\sigma_2 = \sigma_3$ case.

On the other hand Khan and Hoag (1978) found, at any given level of mean normal stress, that values of the octahedral shear stress, τ_{oct},† at failure were higher for the case

---

* Mean normal stress = σ_m = 1/3 $(\sigma_1 + \sigma_2 + \sigma_3)$ and is also referred to as σ_{oct} or the octahedral normal stress.
† Octahedral shear stress: τ_{oct} = 1/3 $[(\sigma_1 - \sigma_2)^2 + (\sigma_2 - \sigma_3)^2 + (\sigma_3 - \sigma_1)^2]^{1/2}$.

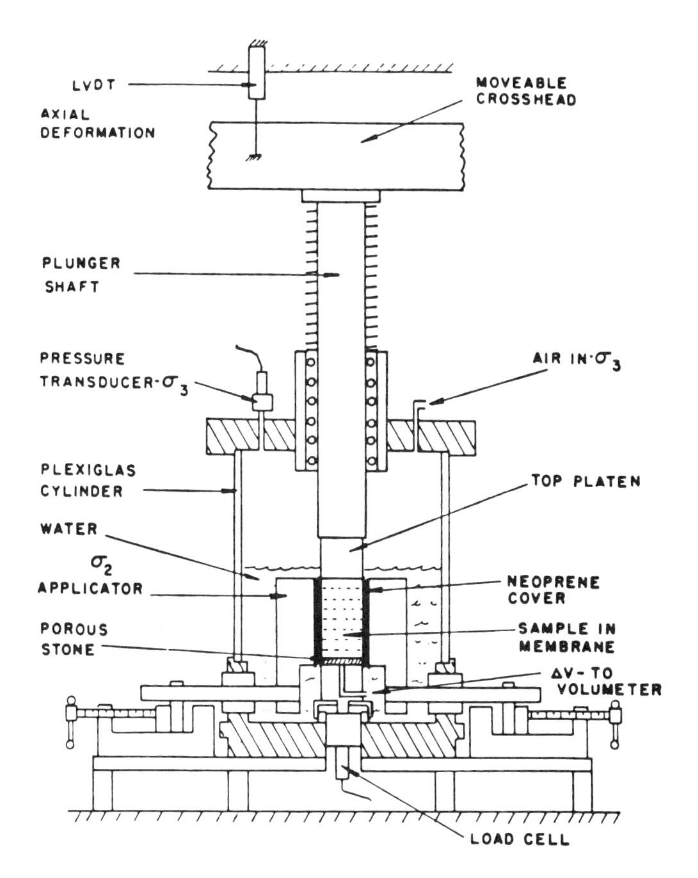

Figure 2.62–Soil-test cell which permits independent control of σ_1, σ_2, and σ_3 as applied to rectangular samples (from Dunlap and Weber, 1971).

of $\Delta\sigma_2 = \Delta\sigma_3$ as would apply in a conventional triaxial cell, as opposed to the case in which $\Delta\sigma_2 = 0.5\ \Delta\sigma_1$ (fig 2.65). However, the differences in τ_{max} are not as great as the differences in τ_{oct} shown in fig 2.65; in the $\Delta\sigma_2 = 0.5\ \Delta\sigma_1$ case, the maximum shear stress, $1/2\ (\sigma_1 - \sigma_3) = \tau_{max}$ is 116% of τ_{oct}, while in the $\Delta\sigma_2 = \Delta\sigma_3$ case, τ_{max} is 106% of τ_{oct}.

Triaxial tests of soil materials to be used for building foundation purposes are frequently done with the soil in a saturated condition and with the top and bottom drainage outlets closed to represent the worst possible condition. Agricultural soils involved in *in situ* soil-machine interactions are frequently loaded for only brief periods but under fully drained conditions. When agricultural soils have a sizable proportion of clay particles, water outflow from the material when loaded is restricted to rates that are very low in comparison to rates of loading. When loaded in drained triaxial tests the material, if near saturation, behaves essentially as it would if the test were conducted in an undrained mode. This gives rise to a positive pressure in the pore water phase. This pressure, which acts undiminished in all directions, is called the *neutral stress*, u, and becomes additive to all intergranular (or "effective") stresses to constitute the total stress

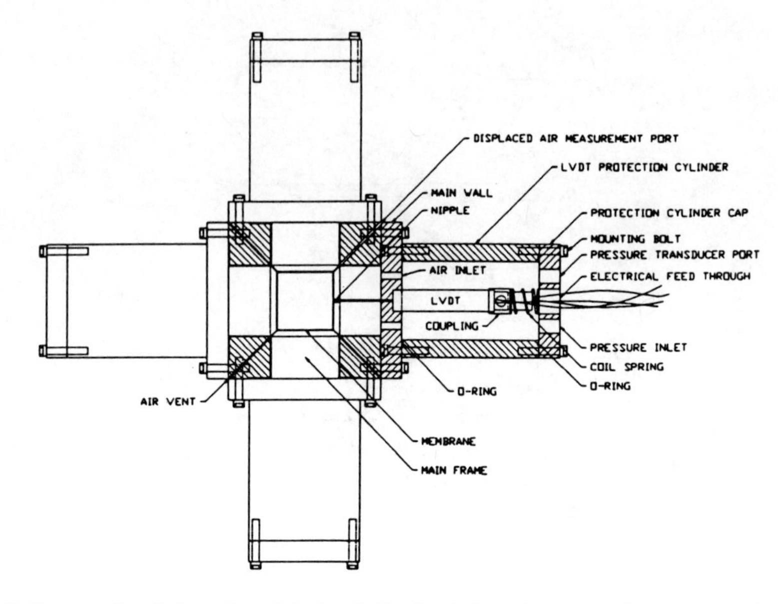

Figure 2.63–Cross-sectional view of a soil-test cell allowing independent control of σ_1, σ_2, and σ_3 (from Gibas et al, 1991).

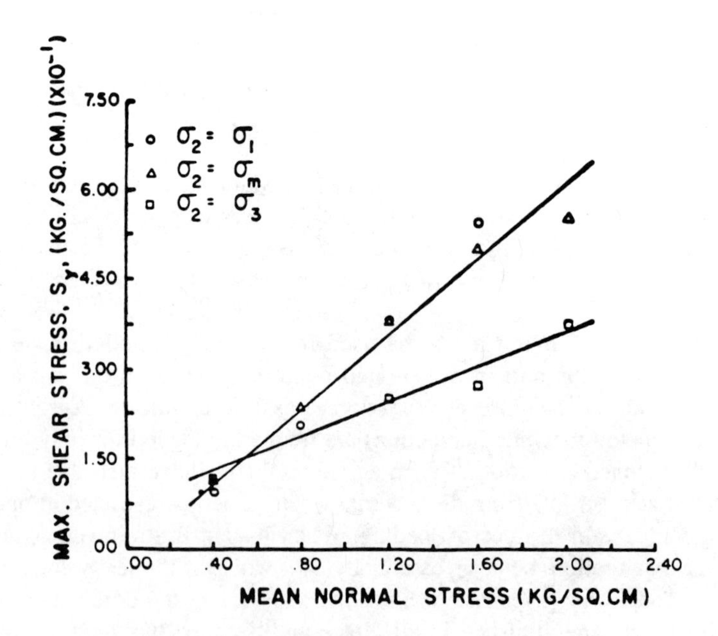

Figure 2.64–Effects of having σ_2 values different from σ_3 values, on maximum shear stress at yield for a silty clay loam soil (from Dunlap and Weber, 1971).

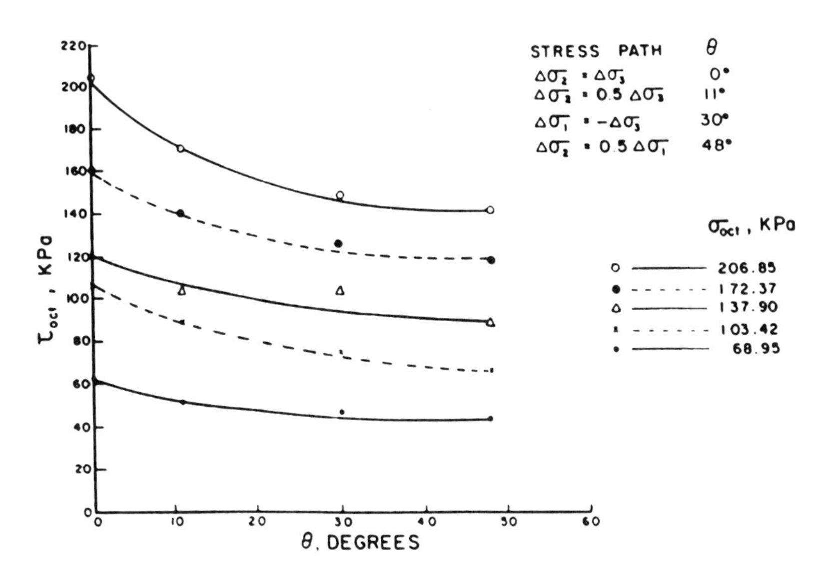

Figure 2.65–Effects of various stress paths on the values of octahedral shear stress at failure for samples of Lloyd clay at specific levels of mean normal stress, σ_{oct} (from Khan and Hoag, 1978).

becomes additive to all intergranular (or "effective") stresses to constitute the total stress (fig 2.66):

$$\sigma = \sigma' + u \quad (2.53)$$

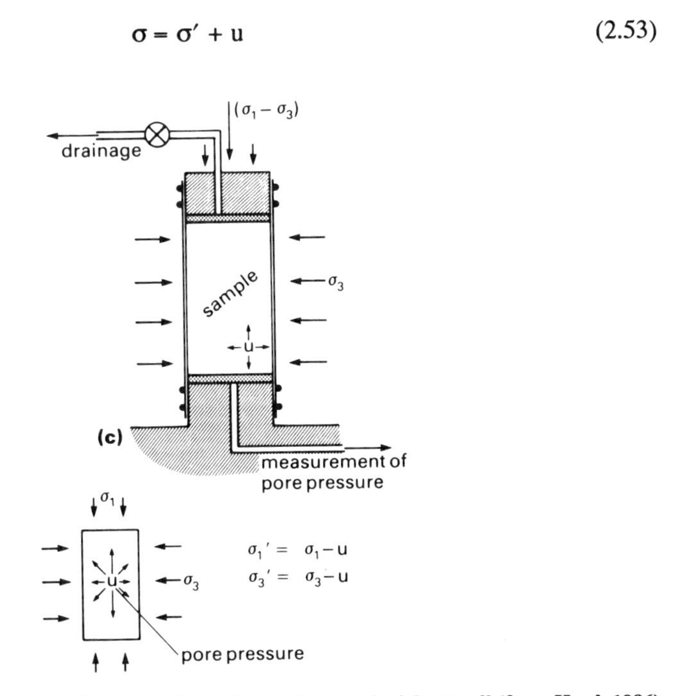

Figure 2.66–Illustration of the neutral stress, u, in a soil sample in a triaxial test cell (from Head, 1986).

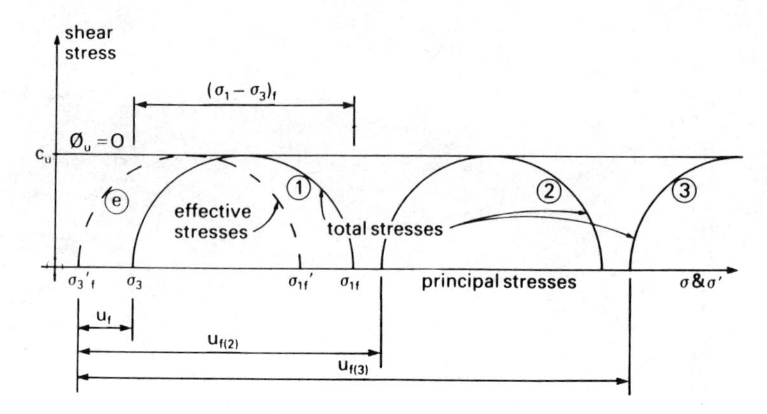

Figure 2.67–The effect of neutral stress in an undrained test. Although the total stresses σ_1 and σ_3 increase, effective stresses $\sigma_1{'}$ and $\sigma_3{'}$ do not change relative to each other. The apparent coefficient of internal friction is zero while apparent cohesion is a constant value (from Head, 1986).

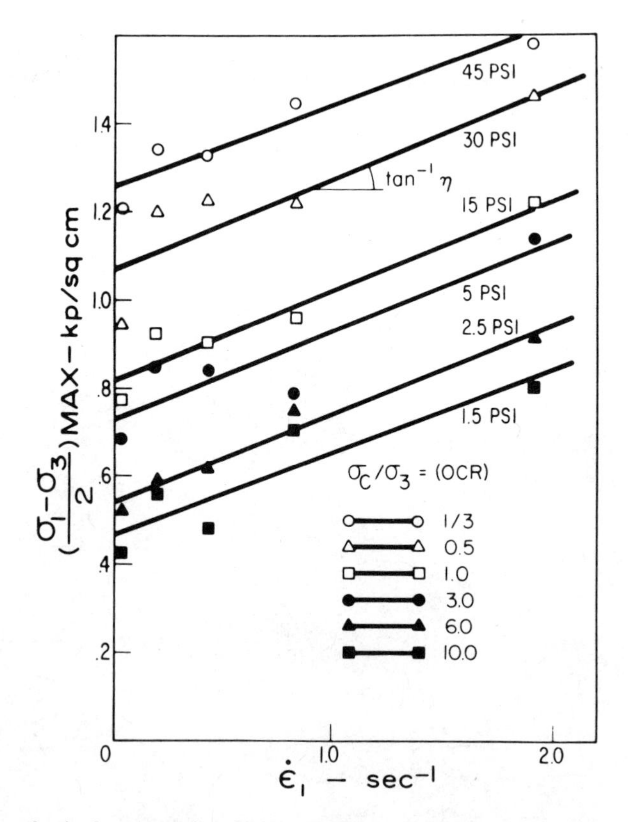

Figure 2.68–Viscoplastic characteristics of a saturated clay soil subjected to various values of axial strain rate, $\dot{\varepsilon}$. The viscosity (slope of lines) appears to be a constant irrespective of preconsolidation stress level, while yield stress increases regularly with preconsolidation stress level (from Hassan and Chancellor, 1970).

in which σ = total stress, σ' = effective or intergranular stress, and u = neutral stress or pore water pressure.

The Mohr failure criterion applies to effective and not total stresses. If the Mohr failure criterion is applied to total stresses of a saturated soil tested either in an undrained triaxial test or in a quick-drained test of a high-clay soil, the results will appear as in figure 2.67, because any increase in σ_3 will result in an increase in u which prevents σ_1 from deviating from σ_3 any further than was the case at lower values of σ_3. Thus the soil appears to have only cohesion as a source of shear failure strength, because the slope of the Mohr envelope is zero. Such soils are called *cohesive* or *frictionless*, although they would exhibit frictional properties if drainage were permitted and the loading rate was sufficiently slow that the neutral stress, u, remained near zero (Sowers and Sowers, 1961).

In some cases, saturated cohesive soils may be described as having visco-plastic characteristics, ie, they do not deform until a certain level of shear stress (called a *yield stress*) is reached. Once that level of shear stress has been reached, any further increase in shear stress is found to be linearly related with the time rate of shear strain (Hassan and Chancellor, 1970). Thus, such a soil could be described by a cohesion or yield stress and by a viscosity in which the shear stress term is the difference between the total shear stress and the yield stress (fig 2.68).

Water in the pore space of unsaturated agricultural soils also affects the strength characteristics of the soil in resisting both shear and tensile loads. The water under tension causes bonds between soil grains (fig 2.69). In many cases this moisture tension produces films of water around the soil grains; these films are so thin that moisture cannot move rapidly between individual bonds. Thus, any rapid loading of the soil causes the soil to act as if it were under the influence of a omni-directional internal

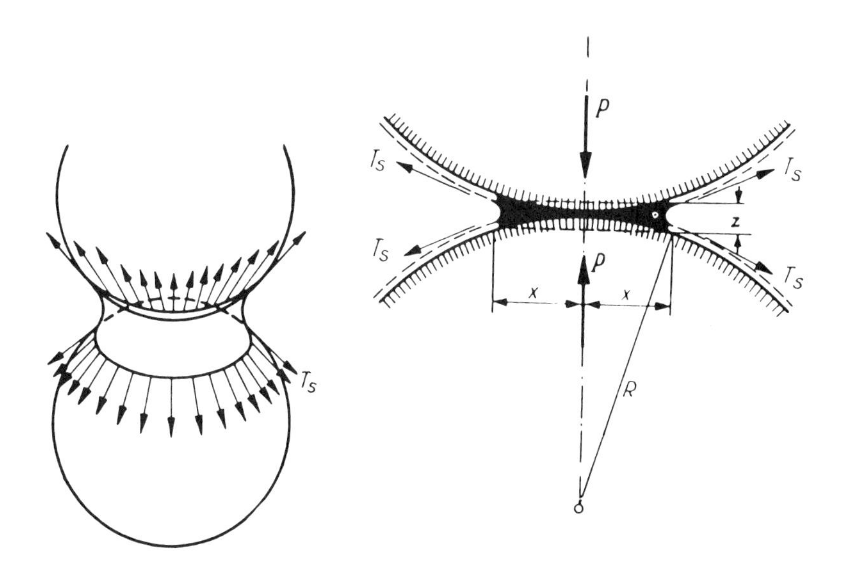

Figure 2.69–Illustration of the bonding effects (intrinsic stress) of moisture surface tension acting between two spheres (from Kézdi, 1974).

tensile stress or intrinsic stress, σ_i. The value of σ_i may be determined by conducting tensile tests and unconfined compression tests on replicate samples of a bonded or monolithic soil material (Chancellor and Vomocil, 1970) (fig 2.70). The resulting Mohr envelope establishes the basis for relating the magnitude of the intrinsic stress, σ_i, to the tensile failure stress. In some cases the intrinsic stress is expressed in terms of the "apparent cohesion" (Kézdi, 1974) for which:

$$\text{apparent cohesion, c,} = \sigma_i \tan\phi \tag{2.54}$$

and conversely the intrinsic stress may be described (McKyes, 1985) by:

$$\sigma_i = c \cot\phi$$

in which σ_i = the intrinsic stress, c = cohesion or "apparent cohesion", and ϕ = angle of internal friction for the soil.

The frictional relations between soil and other materials when the soil is sliding on the material is usually treated in a way similar to that of soil sliding on soil (Gill and VandenBerg, 1968).

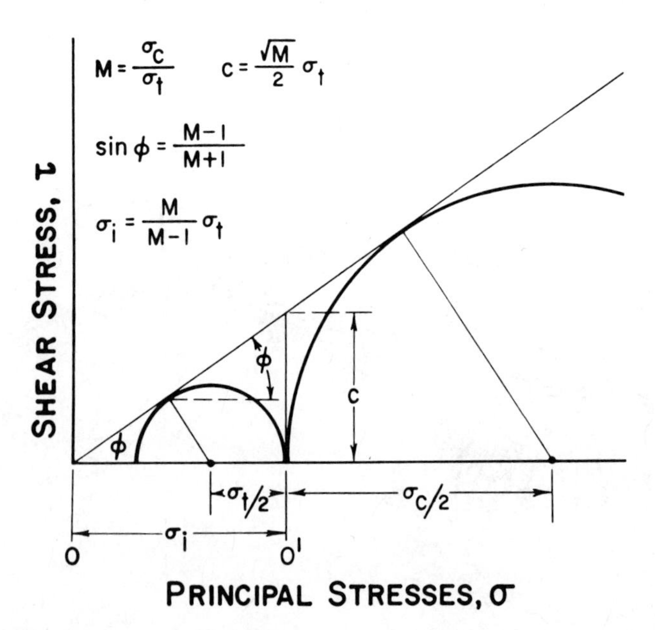

Figure 2.70–Relations between tensile failure stress, σ_t, unconfined compression failure stress, σ_c, intrinsic stress, σ_i, cohesion, c, and the angle of internal friction, ϕ, for a monolithic soil sample (from Chancellor and Vomocil, 1970).

$$S' = C_\alpha + \sigma_n \tan \delta \tag{2.55}$$

in which S' = sliding stress, C_α = adhesion, σ_n = normal stress on the frictional surface, and δ = angle of soil-metal friction. Similarly, the adhesion, C_α, has been related to moisture tension when common soil moisture films wet both the frictional surface and the soil grains in contact with the surface (Gill and VandenBerg, 1968).

Shear Strength of Soils

There are a number of soil testing devices in addition to the triaxial cell that allow the determination of the shear strength parameters of soil, cohesion (c), and the coefficient of internal friction ($\tan\phi$). Many of these have been reviewed by Gill and VandenBerg (1968) and by Johnson et al (1987). Among these are the translational shear box, the parallelogram shear box (Roscoe, 1970), a rectangular grouser plate, and an annular grouser ring. Comparisons among various devices have been made by Wills (1963), Osman (1964), Bailey and Weber (1965), Dunlap et al (1966), and Kuipers and Kroesbergen (1966). In general, it was found that if tests were carefully executed, similar values of c and ϕ would be obtained for a given soil condition irrespective of the device used. One factor which did pose some basis for irregularity entered considerations when a cylindrical test volume (as differentiated from an annular test volume) of material was loaded in torsion. Differences were noticed particularly when the diameter of the cylindrical test volume was small. Various possibilities exist for designating the mean radius at which the applied torque was acting. An assumption that was found to be intermediate among various possible relationships (Johnson et al, 1987; Bailey and Weber, 1965) was that:

$$\tau = \frac{3M}{2\pi \left(r_0^{\,3} - r_i^{\,3}\right)} \tag{2.56}$$

in which τ = shear stress, M = applied moment, r_o = outer radius of the test volume, and r_i = inner radius of the test volume ($r_i = 0$ for cylindrical test volumes).

In a number of cases it was found that the value of ϕ found at a high level of normal stress was somewhat less than that found at low normal stress (Bailey and Weber, 1965); Taylor and VandenBerg, 1966; Chancellor, 1971), while in other cases (Osman, 1964; Wills, 1963) no such characteristic appeared. It was found by Taylor and VandenBerg (1966) that greater displacements of the test apparatus were required to achieve maximum shear stress when normal stresses were high than when they were low (a phenomenon also observable in data obtained by Bailey and Weber, 1965).

The relations between shear stress and shear deformation of soils will be considered in a later section.

Both cohesion (c) and the coefficient of internal friction ($\tan \phi$) are affected by moisture content, porosity, and grain size distribution of the soil material. The interactions among these factors are illustrated in figures 2.71 and 2.72 (Kézdi, 1979). These relations for coarse-grained soils are reviewed by Ayers (1987). In general both

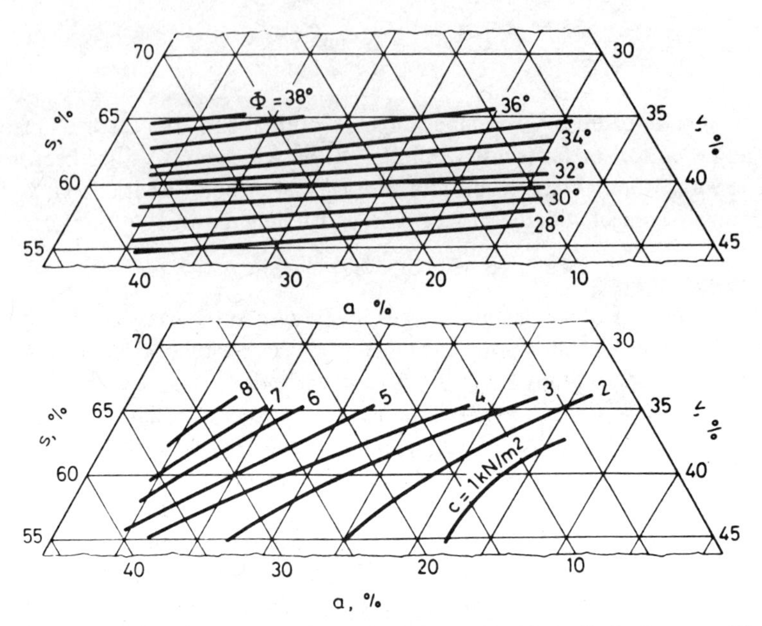

Figure 2.71–Effects of porosity and volumetric moisture content on values of cohesion, c, and the angle of internal friction, ϕ , for a silty sand soil. Air-filled porosity = a, water-filled porosity = v, and volumetric proportion of solids = s (from Kézdi, 1979).

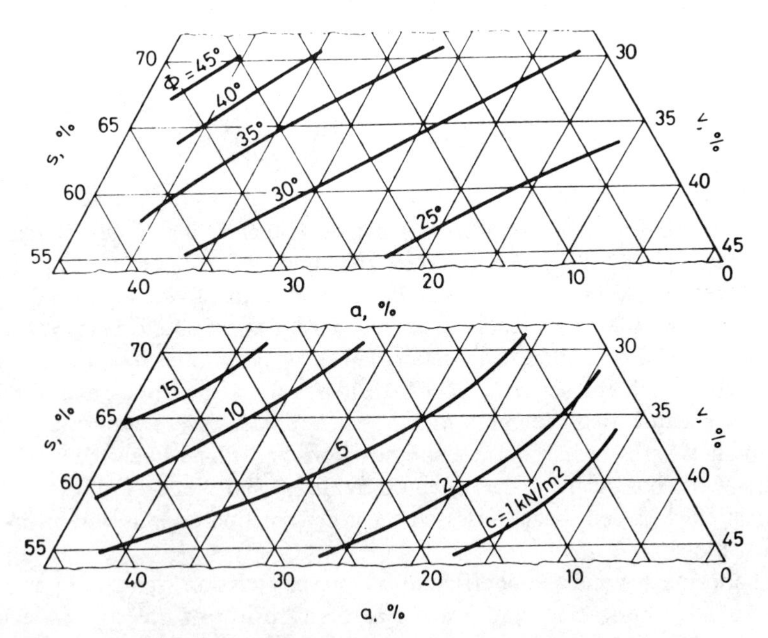

Figure 2.72–Effects of porosity and volumetric moisture content on values of cohesion, c, and the angle of internal friction, ϕ, for a silt soil. Air-filled porosity = a, water-filled porosity = v and volumetric proportion of solids = s (from Kézdi, 1979).

cohesion and angle of internal friction increase with decreases in porosity, whereas cohesion at any given density tends to be maximum at some intermediate level of moisture content. Although a similar characteristic was noted for cohesion by Chancellor (1971), other reports (Kuipers and Kroesbergen, 1966; Kézdi, 1974) and figures 2.71 and 2.72 indicate a general decrease of cohesion with increasing moisture content.

The effects of porosity and normal stress levels on the coefficient of internal friction of soils (tan ϕ) have been reviewed by Hoffmann (1975). Three works are cited linking tan ϕ to normal stress, 10 linking tan ϕ to porosity, and 3 linking tan ϕ to both porosity and normal stress. In general, both higher normal stresses and increased porosities were associated with reduced frictional resistance to shear failure.

One important set of interacting relationships are those linking moisture content, soil porosity, and cohesion based on the compacting or preconsolidation stress to which the soil was subjected prior to the test. The values of cohesion and angle of internal friction as based on effective stresses were found to increase with increased levels of compacting effort (Lambe and Whitman, 1969) or preconsolidation stress (Kézdi, 1974; Hassan and Chancellor, 1970). Kézdi (1974) found that for any given soil, cohesion tended to be in direct proportion to the preconsolidation stress used and the value of the proportionality constant increased linearly with the plasticity index (see the later section on Atterberg limits) of the soil (fig 2.73).

In saturated soil samples, preconsolidation stress increases are associated with lower porosities and reduced moisture contents. For a saturated clay, the viscoplastic yield stress (comparable to the cohesion) increased linearly with decreases in porosity and moisture content (Hassan and Chancellor, 1970). Increasing the time rate of shear strain had no effect on the angle of internal friction but resulted in increased values of cohesion as determined from effective stresses (Hassan and Chancellor, 1970).

In triaxial cells the three principal stresses really hold only two values, σ_1 and $\sigma_2 = \sigma_3$. In translational shear boxes the principal stresses are not measured, but σ_n is measured. Dexter (1981) controlled the tangential stress, σ_t, (stress in a direction mutually perpendicular to the directions of the normal stress and of the shear stress)

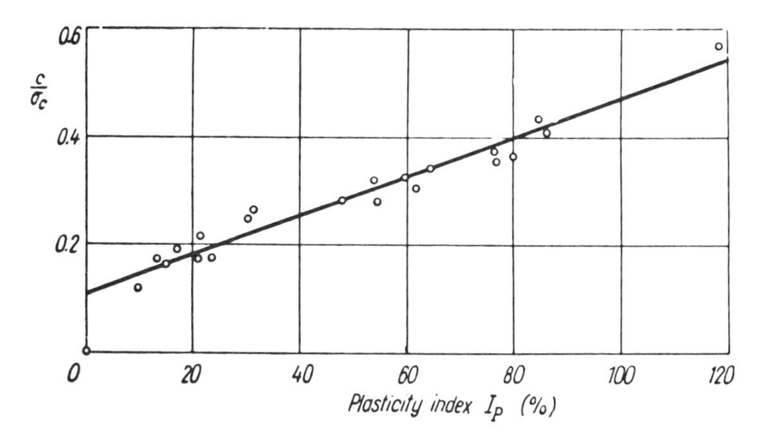

Figure 2.73–Effects of the plasticity index, I_p, on the ratio of cohesion, c, to the consolidation stress, σ_c, for a triaxial test of a cohesive soil (from Kézdi, 1974).

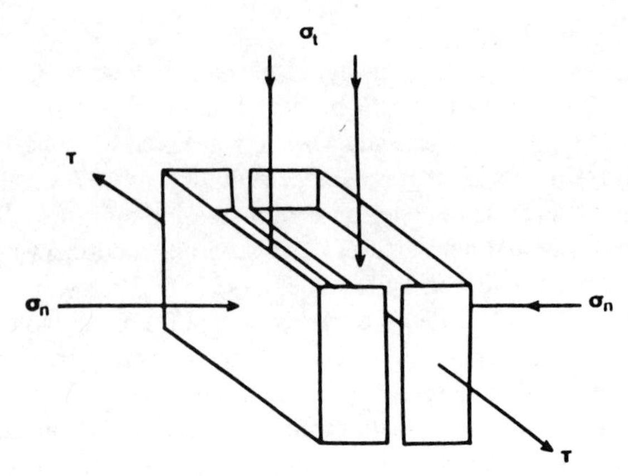

Figure 2.74–Illustration of the tangential stress applied, σ_t, in addition to the normal stress, σ_n, and the shearing force, T, for a shear-box test of the shear strength of soil. Reprinted with permission from *Journal of Terramechanics* 18(4), Dexter, A.R. Soil shear strengths measured with different levels of uniaxial stress acting in a direction tangential to the shear plane. Copyright © 1981, Pergamon Press, Ltd.

(fig 2.74). In tests of three sandy soils in which levels of σ_t up to 2 times σ_n were used, it was found that the shearing stress, τ_{max}, could be represented by:

$$\tau_{max} = 3.7 \text{ kPa} + 1.01\,\sigma_n + 0.18\,\sigma_t \tag{2.57}$$

in which τ_{max} = maximum shearing stress at failure, σ_n = normal stress, and σ_t = tangential stress.

These results indicated that the tangential stress did have some effect on the failure stress in shear but that this effect was comparatively minor.

Tensile Failure Stress of Soil

Soil, and particularly agricultural soil, is considered to be very weak in resisting tensile stresses. Nevertheless, the tensile bonds in soil are responsible for a sizable proportion of the resistance of *in situ* soils to tillage and to horizontal and vertical loads applied in traction and transport processes. Failure stresses during tensile loading can be measured in a number of ways (Kézdi, 1974; Gill and VandenBerg, 1968; Hillel, 1980; Vomocil and Chancellor, 1969; and Farrell et al, 1967):

1. Molding and drying specimens in an hour-glass configuration and pulling them apart using encompassing grips that fit the specimen form.
2. Molding specimens as rectangular beams or cylindrical beams and then loading them in transverse bending (the failure stress calculated from conventional beam theory is sometimes referred to as the *modulus of rupture*).
3. Cementing metal platens to the ends of the test specimen and applying tension to the platens.

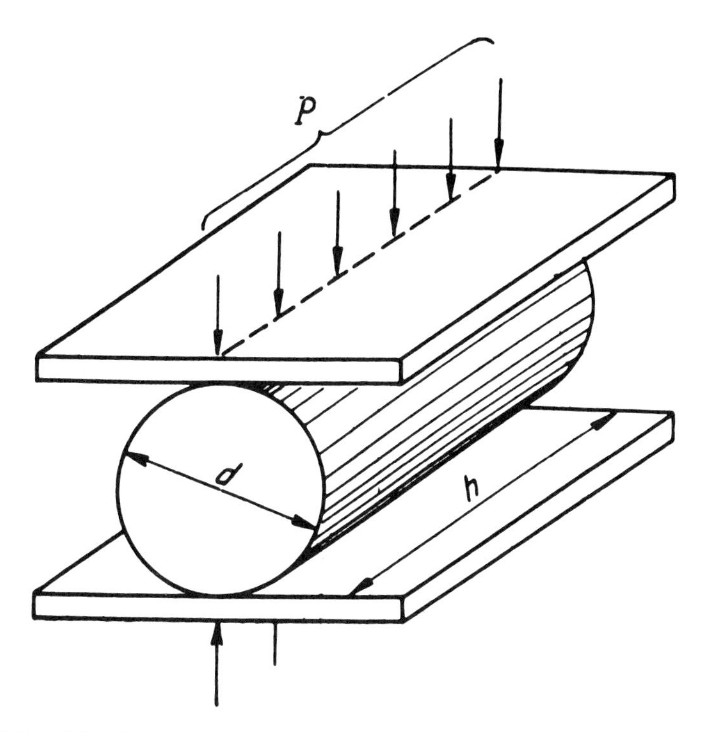

Figure 2.75–Principle of loading a cylindrical, monolithic soil sample to produce tension on a vertical center-plane (from Kézdi, 1974).

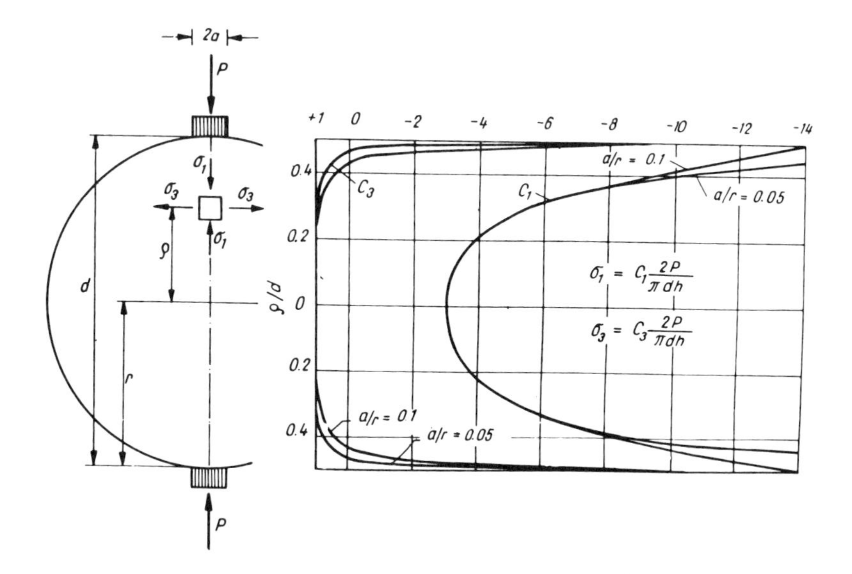

Figure 2.76–Principal stresses induced in a test specimen loaded as shown in figure. 2.75 (from Kézdi, 1974).

4. Laying a cylindrical specimen on its side on a rigid flat surface and exerting a compressive force with a parallel rigid flat surface (the surface compressive stresses gives rise to tensile stresses across the majority of the vertical-axial cross-sectional plane of the specimen (Kézdi, 1974) (figs 2.75 and 2.76).
5. Connecting each end of a cylindrical test specimen to a metal platen of similar diameter with a thin membrane and applying a vacuum to each platen (the maximum tensile stress that could be applied would be slightly less than 1 atmosphere with this type of apparatus).
6. Placing the specimen across the axis of a centrifuge and noting the angular velocity when failure occurred.

Intrinsic Stress. Determination of tensile failure stress, σ_t, by centrifugation, and compressive failure stress, σ_c, by unconfined compression tests provided enough data to determine the Mohr failure envelope for the soil, if it were accepted that the monolithic sample was bonded internally by an active, omnidirectional intrinsic stress, σ_i (see fig 2.70).

It was believed that when monolithic soil samples were cut by the tension in an encircling wire loop radially-moving wedges formed immediately inside the wire-soil interface and generated axial compressive stresses in the surface of the sample. These stresses were resisted by tensile stresses on all but the outermost parts of the sample cross-section and by any externally applied stresses, σ_e, acting in a mutually substitutable manner in resisting wire-loop cutting (see fig 2.77). These results tended to confirm both the existence of an intrinsic stress of a magnitude as determined by figure 2.70 and the active nature of this intrinsic stress. This was also illustrated by Kitani

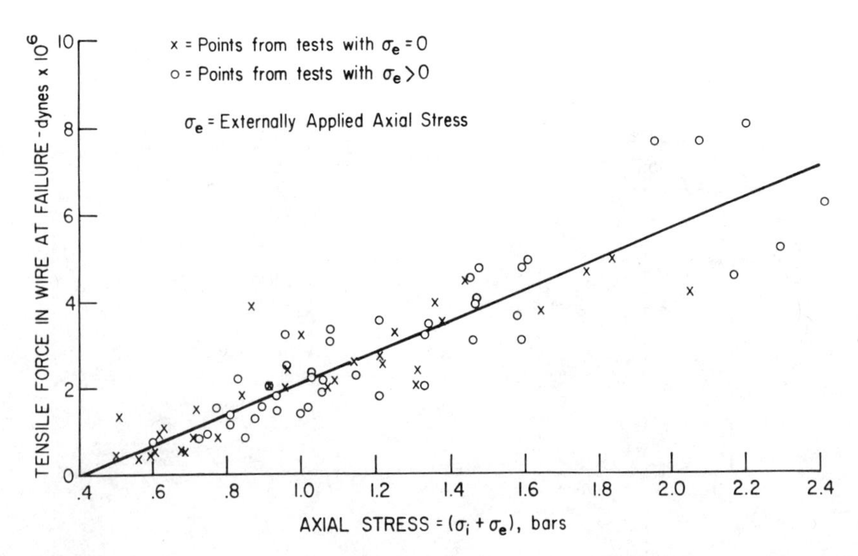

Figure 2.77–Tensile force required for sample failure when a wire was wrapped around a cylindrical, monolithic soil sample, as affected by intrinsic stress, σ_i, (see fig 2.70) or by intrinsic stress plus an externally applied stress, σ_e (from Vomocil and Chancellor, 1969).

(1975), using unconfined compression of a cylindrical sample that was spinning at various angular velocities. There was a near-linear relationship between increased lateral stress and decreased axial failure stress (fig 2.78).

From figure 2.70, it can be inferred that when the equation for shear failure, $\tau_{max} = c + \sigma_n \tan \phi$, is applied to monolithic soil materials, the value of the cohesion, c, is in reality just the product of the intrinsic stress, σ_i, and $\tan \phi$. It can also be inferred that the idea of a residual shear strength at zero normal stress (cohesion, or c) would not be completely realistic in this case because all shear resistance would be truly frictional—only that there was a previously unconsidered source of normal stress, ie, the intrinsic stress, σ_i.

The intrinsic stress is believed to be caused primarily by bonding of soil particles by water bodies existing in two forms. The first of these is water filling in an undrained pore and existing at a suction less than the soil moisture suction just necessary to cause rupture of the air-water interface outside the pore and cause air entry into the pore, with subsequent pore drainage. Such undrained pores are considered, in moist soils, to be interconnected and be at a common suction. The second form of water body is believed to be an isolated volume trapped between an air-water interface film surrounding two

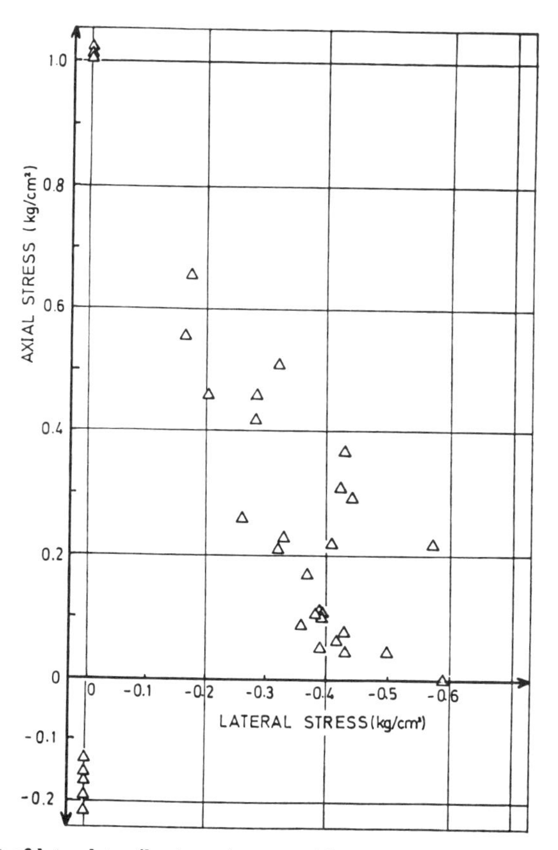

Figure 2.78–The effect of lateral tensile stress (generated by sample centrifugation) in reducing axial stress required for failure of cylindrical monolithic soil samples (from Kitani, 1975).

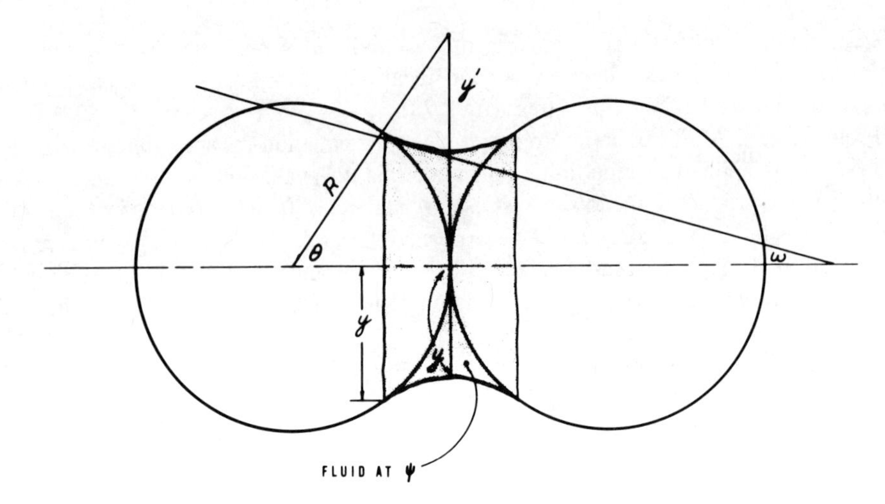

Figure 2.79–Illustration of the moisture tension (negative stress) caused by a water body between two spherical soil particles (from Vomocil et al, 1968).

adjacent soil particles (Kézdi, 1974; Vomocil et al, 1968) (fig 2.79). Suction in the water is considered to be the same as that at which the body was formed when the pore, of which it was once a part, was drained by an increasing level of soil water suction (Morrow and Harris, 1953).

There is a set of theoretical relationships among the size of particle, the form of particle packing, and water surface tension characteristics that determine the soil moisture suction at which a pore will drain (air will enter) and the amount of moisture that will be drained from the pore, as well as the amount of moisture that will remain in isolated water bodies (Vomocil et al, 1968; Snyder and Miller, 1985). These theoretical

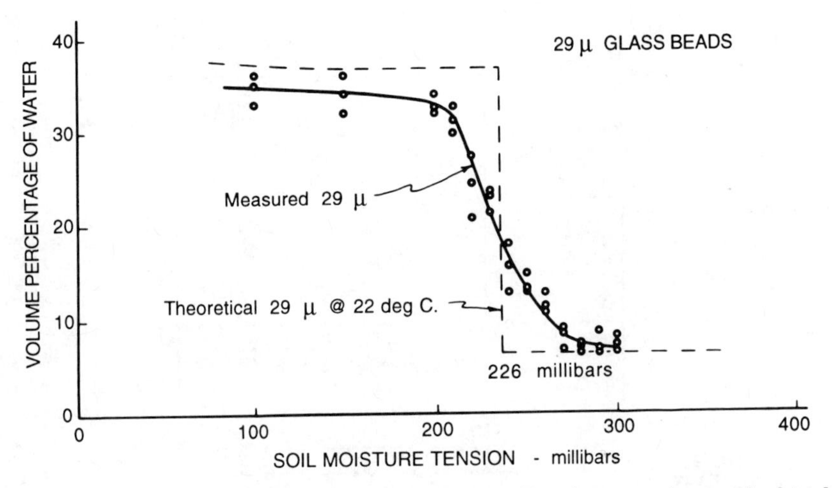

Figure 2.80–Comparison of theoretical and experimental soil moisture tension required to drain a sample made of 29 mm diameter glass beads. (from Vomocil et al, 1968).

relationships have been found to be quantitatively applicable to soil-like materials (see fig 2.80).

When a given volume of saturated soil is subjected to soil moisture suctions slowly increasing from 0 to 15 bars and the volume of water extracted from the soil volume at each level of suction is measured, it is possible to make a computation for the intrinsic stress that would prevail in the soil at each level of soil moisture suction (Chancellor and Vomocil, 1970), based on the following concepts: (1) drained pores have a strength proportional to the soil moisture suction at which they were drained, (2) undrained pores have a strength equal to the soil moisture suction that will just cause additional air entry, and (3) the plane of soil failure caused by applying opposing stresses greater than the intrinsic stress, will be one that passes only through pore space (excluding passage through solid particles) and that passes through low-strength bonds surrounding high-strength material.

In tests applying this method to seven different soils, measured values of intrinsic stress seldom differed from predicted values by more than 50% of the predicted values (see fig 2.81). This indicates that water bonds constitute the major causative factor of the intrinsic stress, although other factors or other types of relationships between soil moisture and intrinsic stress may also exist.

When soil of a given moisture content is forced to assume a more compact state, an increased proportion of the moisture content is held in very small undrained pores associated with high values for soil-moisture suction, while the proportion of drained pore space associated with water bodies isolated at low soil moisture suctions is reduced

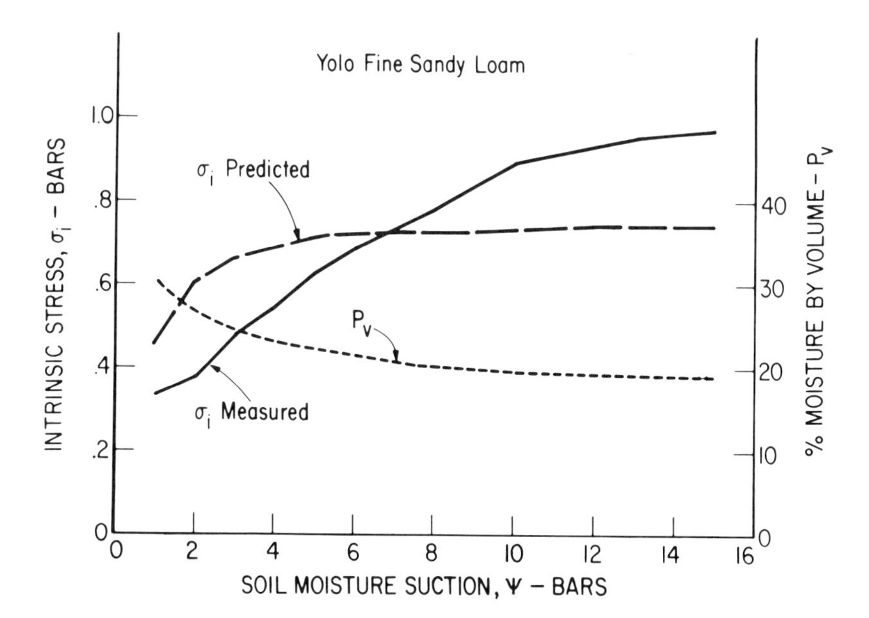

Figure 2.81–Comparison of measured intrinsic stress, σ_i, of a monolithic soil sample with values of σ_i computed using relationships between soil moisture tension and porosity (from Chancellor and Vomocil, 1970).

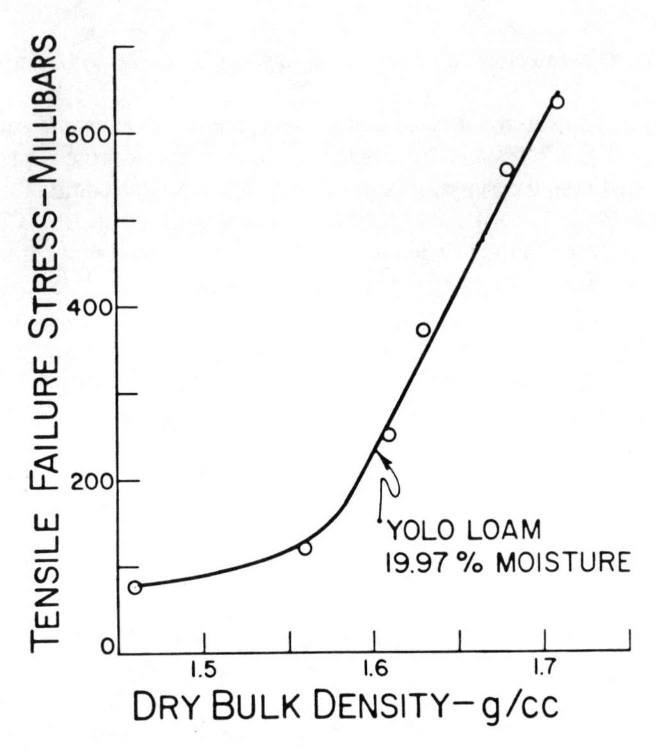

Figure 2.82–Effect of increased dry bulk density on the tensile stress required for failure of a monolithic soil sample (from Chancellor, 1971).

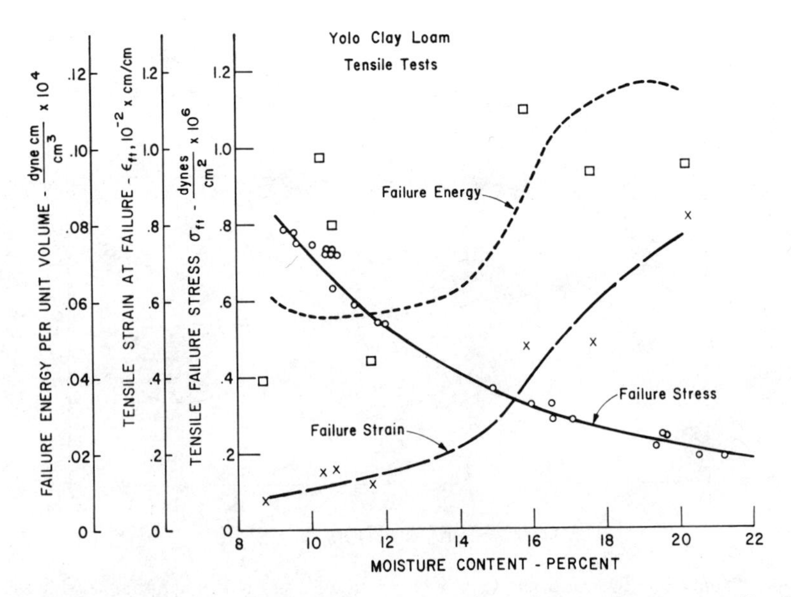

Figure 2.83–Effects of moisture content on the stress, strain, and energy inputs per unit volume for failure of monolithic soil samples in tension (from Vomocil and Chancellor, 1969).

(Jamison, 1953). Thus, the intrinsic stress (and the corresponding tensile failure stress) for a monolithic soil material at a given moisture content, is much higher for this material in a more dense state than in a less dense state (Vomocil et al, 1961; Chancellor, 1971) (see fig 2.82).

Monolithic soil materials subjected to tensile loading or loading in unconfined compression (Vomocil and Chancellor, 1969) require high stresses to produce failure at low moisture contents because of higher intrinsic stresses at the high soil-moisture-suction values associated with low moisture contents. However, because the water bodies responsible for the intrinsic stresses are much smaller at low moisture contents, less strain is necessary before the bonds become broken and the materials fail (see figs 2.83 and 2.84).

As illustrated in figures 2.81 and 2.83, it can be seen that intrinsic stress tends to increase as the soil becomes more dry (except in the case of coarse-grained soils that become very dry). This increased level of intrinsic stress at low moisture is the cause of increased densities of monolithic soil bodies that are allowed to dry (Gill, 1959).

Sliding Friction Between Soil and Other Materials

The most common relationship between the force required to cause one material to slide over another and the normal force between the two materials seems to be of a particularly simple form:

$$F_s = \mu N \tag{2.58}$$

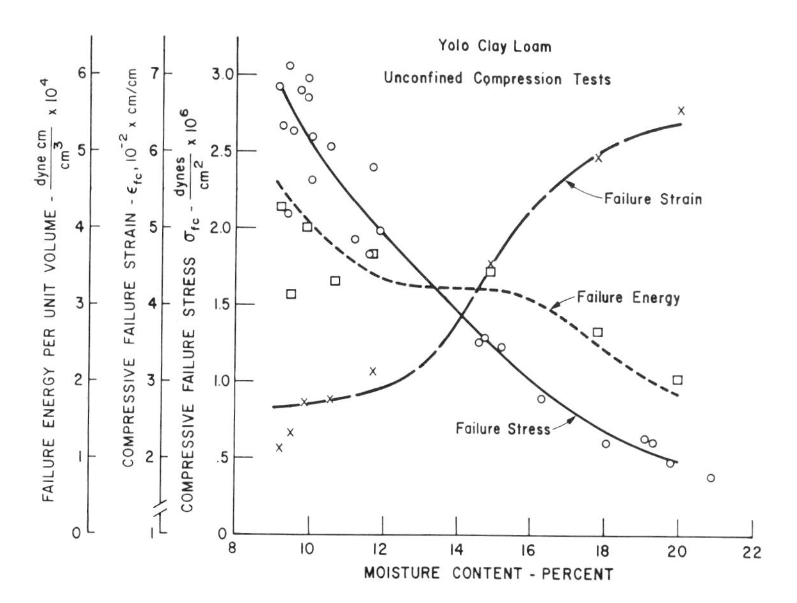

Figure 2.84–Effects of moisture content on the stress, strain, and energy inputs per unit volume for failure of monolithic soil samples in unconfined compression (from Vomocil and Chancellor, 1969).

or in per-unit-area terms:

$$F_s / A = \tau = \mu N / A = \mu \sigma_n \quad (2.59)$$

in which
F_s = force required to produce sliding
A = area of mutual contact
N = normal force between the two materials
τ = shear stress
σ_n = normal stress
μ = coefficient of friction

For agricultural soils sliding against common materials used in tillage and traction devices, such as steel and rubber, the relationships found by numerous investigators appear to be neither simple nor consistent across soil type and conditions. Three comprehensive reviews of this situation are Gill and VandenBerg (1968), Hendrick and Bailey (1982), and Koolen and Kuipers (1983).

In order to deal with the numerous sources of complication, the basic relationship given above has been modified to the following:

$$\tau = \mu \sigma_n + C_\alpha \quad (2.60)$$

in which C_α is the adhesion, or residual between-material shear resistance per unit area when no normal force between the materials exists. Some investigators have chosen a slightly different form of this expression:

$$\tau = \mu (A_N + \sigma_n) \quad (2.61)$$

in which $A_N = C_\alpha/\mu$. Implicit in this latter form (Robbins et al, 1987b) is the idea that shear resistance (adhesion) is the result of bonding forces between the soil and the other material (most likely caused by moisture tension), which should then be added to the applied normal stress, σ_n, to obtain the total normal stress. These total normal stresses, when multiplied by the coefficient of friction, then result in the total sliding resistance. Thus, all the factors responsible for variations in the final sliding force are factors that modify the coefficient of friction.

Factors that have been found to affect the forces required to slide soil on other materials are, in part:

1. Soil texture
2. Soil moisture content and/or tension
3. Soil porosity
4. Material hardness

5. Sliding path length—the displacement of a given spot on the soil surface relative to the material in question
6. Sliding velocity
7. Material type
8. Level of normal stress
9. Stiffness of loading system and rigidity of the soil materials
10. Quasi-static loading vs fully kinetic sliding
11. Maximum values of the normal stress during the course of the test history
12. Sudden increases or decreases of normal stress

In addition, it was found that a number of interactions between two or more of the above factors also affect the force levels to produce sliding of soil on another material.

There are two main categories of apparatus types used to measure the sliding friction forces of soil on other materials. The first of these apparatus types is the simple displacement of a piece of the material across the soil surface or vice-versa. In one case (Butterfield and Andrawes, 1972) the lower half of a conventional translational shear box was filled with a piece of the material in question while the upper half of the box was filled with soil. The other category of test device involves the use of an annular piece of the material that is caused to rotate on either an annular sample of soil (fig 2.85) or on an *in situ* soil surface.

A third type of device used in one instance was a special transducer that measured normal and frictional forces simultaneously (Spektor et al, 1985). Data were obtained when this device was pressed into preformed cavities in a soil mass.

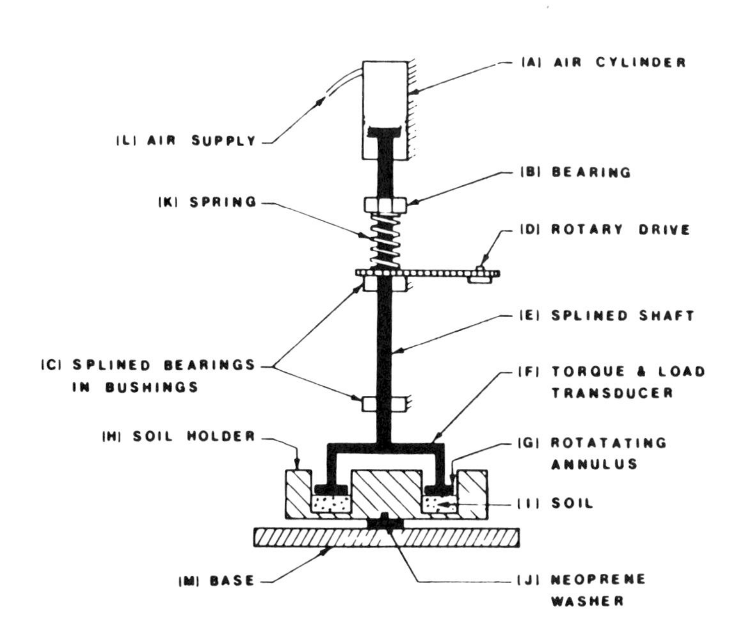

Figure 2.85–Apparatus for measuring soil-metal sliding friction forces (from Hendrick and Bailey, 1982).

One comprehensive set of measurements using an annular apparatus (Soehne, 1953) involved soil variables of texture, moisture content, and porosity. Test variables included normal stress, sliding path length, rubber vs steel materials, as well as hardness and degree of polish of the steel. Results in which the coefficient of friction, μ, was the only dependent variable yielded values for μ ranging from 0.2 to 0.75 for steel sliding on soil, and from 0.3 to 1.1 for rubber sliding on soil.

Another study (Robbins et al, 1987a and b; Robbins et al, 1988) included variables of soil type, soil moisture, adhesive bonding forces, sliding path length, normal stress levels, and the sequence of normal stress applications. It was found that a set of relationships involving seven constants (some of which reflected interactions among several variables) was required to reflect the results. When used to predict the force required to drag a flat steel plate over the soil surface these results yielded values ranging from 0.35 to 1.31 times the normal load on the plate.

Sliding Path Length. In general, with a comparatively rigid material sliding against soil, a certain amount of relative travel was required before the full frictional force would be developed. For annular test devices any sliding path length can be obtained. However, with a plate being dragged over the soil surface, the maximum sliding path length is equal to the length of the plate, and points toward the forward edge of the plate will have shorter sliding path lengths than points toward the rear of the plate. In some cases

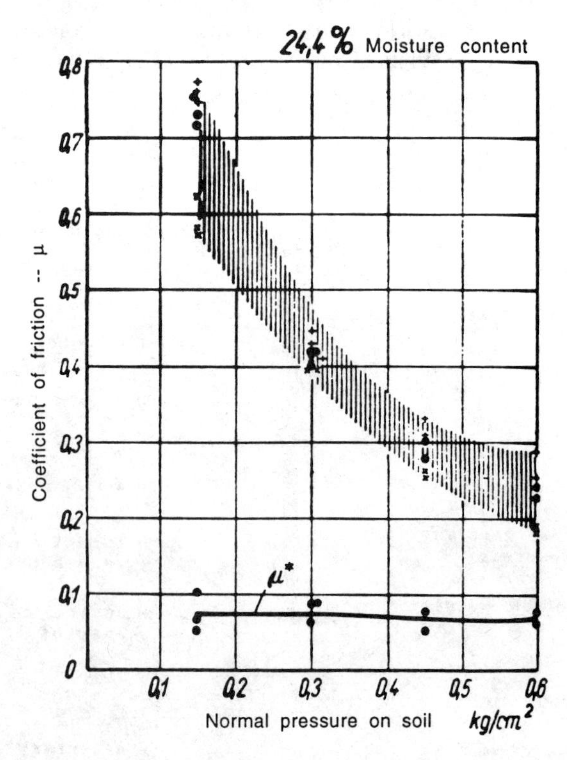

Figure 2.86–Effect of normal stress on the coefficient of friction of steel sliding on a loamy clay soil at 24.4% moisture (hatched zone). When frictional forces were divided into components for adhesion and for that in proportion to normal load, μ*, the resulting low values of μ* are shown (from Soehne, 1953).

frictional forces decrease once a certain sliding path length has been achieved (Yao and Zeng, 1988), but generally this is not the case. The sliding path length required to achieve near-maximum friction forces ranges from about 2 mm (Yao and Zeng, 1988) to as much as 1 m (NeSmith et al, 1986). This distance appears to be related to the physical nature of the test apparatus and to the characteristics of the soil — dry soils require longer distances (Soehne, 1953). For the case of rubber against soil, sliding path lengths required to achieve near-maximum frictional forces tended to be close to zero for many soil conditions (Soehne, 1953), and there was a common tendency for frictional forces to decrease with increased sliding path lengths.

Normal Stress. When adhesion, C_α, between soil and the other material affects the frictional force, but attempts are made to account for all variation by changes in μ alone, increasing normal stress levels will cause an apparent decrease in μ, the coefficient of friction (fig 2.86) (Soehne, 1953). In such cases this apparent characteristic will disappear if adhesion is accounted for independently. This characteristic was found to apply for both steel and rubber when sliding on soil. For most soils, however, friction stresses tend to change linearly with normal stress level, although cases in which small increases of the coefficient of friction that occurred with increases in normal stress for one soil/moisture condition have been reported in the same study in which small coefficient of friction decreases have occurred with normal stress increases for another soil/moisture condition (Soehne, 1953). The value of adhesion, C_α, in some cases has been shown to increase with increases in the normal stress level (Stafford and Tanner, 1976).

Soil Moisture Content. For many soils the coefficient of friction tends to be maximum at some intermediate moisture content, with lower values at higher and lower moisture contents (Soehne, 1953; Stafford and Tanner, 1976; Yao and Zeng, 1988; Zhang et al, 1986). This effect is more pronounced with fine-grained soils than with coarse-grained soils (Soehne, 1953; Robbins et al, 1988; Yao and Zeng, 1988). This same characteristic also appears to be the case for rubber sliding on soil, but on some soils reduced friction at low moisture content is not evident (Soehne, 1953; Neal, 1966). Moisture content has a significant effect on adhesion, C_α, with adhesion increasing as moisture goes from low to intermediate values (Zhang et al, 1986). The soil-to-material bonding stress will depend on the soil moisture tension, and so it could be expected that more fine-grained soils (fig 2.86) would exhibit greater adhesion at certain moisture contents than would be the case for coarse-grained soils. At very high moisture contents that would produce positive pore water pressures under load, tension would decrease reducing adhesion (Neal, 1966), and may even increase to effect lubrication.

Soil Porosity. Two studies (Stafford and Tanner, 1976; Soehne, 1953) provide data to indicate that the coefficient of friction is not much affected by soil porosity, but that C_α tends to increase as porosity is reduced (the soil becomes more dense). Values of C_α for soils of 30 to 35% porosity may be in the range of 5 to 6 kPa. In another study with pure sand at 36 and 44% porosity (Butterfield and Andrawes, 1972), the coefficients of friction on steel, roughened steel, glass, and methyl methacrylate (Plexiglas*, Lucite*,

---

* Trade names are used solely to provide specific information. Mention of a trade name does not constitute an endorsement of the product to the exclusion of other products not mentioned.

Perspex*) were, at the higher porosity, for all materials, 66% of that at the lower porosity.

Quasistatic vs Sliding Friction. When the frictional load is supplied via a soft spring structure, if the static coefficient of friction is greater than the kinetic coefficient of friction, a "stick-slip" sequence of events will take place (fig 2.87) (Butterfield and Andrawes, 1972). When a pure sand was tested against steel, glass, and methyl methacrylate, the value of the coefficient of friction found for steel in kinetic friction was 87% of that for static friction. For glass and methyl methacrylate the values were 83 and 72%, respectively.

Change of Normal Stress. The value of adhesion, C_α, is especially sensitive to the highest value of the normal stress, σ_n, that ever occurred on the surface element. The higher σ_n will give a closer contact. In one study (Soehne, 1953), up to a fivefold increase in C_α was found with a tenfold increase in σ_n (Koolen and Kuipers, 1983). In cases in which the normal stress underwent a step decrease, values of C_α, found were greater than when, σ_n, underwent a step increase (Robbins et al, 1987b). In both cases, however, the level of adhesion measured tended to decrease with increases of the magnitude of the step in σ_n both for step increases and step decreases. The coefficient of friction, μ, in these tests increased with the magnitude of the step change in σ_n. In some cases with a moist clay soil, the sliding resistance was less than the normal load at a higher initial level of σ_n. When σ_n was sharply reduced, sliding resistance decreased, but only a relatively small amount so that it was higher than the subsequent normal load (fig 2.88) (Robbins et al, 1987a).

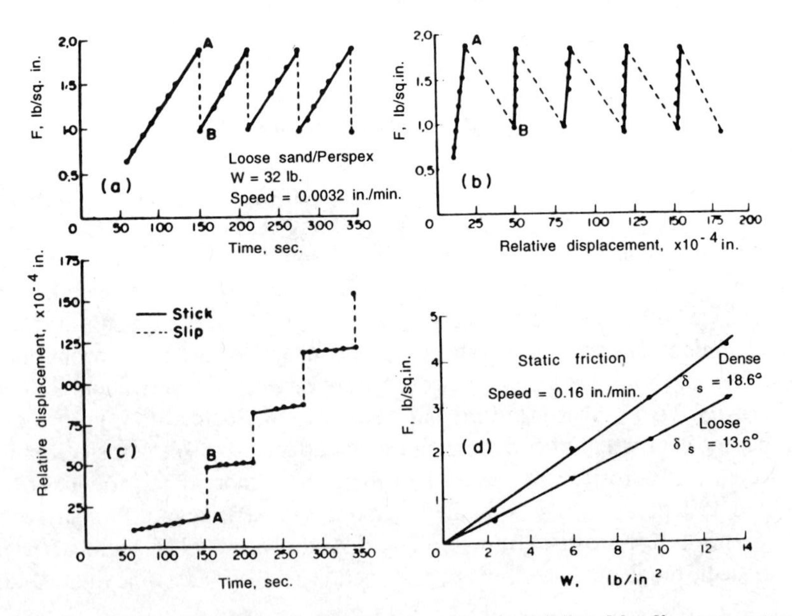

Figure 2.87–Illustration of the stick-slip phenomena when the frictional loading apparatus was not particularly rigid. The effects of sand density can be seen. Reprinted with permission from *Journal of Terramechanics* 8(4), Butterfield, R. and K. Z. Andrawes, On the angles of friction between sand and plane surfaces. Copyright © 1972, Pergamon Press, Ltd.

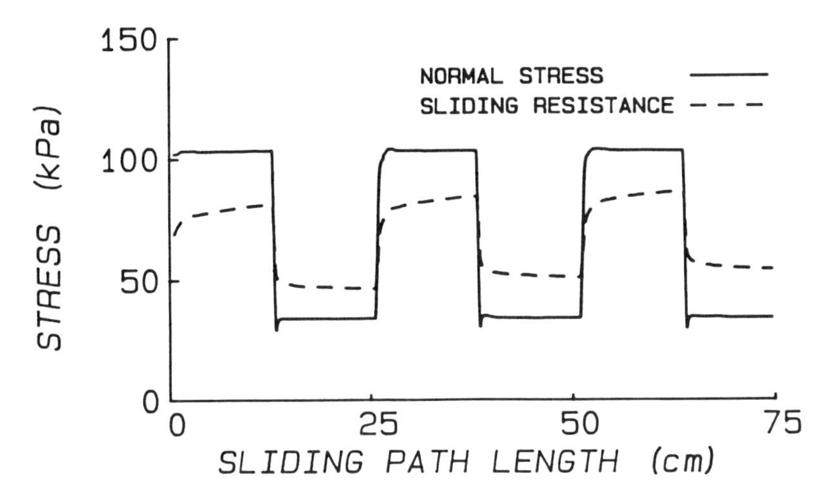

Figure 2.88–Effects of sudden application and release of normal stresses when an annular steel ring was forced to slide on a soil surface. At high normal stress, the coefficient of friction was less than 1.0, but when the normal stress was suddenly reduced, the coefficient of friction became greater than 1.0 (from Robbins et al, 1987a).

Type of Material. In a study with a pure sand (Butterfield and Andrawes, 1972) coefficients of friction for smooth glass, methyl methacrylate (a plastic), and steel were 0.166, 0.210, and 0.243, respectively. In a case where a polished tempered steel ring (Vickers Hardness, 550 kg/cm^2) was compared with a hardened ground steel ring (Vickers Hardness, 900 kg/cm^2), the coefficients of friction were nearly the same initially, but at higher levels of sliding path length the values for the ground ring were somewhat higher. For soils at high moisture content, there was no difference between the two rings (Soehne, 1953). The coefficient of friction measured with steel of various levels of hardness against soil (Koolen and Kuipers, 1983) was generalized by:

$$\mu = 0.37 - 0.00015H \tag{2.62}$$

in which H = Brinell hardness number (650 for very hard tempered steel; 125 for low carbon steel).

In a study with a wet clay soil (cohesion = 10 kPa, coefficient of internal friction = 0.32, moisture content = 49% by weight) Salokhe and Gee-Clough (1988) determined values of μ and C_∞ for steel wheel lugs coated with the various materials (table 2.8).

Teflon*-coated steel and a comparison steel surface were used in tests with an annular ring device and with an inclined blade moved through a soil bin. In both cases involving three different soils (Fox and Bockhop, 1965), the Teflon-coated surface displayed negligible adhesion, although significant adhesion was noted for the steel at high

* Trade names are used solely to provide specific information. Mention of a trade name does not constitute an endorsement of the product to the exclusion of other products not mentioned.

Table 2.8. Coefficient of friction and adhesion values for coated steel wheel lugs

Coating	Coefficient of Friction (μ)	Adhesion (C_α) kPa
Lead oxide paint	0.51	4.33
Ceramic tile	0.23	0
Teflon* sheet	0.36	0.1
Teflon tape	0.70	0.27
Enamel	0.38	0

* Trade names are used solely to provide specific information. Mention of a trade name does not constitute an endorsement of the product to the exclusion of other products not mentioned.

moisture contents. At high moisture contents there were negligible differences in the coefficient of friction due to the Teflon in the annular ring tests. However, at intermediate moisture contents the average coefficients of friction for steel and Teflonwere 0.583 and 0.322, respectively. At low moisture contents the respective values were 0.42 and 0.282. In tests with the inclined blade, draft was reduced by 27% at low moisture contents but by 31% at high moisture contents. The Teflon material exhibited rates of wear which were 8 to 10 times that of steel.

Surface roughness has a large influence on friction. For a rusted steel surface the coefficient of friction, μ, may be as high as the coefficient of internal friction of the soil, tan ϕ, and thus may exceed 0.8. If the rust is removed, friction will be considerably less, but a high degree of surface polish will result in only a minor decrease in the coefficient of friction (Koolen and Kuipers, 1983).

The coefficients of friction of steel on soil, rubber on soil, and soil on soil were compared at various sliding path lengths for two soils (Soehne, 1953) (fig 2.89). The coefficient of friction for soil on soil did not change with path length and was consistently higher than μ values for the other materials on soil. At initial levels of sliding path length, the coefficients of friction for rubber on soil were considerably higher than those for steel on soil. As the sliding path length exceeded 1 m, the values for the two materials tended to converge.

Soil Texture. For low levels of sliding path length for steel sliding on soil, Nichols (1931) found that the coefficient of friction, μ, could be represented by:

$$\mu = U + V \times (\text{percent of clay content in soil}) \tag{2.63}$$

in which U and V are arbitrary constants. U = 0.24 and V = 0.005 when clay percentage is less than 32; U = 0.32 and V = 0.0 when clay percentage is equal to or greater than 32. However, a study with clays with up to 45% clay indicated that the coefficient of friction continued to increase with clay content even at these high clay contents (Zhang et al, 1986). This indicates that fine textured soils tend to have a higher coefficient of friction than coarse-textured soils (Koolen and Kuipers, 1983).

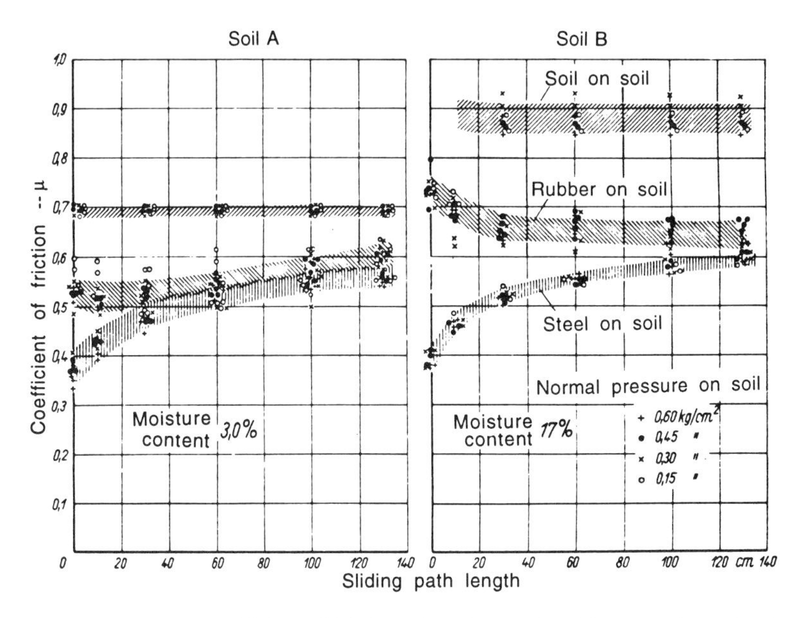

Figure 2.89–Comparison of coefficient of friction of steel on soil, rubber on soil, and soil on soil for various levels of sliding path length: Soil A = sandy soil. Soil B = fine-grained loam soil (from Soehne, 1953).

In a study with a sand, a loam, and a clay soil (Soehne, 1953), while the coefficients of friction of the loam and clay soils tended to decrease considerably at high moisture contents (which was not the case for the sand soil) the maximum values for the coefficient of friction for the loam and clay soils were slightly higher than that for the sand (fig 2.89). An identical pattern of differences between a sandy loam and a clay loam soil were noted when sliding friction tests for soil against steel were conducted at sliding velocities of 1.7 m/s, while at velocities of 0.45 m/s there were few differences between the two soils (Yao and Zeng, 1988). In general, coarse-grained soils exhibit ratios of sliding resistance to normal load which change much less with moisture content than is the case for fine-grained soils (Soehne, 1953; Yao and Zeng, 1988; Robbins et al, 1988).

Sliding Velocity. It is with regard to the effects of sliding velocity on the friction between soil and other materials that there is the most disagreement among various sources of information. On one hand, Koolen and Kuipers (1983) cite data indicating that increases in the coefficient of friction on the order of 0.1 to 0.15 are associated with increases in sliding velocity above the quasi-static case. It is believed that this characteristic applies to wet soil and/or for speed variations over one or more orders of magnitude. In a study aimed specifically at the effects of sliding speed, Yao and Zeng (1988) found that the coefficient of friction increased linearly with the log of the sliding velocity when velocities ranged from 0.45 m/s to 1.7 m/s. The increase in shear stress between these two velocities was 32 kPa, irrespective of whether the test area carried a normal stress at 25 kPa or 250 kPa, indicating that the effect of velocity was primarily in the value of adhesion, C_α.

On the other hand, Gerlach (1953) cites data showing a linear decrease of the coefficient of friction from 0.83 at 0.2 m/s to 0.72 at 1 m/s (fig 2.90). Stafford and Tanner (1976) presented data to indicate that between speeds 0.017 mm/min and 0.29 mm/min the coefficient of friction tended to decrease linearly with the log of the speed. The decrease in the coefficient of friction between those speeds was approximately 0.1 irrespective of the general level of the coefficient of friction (coefficients at 0.017 mm/min ranged from 0.2 to 0.33 due to differences in moisture content; higher friction coefficients at higher moisture contents).

Reduction of Frictional Resistance. Much of the energy used to operate tillage tools goes to overcome frictional sliding resistance as soil moves over tillage tool surfaces. One approach to reducing tillage energy requirements has been to use surfaces with low frictional properties (see above). Another approach is to use a lubricating fluid to reduce soil-metal friction. Three procedures will be reported here, lubrication with a polymer, lubrication with air, and lubrication with water attracted to the metal surface by electro-osmosis.

Tests were conducted with a clay loam and a sandy loam soil and a steel annular friction testing apparatus in which Teflon coatings and polymer application were used (Schafer et al, 1975). Then a polymer lubricated blade was tested at different rake angles. The Teflon coating caused an average 47% reduction in the coefficient of friction from that with steel while with the application of 1, 3, and 5% polymer materials to the sliding surface resulted in percentage reductions in the coefficient of friction of 38, 56, and 61, respectively. With the blade lubricated by a 3% solution of polymer, the average draft reduction was 16%. In this case the application rate of polymer-water solution was at a rate equivalent to 103 L/ha. When a 3% polymer solution was injected at the base of the moldboard on a moldboard plow, plow operation was made less troublesome in very sticky soils (Schafer et al, 1977). Average draft reduction in 15 trials on widely varying soils was 22% with an average application rate of 140 L/ha, although most of the draft reduction was obtained with application rates of 50 to 100 L/ha.

The use of air-flow through or adjacent to tillage tool surfaces has been used to reduce frictional forces. When a perforated steel plate with 0.71-mm-diameter holes arrayed on

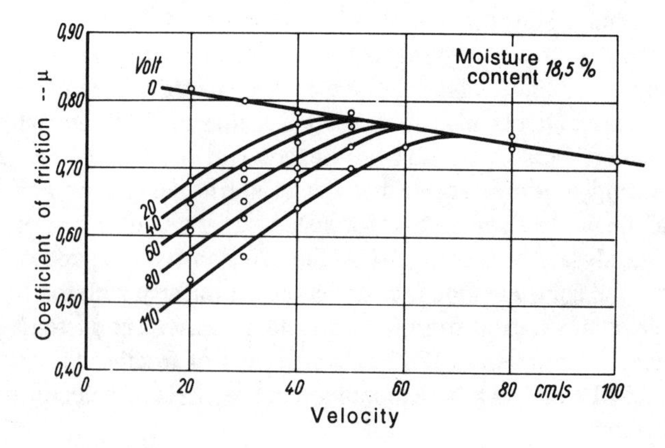

Figure 2.90–Effect of various electro-osmosis voltages on the coefficient of friction of soil sliding at various speeds against a steel plow share (from Weber, 1932, as reproduced in Gerlach, 1953).

a 2-mm grid spacing was compared with a plain steel plate, it was found that even when air flows through the perforated plate consumed 1½ times the draft power for compression, the perforated plate required higher draft forces under all test conditions. Many of the perforations near the cutting edge became plugged with soil (Bigsby and Bockhop, 1964). In another study (Mink et al, 1964), air was injected in longitudinal slots milled in the sides of a vertical tool shank. The maximum air-flow rate used was 21 m^3/min (at standard conditions)/m^2 of tool side surface area at a pressure of 172 kPa. (This pressure was applied across 30, 2.8 mm diameter holes, tool surface area was 0.016 m^2). An approximate 30% draft reduction (from an unaided draft force of 84 N) under these maximum conditions was obtained when the tool moved forward at 0.25 m/s.

In a third study (Kitani, 1978), air nozzles were located at the base of the moldboard plow. When the nozzles were aimed 60° forward of the normal (in the longitudinal-vertical plane) to the moldboard surface and 60° to the right of the direction of motion, draft reduction was 18% at a gage air pressure of 80 kPa and 37% at a gage air pressure of 200 kPa (a combination of angles of 30 and 90°, respectively, was equally effective). Further increase in air pressure did not have much effect (plowing speed was 1.62 km/h). The vertical force on the plow was not affected much at all by the air flow application.

In an attempt to produce low-friction soil failure ahead of a subsoiler (Araya and Kawanishi, 1984), the subsoiler was equipped with a 14.5-mm-diameter air nozzle in the blade tip and, in some cases, with a second nozzle, about 15 cm above the tip. When operating at 1.4 cm/s in soil at 16% moisture, the application of 290 kPa air pressure to the nozzle caused a 50% reduction in draft force. Draft power savings amounted to 18 W, while the power required for air application was 880 W. In soil at 30% moisture content, it was found that a short pulse of air pressure could crack the soil about 30 cm ahead of the tool. A 2-s pulse of air at 98 kPa resulted in a 25% draft reduction lasting for a period of 30 s (forward velocity = 1.13 cm/s). The energy to compress the air was 967 J, while the draft energy saved was 1020 J.

A third method used to reduce friction of tools sliding against soil surfaces is that of electro-osmosis in which the negative pole of a direct-current voltage source is attached to the tool, and the positive pole is attached to a large, nearby electrode in contact with the soil. If the voltage is high enough and the speed of sliding is sufficiently slow, additional moisture will be deposited at the tool surface, which will tend to lubricate the surface and reduce sliding friction. Figure 2.90 shows the effect that electro-osmosis can have on the coefficient of friction (Weber, 1932, as cited by Gerlach, 1953). When this voltage was applied to the share of a plow, draft was reduced as much as 19% with a current flow of 1 A (Weber, 1932, as cited by Gerlach, 1953).

Clyma and Larson (1991) tested the effects of electro-osmosis on the draft of a cultivator sweep 260 mm wide and 203.2 mm long while operating in a soil bin. Two coulters 203 mm in diameter were used as anodes and located 102 mm behind and 13 mm to either side of the sweep. Various voltages were applied, and various speeds were used in different soils at different moisture contents. The largest draft reduction, 15.9%, (for the sweep alone), was found while operating with a 45-V DC potential (highest voltage used) and a forward speed of 3.3 km/h (lowest speed used). The draft power (sweep alone) was 361.2 W (down from 429.5 W without electro-osmosis), while the electric power consumed was 15.1 W (current = 0.33 A) in a clay loam soil at 10% moisture content. The draft of the coulters themselves required a power input of 265.6 W, so that such a system would have a net advantage only in cases only where the coulters were already part of the operating system. Under the above test conditions, removal of one of the coulters allowed a draft reduction (for the sweep alone) of 11.4%, somewhat less than with both coulters.

Empirical Dynamic Properties of Soil

Because agricultural soils are so variable and because of difficulties both in measuring soil properties that are adaptable to a system of mechanics and in applying these measures in the design process, many engineers have made use of empirical measures of soil properties and have then correlated these properties with the physical performance of the soil under operating circumstances. It is only through the correlation process that such empirically measured properties have derived significance.

Abrasiveness

There are no well-known measures of abrasiveness for agricultural soils. Generally, designers are concerned about wear of soil-engaging tools. The wear process depends on properties of the soil, properties of the tool, and features of the process in which the two are engaged. To categorize soil abrasiveness, usually a common steel tool is used, and the rates of wear are compared among various soils. In some cases, a standard abrasive material may be used in place of the soil for comparison purposes. Similarly, it is difficult to specify the wear resisting properties of various materials. Thus the comparison process is again used to compare (in a given soil or abrasive material) the rate of wear of the material in question with the rate of wear of a common steel material.

The hardness of soil materials such as quartz is several times that of common steel. A number of reports specifying the hardness of steel and other materials have used a Vickers diamond pyramid indenter having an angle of 136° between opposite faces. The resulting readings are given in kg (force units)/mm^2. Mild steel has hardness numbers of about 200 kgf/mm^2, and hardened die steel may have readings as high as 900 kgf/mm^2, (Richardson, 1967). The hardness of quartz and flint, however, may be in the range of 1000 to 1140 kgf/mm^2 (Richardson, 1967). Hardness values of sand grains from several soils ranged from 813 to 1088 kgf/mm^2 (Zhou, 1986a). Hardness values of various abrasive materials are as follows (Richardson, 1967): Silicon carbide, 3000 kgf/mm^2; Corundum,2160 kgf/mm^2; Alumina, 1200-1500 kfg/mm^2; Garnet, 1360 kgf/mm^2; Flint, 1060 kgf/mm^2; Glass, 590 kgf/mm^2.

One study of abrasive wear on plow shares concluded that if the hardness of the abrasive is less than 70 to 100% of the hardness of the tool material (depending on various factors such as abrasive grain size), that wear rates will not be a practical problem. If the hardness of the abrasive was more than 130 to 170% of the hardness of the tool (depending on various conditions), then wear would be a serious practical problem. When the ratio of hardnesses would fall between these two general ranges, then there was some potential for taking practical steps to increase the hardness of the tool material (Zhou, 1986b).

In wear tests, the results may be expressed in terms which can be normalized by dividing wear volume by common factors. In some cases (Zhou, 1986b) wear results were given in terms of mm^2/GN, which is the result of:

$$\frac{\text{volume of wear}\left(mm^3\right)}{\text{contact area}\left(mm^2\right)} \times \frac{1}{\text{relative movement (mm)}} \times \frac{1}{\text{normal pressure}\left(N/mm^2\right)} \quad (2.64)$$

This equation form indicates that the abrasive effect is expected to be in proportion to the relative movement and to the normal pressure applied. The factors of relative movement and normal pressure have been found in both laboratory and field tests to be unaffected by differences in friction when the tool material hardness was greater than 400 kgf/mm^2 (Richardson, 1967).

Although the relative rate of abrasive wear of various tool materials in contact with a given soil is usually linearly correlated with the hardness of the material (Richardson, 1967), some materials with hard inclusions of comparatively large size in a softer matrix tend to have less wear than those in which the hard inclusions are smaller. Similarly, the size of the hard particles in the soil plays an important role in the rate of wear. In one case in which steel materials were tested in flint grit, (Moore and McLees, 1980), the wear rate when flint particles diameter averaged 357 μm was approximately 30% greater than when the average diameter was 92 μm. When wear rates were compared in two soils in which the fine fractions had the same particle size distribution, same moisture content and same *in situ* shear strength, the wear rate in the soil containing stones (particles greater than 2 mm in diameter) was about 10 times that in the soil without stones (Richardson, 1967). When four different plowshare materials were tested in three soils (Zhou, 1986a), the relationships between particle size distributions and mean wear rates were as shown in table 2.9.

When mild steel, low-alloy steel, and spring steel were tested in two soils (a clay loam and a sandy clay loam), wear rates for the mild steel were greater in the clay loam (by a factor of about 60% over those in the sandy clay loam), while the wear rates for the other two steels were greater in the sandy clay loam (by factors of 85 and 175%, respectively, over those in the clay loam) (Moore and McLees, 1980).

Table 2.9. Mean wear rate of plowshare materials with regard to soil particle size distribution (Data from Zhou, 1986a)

Particle Size Distribution (%)			
> 1 mm	0.25 – 1 mm	0.05 – 0.25 mm	Mean Wear Rate*
0.0	0.58	25.17	0.12
1.33	3.44	57.54	0.520
23.95	29.10	28.38	2.437

* Designated in units of loss of length of a plowshare tip per unit distance traveled, mm/km.

Wear rates on a common steel tool were compared in different soils (Foley et al, 1984; Lawton and Foley, 1986) for several years. In a similar study, a standard steel seed-drill opener was used (Foley and McLees, 1986). The average wear rates determined by these studies are given in table 2.10.

Soil Consistency and Atterberg Limits

The term *soil consistency* (not exactly definable) refers to the characteristics of a soil that cause it to remain "consistent" under stress or maintain its shape when subjected to deformative forces (Hillel, 1980). Some empirical criteria aimed at making some measure of consistency are known as the Atterberg limits and comprise the following:

The *liquid limit* (or *upper plastic limit*) is the soil moisture content (dry basis by weight) at which the soil-water system changes from a viscous liquid to a plastic body. It is measured with a special apparatus (fig 2.91) (Sowers, 1965). A cup is filled with soil at one of several different moisture contents; a groove is then made in the soil with a grooving tool. The cup is lifted and dropped from a standard height of 1 cm until the soil has flowed sufficiently to close the groove along a 12 mm length. The number of impacts and the corresponding water content are plotted. The liquid limit is interpreted as the moisture content for which the groove is closed by 25 impacts (fig 2.92). The *plastic limit* (or *lower plastic limit*) is the moisture content at which the soil stiffens from a plastic to a semi-rigid friable state. In practice it is defined as the moisture content at which a sample of soil can just be rolled into a thread of 3 mm diameter without breaking (Sowers, 1965). The *shrinkage limit* is the moisture content at which the soil changes from a semi-rigid to a rigid solid with no additional change in specific volume as drying continues (fig 2.93). The *sticky limit* is the minimum moisture content at which a soil paste will adhere to a steel spatula drawn over its surface (Sowers, 1965).

Table 2.10. Effect of various soils on wear rates of steel agricultural tools

Mean Wear Rates*	Soil Description
Data from Foley et al (1984) and Lawton and Foley (1986)	
0.2769	Fine silty over clayey and fine loamy over clayey - very flinty
0.1787	Fine loamy over clayey and clayey soils - with flint and gravel
0.1287	Coarse and fine loamy soils, some calcareous coarse loamy soils over chalk
0.0982	Coarse loamy and sandy soils over sand or sandstone, and fine loamy over clayey soils
0.0650	Calcareous silty soils over chalk
0.0609	Calcareous clayey soils
0.0586	Coarse loamy and sandy soils over sand or sandstone
Data from Foley and McLees (1986)	
0.062	Deep, well-drained coarse loamy, coarse loamy over clayey and sandy soils, variable flint
0.047	Shallow, well-drained calcareous sandy and coarse loamy soils over chalk
0.033	Slowly permeable fine and coarse loamy over clayey soils
0.028	Well-drained sandy and coarse loamy soils over soft sandstone
0.024	Shallow, well-drained brushy calcareous fine loamy soils over limestone

* Designated in units of loss of length of a cultivator point per unit distance traveled, mm/km.

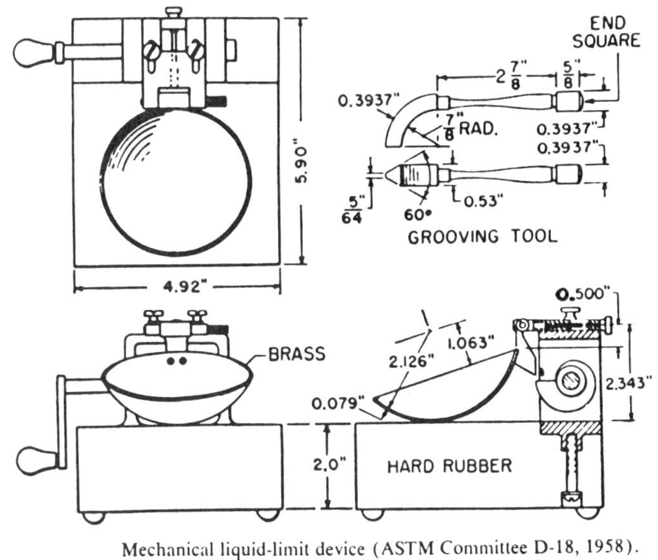

Mechanical liquid-limit device (ASTM Committee D-18, 1958).

Figure 2.91–Apparatus for determination of the Atterberg Liquid Limit (from Sowers, 1965).

From the liquid and plastic limits, it is possible to compute a plasticity index. The plasticity index is the difference between the liquid and plastic limits:

$$\text{plasticity index} = I_p = W_L - W_p \tag{2.65}$$

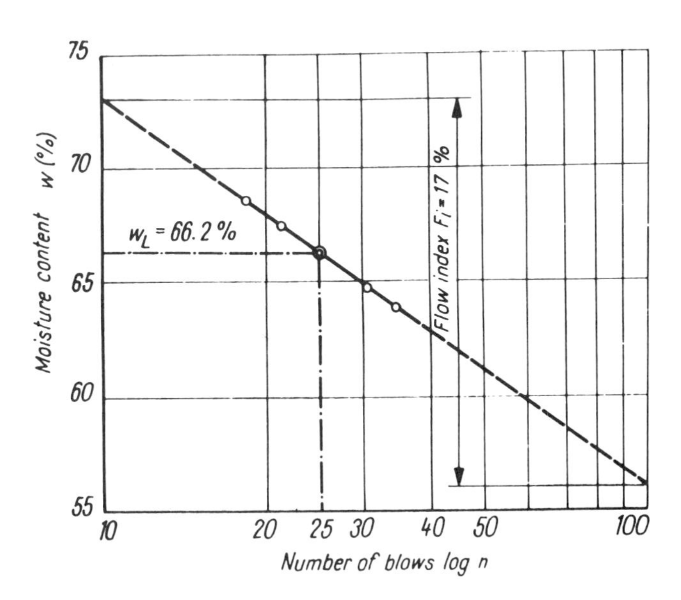

Figure 2.92–Interpolation diagram for obtaining the Atterberg Liquid Limit using a limited number of moisture content levels (from Kézdi, 1974).

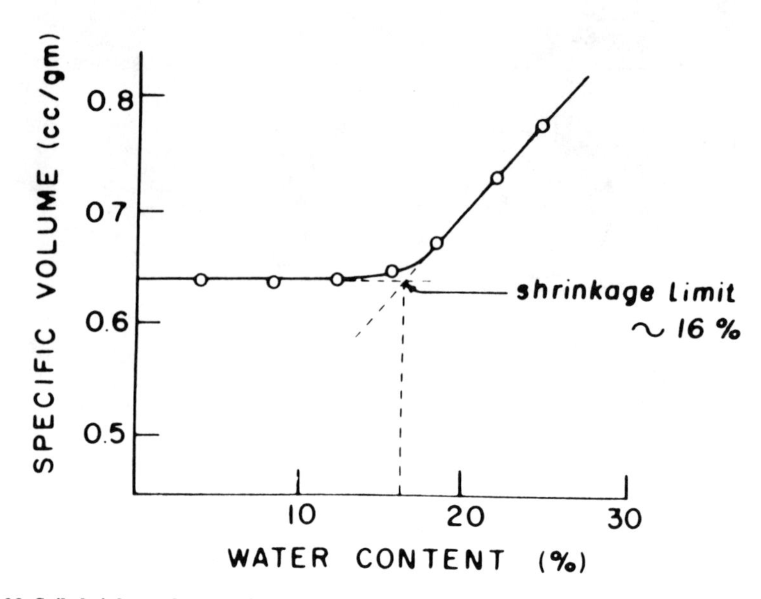

Figure 2.93–Soil shrinkage characteristics upon drying, showing how the Shrinkage Limit is determined. Reprinted with permission of Academic Press, Inc. from Hillel, 1980. Copyright © 1980, Academic Press, Inc.

in which W_L = liquid limit in percentage points, and W_p = plastic limit in percentage points. For clay soils, it is then possible to use the plasticity index to compute the activity of the clay (see fig 2.94) (Lambe and Whiteman, 1969):

$$\text{activity of clay} = \frac{I_p}{\text{\% of dry soil by weight finer than 2 } \mu\text{m}} \quad (2.66)$$

The activity of sodium montmorilirite is 7.2; of illite, 0.9; and of kaolinite, 0.38 (Lambe and Whitman, 1969).

The Atterberg limits can also be used in computing a consistency index (Kézdi, 1974):

$$\text{cosistency index} = I_c = \frac{W_L - W}{W_L - W_p} \quad (2.67)$$

in which W_L = liquid limit in percentage points, W_p = plastic limit in percentage points, and W = dry basis moisture content of the soil in percentage points.

Relationships between I_c and various physical soil conditions are illustrated in figure 2.95 (Kézdi, 1974), along with typical values for unconfined compression strength levels for each condition.

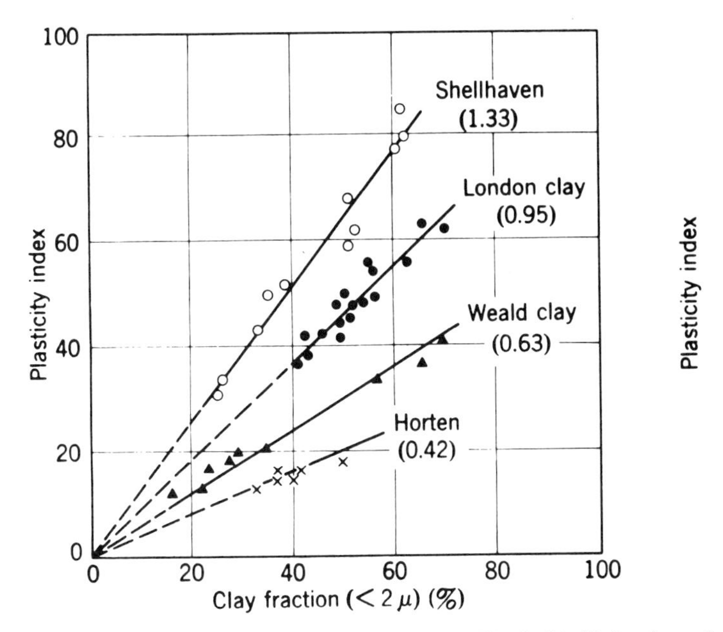

Figure 2.94–Relations between Plasticity Index (Liquid Limit - Plastic Limit) for three clay soils with various clay activity values shown in parentheses. Reprinted with permission of John Wiley & Sons, Inc. from Skempton, 1953 as reproduced in *Soil Mechanics*, Lambe, T. W. and R. V. Whitman, copyright © 1969.

The liquid limit and the plasticity index can be used to classify soils (fig 2.96) (Kézdi, 1974).

The coefficient of internal friction (tan ϕ) of soils may also be related to the plasticity index (fig 2.97) (Lambe and Whitman, 1969). The cohesion of soil can be related to the consolidating stress by a factor that is linearly correlated with the plasticity index (Kézdi, 1974) (see fig 2.73).

Compressibility characteristics of soils in drained tests can be expressed by the Compression Index, C_c (Lambe and Whitman, 1969):

$$\text{compression index} = C_c = -\frac{\Delta e}{\Delta \log \sigma_v} \tag{2.68}$$

in which e = void ratio, and σ_v = consolidating stress.

For compression of soils not previously compressed by other than natural processes, values of C_c may be estimated from:

$$C_c = 0.007\,(W_L - 10\%) \text{ for remolded soil}$$
$$C_c = 0.009\,(W_L - 10\%) \text{ for undisturbed soil}$$

in which W_L = liquid limit in percentage points.

The ability of a lugged wheel to develop tractive thrust has been examined relative to plastic and liquid limits of a sandy loam and a clay soil (Ali and McKyes, 1979). It was found for the sandy loam that as moisture content increased, approaching the plastic limit, tractive thrust increased regularly. As moisture content increased beyond the plastic limit, thrust decreased sharply to very low levels at the liquid limit. In contrast, with the clay soil, as moisture content increased from low levels, tractive thrust values increased regularly until the liquid limit was reached. As moisture content increases beyond the liquid limit, thrust values fell rapidly (fig 2.98).

There exist few effective means of estimating the Atterberg limits from more basic measurements such as moisture content, porosity, clay content, etc. Thus, it appears that these Atterberg limits are able to describe soil properties that are seemingly unique because the physics of soil strength is not yet a fully developed body of knowledge.

Penetrometer Pressures and Depth of Penetration

The use of penetrometers for evaluating soil physical properties has been practiced for many years and has a great deal of appeal because of the ease with which measurements can be made over extensive areas and depths of field conditions (Perumpral, 1987). Three general types of penetrometers have received common use: (1) a pointed device of a given mass that is allowed to fall a specified distance, after which the depth of penetration is measured, (2) a pointed device that is subjected to blows of a weight of given mass being dropped repeatedly from a given height onto a stop attached to the

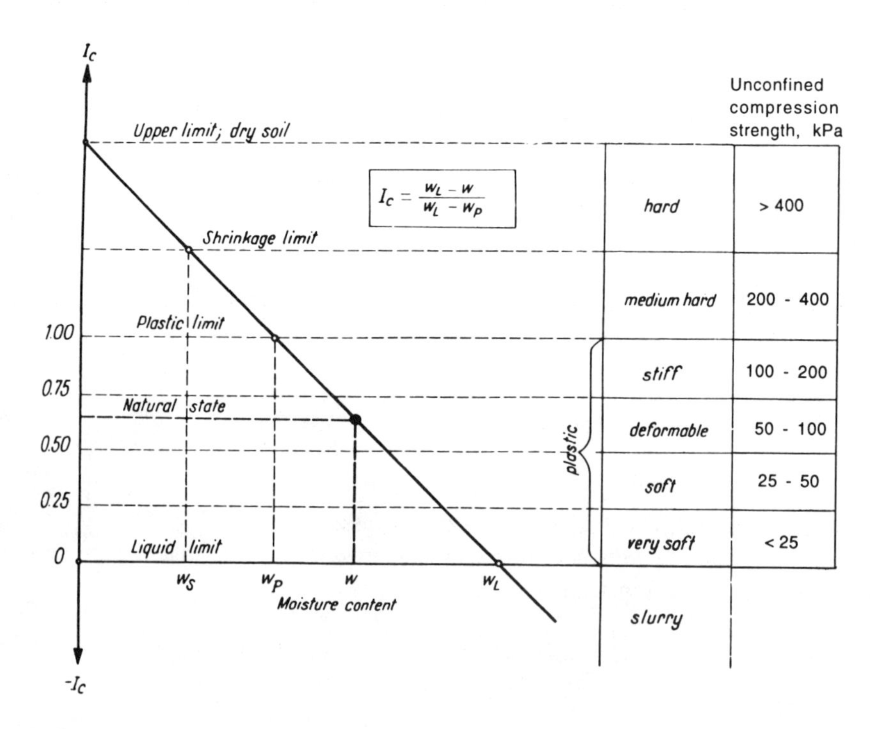

Figure 2.95–Relations among the Consistency Index (I_c), Atterberg Limits, and general categories of soil strength (from Kézdi, 1974).

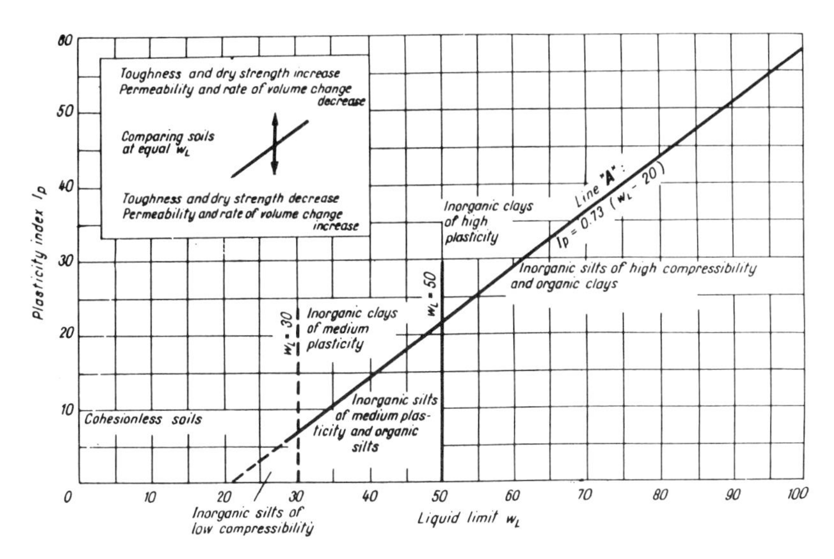

Figure 2.96–Relations of various soil categories to the "A" Line based on Atterberg Limits (from Kézdi, 1974).

shaft of the device, with this number of drops per unit depth traversed measured, and (3) a pointed device that is pushed into the soil; the force required is measured and usually normalized by dividing by the base area of the cone or pyramid which forms the point.

In agricultural operations, the third type is the most frequently used, and the most common design is that specified in ASAE Standard: ASAE S313.1 (ASAE, 1983-1984)

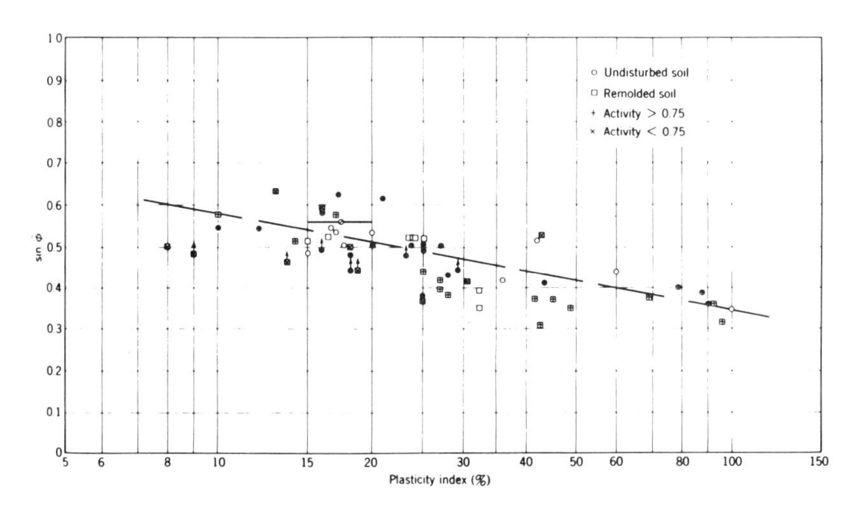

Figure 2.97–Relations between the Plasticity Index and the friction angle, φ , obtained in fully drained tests (u = 0.0) with cohesive soils. Reprinted with permission of John Wiley & Sons, Inc. from Kenny, 1959, as reproduced in *Soil Mechanics*, Lambe, T. W. and R. V. Whitman, copyright © 1969.

(fig 2.99). It is the same as that used by the US Army Waterways Experiment Station in Vicksburg, Mississippi (Knight and Freitag, 1962). The penetrometer has a polished steel cone with a 30-degree included angle at the tip and a 3.23 cm^2 (0.5 $in.^2$) base area. Tests with 30-degree cones have shown that different sizes of base area are unlikely to affect the force-per-unit-area values registered (fig 2.100) (Schafer et al, 1969). However, higher values of cone index have been found with larger diameter cones in some cases (Hassan, 1983). The readings obtained are, in some cases, affected by the speed of insertion (Freitag, 1965), and the standard rate of insertion (ASAE, 1983-1984) is 1.83 m/min (72 in./min). Readings are also affected by the general mechanism of insertion, with hand-powered insertion giving somewhat lower readings than hydraulically powered insertion (Sirois et al, 1989). One study in which the penetrometer was inserted at the standard speed and also inserted by striking the penetrometer with falling weights of varying mass (4 to 24 kg) and velocity (2 to 5 m/s) (Womac et al,

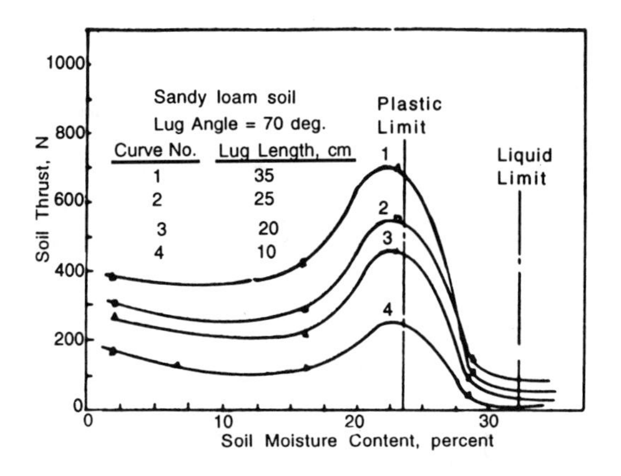

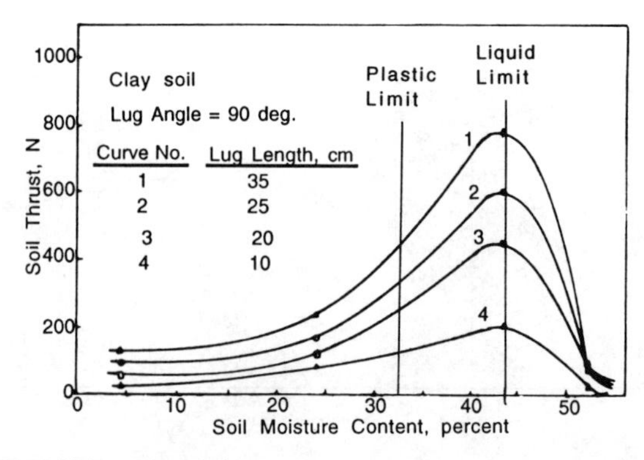

Figure 2.98–Tractive capabilities of a single wheel lug in relation to the Atterberg Limits for two soils (from Ali and McKyes, 1979). Normal load was 623N in all cases. Lug depth was 5 cm in all cases, and lug angle was was measured relative to the direction of forward motion.

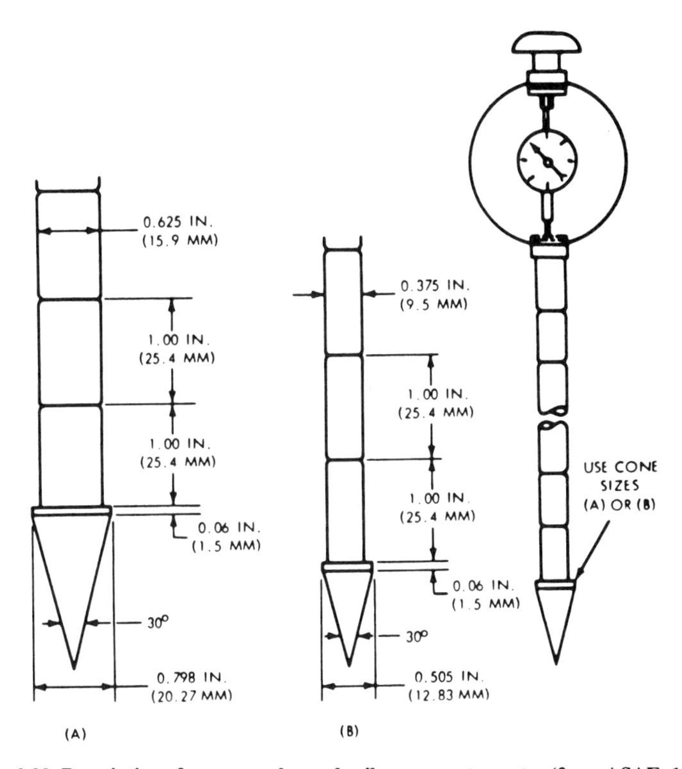

Figure 2.99–Description of a commonly used soil cone penetrometer (from ASAE, 1983).

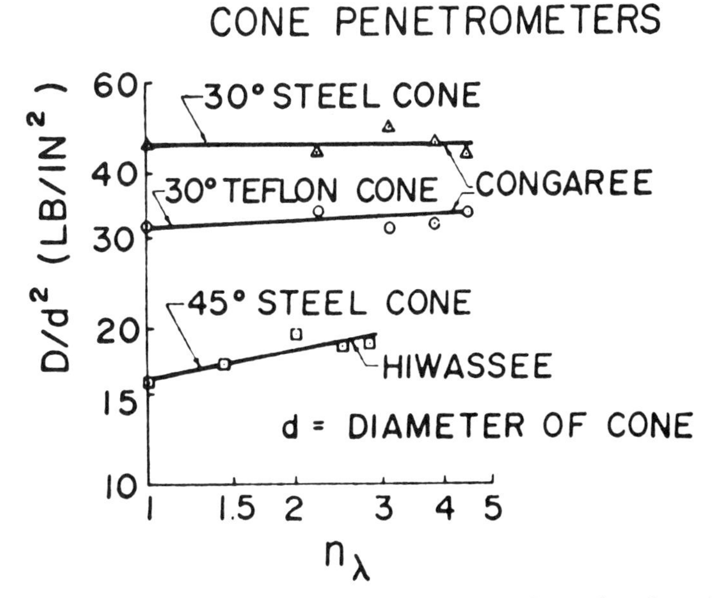

Figure 2.100–Results of tests with soil cone penetrometers constructed at various size scales (n_λ). It is illustrated that size scale has little effect on the values of force (D) per unit cross-sectional area ($0.785d^2$) for the 30° steel cone (from Schafer et al, 1969).

1987), indicated that the average peak cone index under the high speed conditions was 2.23 times that at normal speed. An equation to predict penetration depth from mass and falling velocity for a penetrometer cone of given geometry depended on the use of a soil penetrability index. Even with this inclusion, predictions were not completely representative of actual values.

The literature contains reports on factors that affect the penetrometer reading. The factors cited include soil moisture content, porosity (bulk density), coefficients of friction of soil against steel and against soil, soil grain density, shear and bulk moduli, vane shear strength, Atterberg limits, organic matter content, clay content, etc. (Freitag, 1968; Rohani and Baladi, 1981; Upadhyaya et al, 1982; Ohu et al, 1986; Ohu et al, 1988; Ayers and Bowen, 1987; Gameda et al, 1988, 1989). Even with all these variables, as many as three or more special coefficients per soil were required to fit results to analytic models. Most of the efforts to interpret correlations between penetrometer readings and soil properties have been made using data from remolded soil in laboratory containers. One such study used the laboratory-data-based model used to predict *in situ* penetrometer readings based on moisture and porosity values of the same soil (Chesness et al, 1972). In that study, actual *in situ* values averaged 12.9 times those predicted, indicating that remolded soils in laboratory tests may not be able to provide the sort of experimental medium to allow interpretation of field soil properties. Another study (Womac et al, 1987) also found distinct differences between remolded and *in situ* results, although moisture and porosity were the same. Efforts to associate cone penetrometer data with soil classification parameters also have not been particularly definitive (fig 2.101) (Meyer and Knight, 1961). Consequently, it is extremely difficult to interpret penetrometer readings in terms of more basic physical properties of soil.

The empirical correlation of penetrometer data with whether or not a particular vehicle could traverse a given soil condition (US Army, 1959; Knight and Freitag, 1962; Rush, 1968), has been found to be a workable system. This system has been further developed to the point that general estimates can be made reliably for the motion resistance of a given soil-vehicle combination and for the maximum drawbar pull that could be developed. The system was extended even further to cover the performance of pneumatic tires at 20% slip on soft soils (Freitag, 1965). This last development served as a base for a system for predicting off-road traction of wheeled vehicles for a full range of slip values (Wismer and Luth, 1974). In this last system, the "cone index" (cone penetrometer resistance force per unit cone base area at the depth to which drive wheels usually sink) is the only soil parameter measured.

Penetrometer resistance values have thus become established as a basis for predicting traction performance. Because of the great variability in soil strength characteristics that affect tractive performance, it is not possible for the full range of these characteristic values to be represented in one value, and consequently deficiencies in tractive performance predictions based on cone index alone have been found (Upadhyaya et al, 1989). However, because the cone index has become such a well-established parameter in this field, the role of soil physical characteristics in traction systems has been described in the form of a value of cone index imputed from actual tractive performance data (Wang and Zoerb, 1988). This value, called the *tractive index* has been found to be different in quantity than that derived from the actual cone penetrometer readings, and found to vary considerably under conditions in which the penetrometer readings did not change appreciably (Wang and Zoerb, 1988). It is therefore anticipated that, in addition to the established usefulness of cone penetrometer values in predicting tractive performance, measurement of other soil parameters will be required if improved tractive performance prediction is to be achieved.

Efforts have been made to relate cone penetrometer readings to the draft force required by implements. When procedures were used that were similar to those used with traction devices, it was possible to make very general draft force predictions for simple laboratory tools (Wismer and Luth, 1972; Luth and Wismer, 1971). In other tests with penetrometer cones, single discs, or simple chisels of varying size scales, correlations between the effect of size scale on cone penetrometer resistance and on disk draft were found, although similar correlations with chisel draft were not suitable for prediction purposes (Johnson et al, 1978). In one study in which cone index and soil wet bulk density were used as the soil parameters in a system for predicting the draft of moldboard

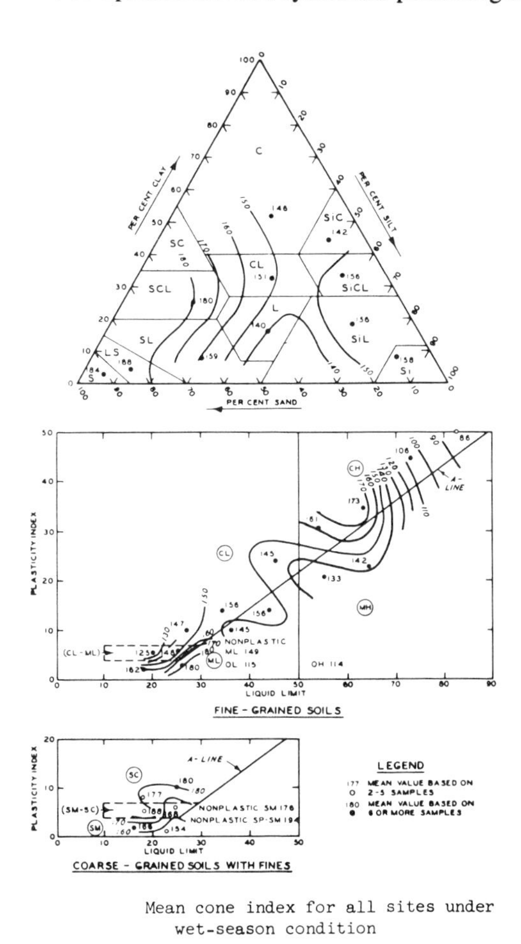

Figure 2.101–Generalized relations between soil texture, Atterberg Limits (as reflected in the A-line) and cone penetrometer readings (Meyer and Knight, 1961, US Army Corps of Engineers).

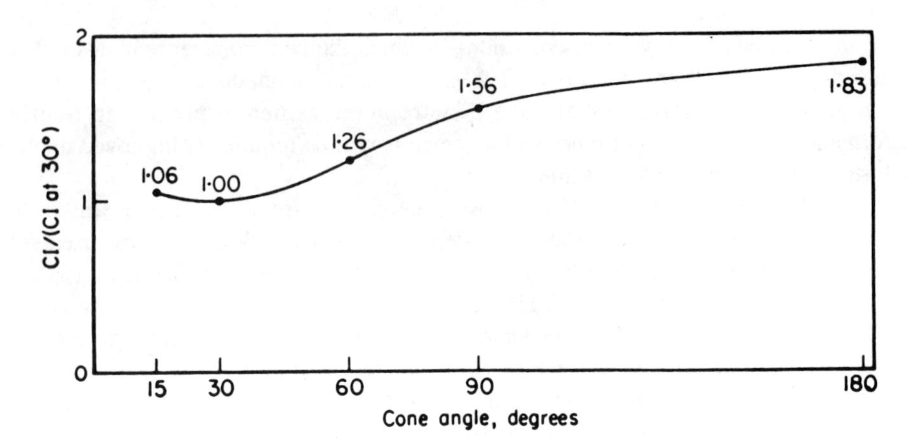

Figure 2.102–Effect of various cone included angles on the cone index (as compared to the cone index with a 30° cone) (from Koolen and Vaandrager, 1984).

plows; in general good correlations were found between predicted and measured values (Oskoui et al, 1982). The subject-matter field in which tillage forces are related to penetrometer readings is less well developed than that for tractive force.

A number of variants of the cone penetrometer have been investigated. In addition to cone size, mentioned above, various cone angles have been explored (Schafer et al, 1969; Koolen and Vaandrager, 1984; Gill, 1968; Tollner and Verma, 1984). A general result of one such investigation is presented in figure 2.102 (Koolen and Vaandrager, 1984).

Another variant was the use of a rotating cone penetrometer (Leviticus, 1973). A linear correlation was found between shear stress on the cone surface and normal stress per unit cone base area.

A further variant was the use of a cone penetrometer for which a lubricating polymer could be forced out along the cone surface (Tollner, 1983; Tollner and Verma, 1984). It was believed that such a penetrometer would more closely represent root motion through soil than would an unlubricated penetrometer. In general, the forces on the lubricated penetrometer responded similarly to variations in soil conditions, although resistance values were lower than those of the unlubricated one. In some of these tests, the penetrometer was stopped in position and the pattern of force relaxation with time recorded (fig 2.103) (Tollner, 1983).

In an effort to devise a soil measurement that might parallel motion resistance of a vehicle, a penetrometer-like device was constructed in which a 10-kg cylindrical weight 99 mm in diameter was allowed to drop 1 m onto the soil surface, and the depth of the impression was measured (Tijink and Koolen, 1985). Values measured were not particularly well correlated with cone index. When both cone index and impression depth were used, good predictions of rut depth were obtained. In this same experiment, cone index and shear vane torque were found to be highly correlated.

Shear-Vane Torsion

The shear vane has for many years been regarded as a device for measuring strength properties of soils *in situ* (Gill and VandenBerg, 1968) because of the ease and

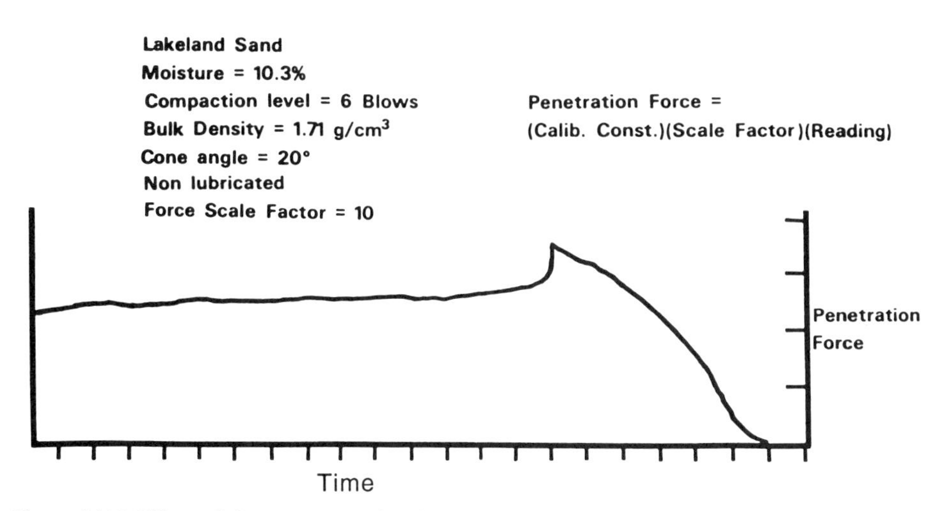

Figure 2.103–Effects of time on penetration force relaxation. The time axis starts on the far right and progresses to the left (from Tollner, 1983).

convenience of such measurements. A schematic description of the shear-vane apparatus is shown in figure 2.104 (Hillel, 1980). Because it is assumed that the final form of the soil failure surface is that of a cylinder with the length and diameter of the vane, and because no normal forces are applied to this surface, torsional resistance is caused by cohesion acting on all surfaces of the cylinder (including the top if the instrument is pressed well below the soil surface):

$$T = c\,\pi\left(\frac{1}{2}\,d^2h + \frac{1}{6}\,d^3\right) \qquad (2.69)$$

in which T = torque applied to cause failure, c = cohesion, d = diameter of shear-vane apparatus, and h = length of vane assembly.

Common ratios of length to diameter used range from h = 1.5d to h = 2d (Koolen and Kuipers, 1983). Many different values of d are used, but common diameters vary from 18 to 64 mm. Some commercial shear-vanes have the value of h at the outer periphery of the vane somewhat less than the value of h at the inner radius. This tapering effect, shared equally between top and bottom of the vane assembly, makes it easier to insert and remove the vane.

Shear vane use is generally now compatible with soils that are very dry and hard or soils with stones. Shear-vane readings are very dependent on soil moisture content, so this should also be recorded where comparative data are required. Normally 10 to 15 readings are required to provide a reliable estimate of shear strength (Archer and Marks, 1985).

By using shear vanes of different lengths, it is possible to estimate the torque required to shear only one end surface of the failure cylinder (Schafer et al, 1963):

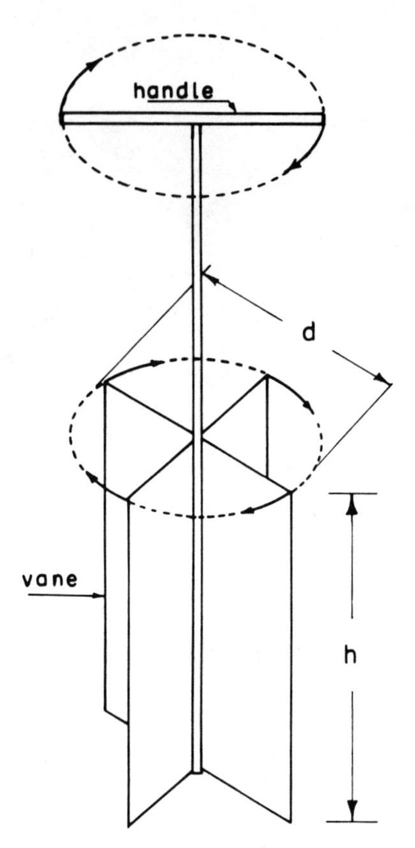

Figure 2.104–Schematic description of a shear-vane apparatus. Reprinted with permission of Academic Press, Inc. from Hillel, 1980. Copyright © 1980, Academic Press, Inc.

$$T_E = (1/2)\left(T_1 - L_1\, T_3 / L_3\right) \qquad (2.70)$$

in which

T_E = torque required to shear one end surface of the short-length vane, V_1
T_1 = total torque required to shear the soil with the short-length vane, V_1
T_3 = total torque required to shear the soil with the long-length vane, V_3
L_1 = length of the short-length vane, V_1
L_3 = length of the long-length vane, V_3

Cohesion, assumed to be the only form of shear resistance acting on the ends of the vane, can be computed by:

$$c = \frac{3\, T_E}{2\, \pi \left(r_o^3 - r_i^3\right)} \qquad (2.71)$$

in which c = cohesion, r_i = inside radius of vane apparatus, and r_o = outside radius of vane apparatus.

Estimates of tan ø (coefficient of internal friction) may also be made using the values of cohesion obtained above (Schafer et al, 1963):

$$T = \frac{4\pi c}{3}\left(r_0^3 - r_i^3\right) + 2P\,L\,r_o^2\,c + 4\,L\,r_o^2\,c\left[\frac{e^{[(\pi/2)(\tan\phi)]} - 1}{\tan\phi} - \frac{\pi}{2}\right] \quad (2.72)$$

in which T = total torque required to shear the soil and L = length of the vane assembly.

Field tests for cohesion using the above method and using a torsional shear box showed shear-vane cohesion values several times higher than those obtained with the shear box (Schafer et al, 1963). Subsurface test data for a construction site are shown in figure 2.105 (Lambe and Whitman, 1969), indicating that shear stress values obtained with the shear-vane were approximately twice as high as those obtained with triaxial tests.

Tests conducted with both a cone penetrometer (30°, 3.23 cm^2 base area) and a shear vane on a sizable range of soils at various conditions (Tijink and Koolen, 1985) showed that the result data were highly correlated (R = 0.96).

Efforts to obtain a system for predicting cone penetrometer resistance using laboratory tests on remolded soils (Ohu et al, 1986; Ohu et al, 1988) found that:

$$CI / P_c = A\left(c / Q_p\right)^{n\theta} \quad (2.73)$$

in which

CI	=	cone penetrometer resistance per unit cone base area
P_c	=	pressure at which the laboratory sample was compacted
c	=	soil shear strength (cohesion) as measured with a shear vane
Q_p	=	overburden pressure
θ	=	volumetric fraction occupied by water
A, n	=	constants depending on soil type

These results indicate that there is a considerable similarity between the soil strength properties measured by the shear vane and those measured by the cone penetrometer. Although the soil properties of cohesion and coefficient of friction affect the readings obtained with both instruments, the failure mechanism of the two instruments are distinctly different.

Remolding Index

The remolding index is a parameter used to describe the loss of strength (for fine-grained [clay] soils) or for poorly drained sands with many fine particles due to agitation while in a high moisture state. This characteristic is sometimes called *thixotropy*. The remolding index is a factor used as a coefficient of the cone index (cone penetrometer resistance force per unit cone base area) to yield the rating cone index (US Army, 1959).

$$(\text{cone index}) \times (\text{remodeling index}) = \text{rating cone index} \tag{2.74}$$

The rating cone index is the soil parameter that is used to interface with vehicle parameters in the US Army soil-vehicle mobility system (US Army, 1959; Knight and Freitag, 1962; Rush, 1968).

To conduct the remolding index test, a core sampler is used to obtain a field sample slightly less than 50 mm in diameter and about 150 mm long. The sample is placed in a vertical cylinder (mounted on a base plate) which is 50 mm inside diameter and 200 mm long. For clay soils, a cone penetrometer reading (using the 3.23 cm^2 base area cone) is made to a depth of 100 mm. For sands with fine-grained particles, the 1.29 cm^2 base area cone is used to obtain the cone index. Then 100 blows of a drop hammer falling 305 mm to strike a foot of slightly less than 50 mm diameter resting on the top surface of the fine-

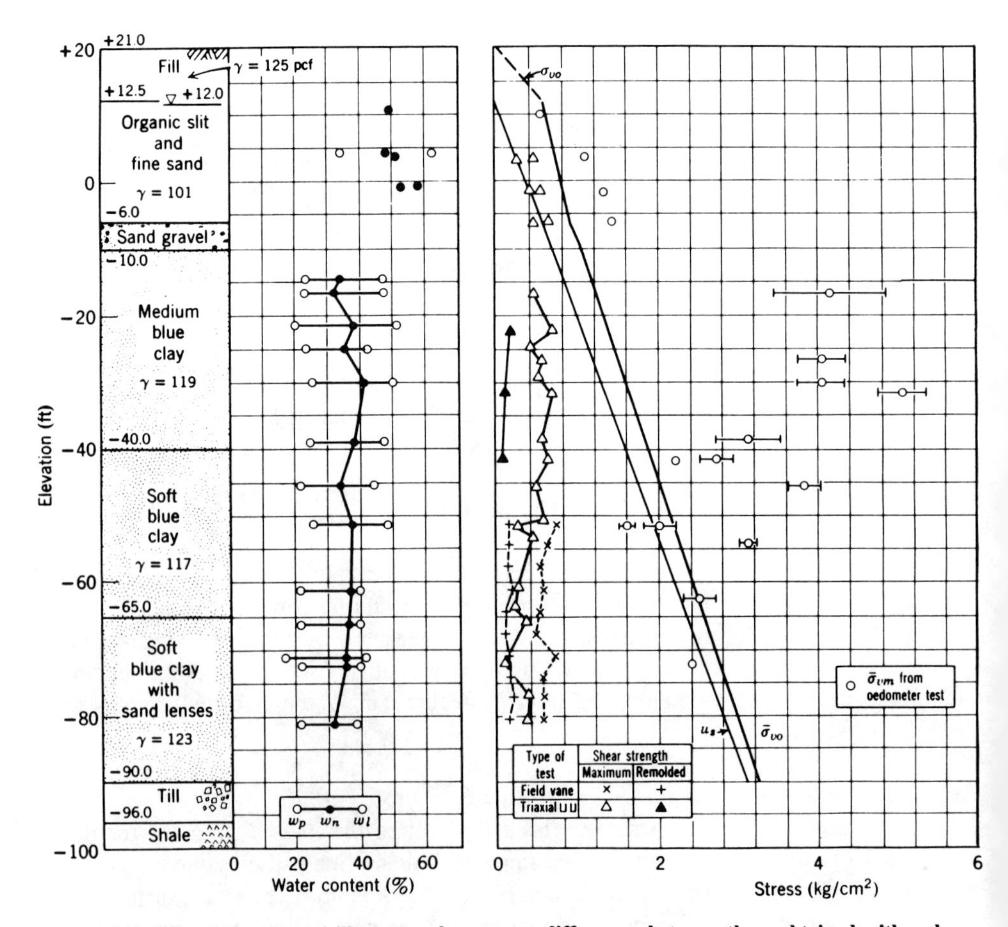

Figure 2.105–Right-hand panel illustrates shear stress differences between those obtained with a shear-vane apparatus and those obtained in triaxial cell tests. Reprinted with permission of John Wiley & Sons, Inc. from *Soil Mechanics*, Lambe, T. W. and R. V. Whitman, copyright © 1969.

grained soil in the cylinder are applied. For sands containing a sizable proportion of fine silt and clay, the drop hammer is not used, but the sample, along with the cylinder and cylinder base, are dropped 25 times onto a hard surface from a height of 150 mm.

After the remolding (by drop hammer or sample dropping), another cone index reading is obtained using the appropriate cone size. The remolding index is computed by:

$$\text{remodeling index} = \frac{\text{cone index after remolding}}{\text{cone index before remolding}} \tag{2.75}$$

Typical values for the remolding index, classified according to soil texture and soil consistency parameters, are given in figure 2.106 (Meyer and Knight, 1961).

The thixotropic characteristics of a rice field soil were defined for a high clay (37% of particles smaller than 1 μm and 50% smaller than 5 μm, plastic limit = 23%, liquid limit

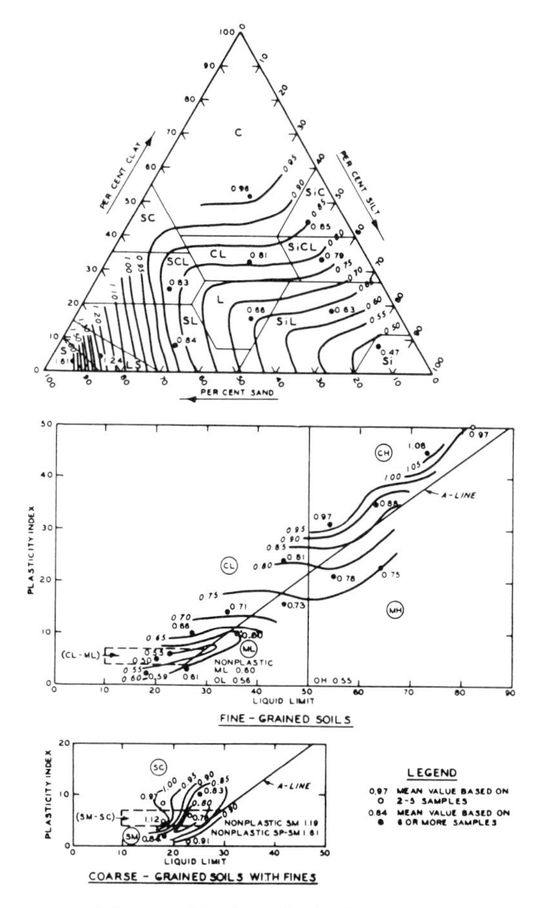

Figure 2.106–Relations among soil texture, Atterberg Limits (as represented by the A-line) and values of the remolding index (from Meyer and Knight, 1961, US Army Corps of Engineers).

= 47%) soil in terms of the change in viscosity upon forcing the soil through an orifice (Qian et al, 1982):

$$T = \frac{\eta_1 - \eta_2}{\eta_1}$$

in which T = thixotropy parameter, η_1 = viscosity before agitation, and η_2 = viscosity after agitation. It was found that at a moisture content slightly above the liquid limit, the thixotropic effect was maximum (fig 2.107) (Qian et al., 1982).

Torsional Shear Devices

Torsional shear strength measuring devices which use a circular rather than an annular test section, have been in use for some time (Gill and VandenBerg, 1968). There has always been some concern that when the torsional device began to rotate relative to the soil, the shear deformation of the soil near the periphery of the test section was greater than that near the center of the test section. Generally, the shear stress has been computed by:

$$S = \frac{3M}{2\,\pi\,R^3} \tag{2.76}$$

in which S = maximum shear stress, M = maximum torque applied to the device, and R = outer radius of the test section. A second concern has been that with most of these

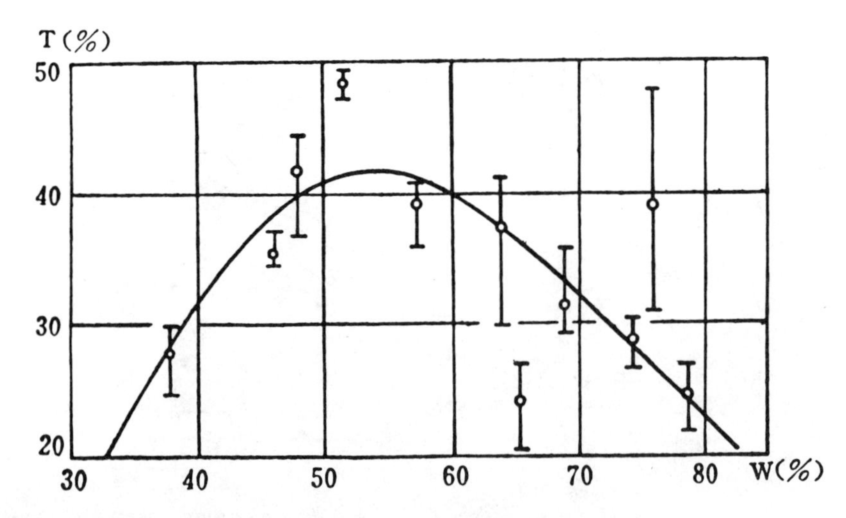

Figure 2.107–Effect of moisture content (W) on the loss of viscosity of a clay soil due to agitation (thixotropy) (from Qian et al, 1982).

$$T = \frac{\text{viscosity before agitation} - \text{viscosity after agitation}}{\text{viscosity before agitation}}$$

devices, the normal stress levels used are much lower than those experienced by field soils during tillage and traction operations (O'Callaghan et al, 1965). Thus, with such shear devices, pore water pressure in the soil would not be fully mobilized, while in actual operations, particularly in clays at higher moisture levels, pore water pressure might play a large part in determining the apparent shearing resistance of the soil.

The normal stress for one such torsional shear device (Schafer et al, 1963) tended to be less than 24 kPa while that for the Cohron Sheargraph was generally less than about 100 kPa (Cohron, 1962; Cohron, 1963). The Cohron Sheargraph (fig. 2.108) is notable because it can be manually carried and operated. Operation is by placing the torsional shear head on the soil surface and applying a normal force by pushing the handle down against a calibrated spring. Once the normal force has been applied, the handle is rotated to apply a torsion (via the same spring) to the shear head. After the shear head begins to rotate, the handle is rotated continuously while the normal force is gradually reduced. A plot of torsional moment vs normal force is made by the device.

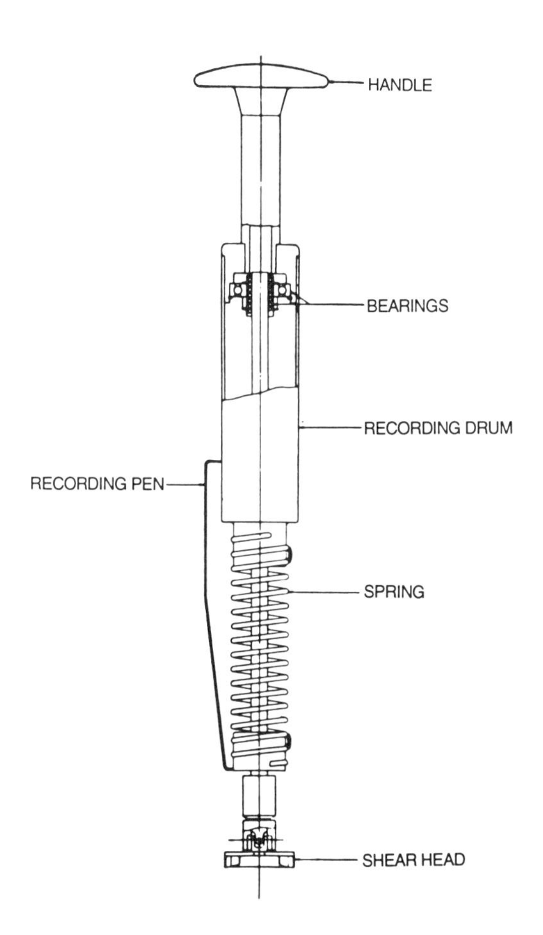

Figure 2.108–Description of the Sheargraph (from Cohron, 1963).

Because of the convenience of use of the Cohron Sheargraph, several researchers have made tests in which results of this and other shear-strength measuring devices have been compared (Bailey and Weber, 1965; Dunlap et al, 1966). Examples of these comparative results are shown in figure 2.109 (Bailey and Weber, 1965) and figure 2.110 (Dunlap et al, 1966). The comparative differences varied with type of soil (fig 2.111) (Bailey and Weber, 1965), but generally the Cohron Sheargraph indicated higher levels of cohesion than other methods. This may have been due to the small diameter of the circular zone loaded in torsion.

The Cohron Sheargraph was used in the development of a model which linked soil moisture and bulk density values to those of cohesion and internal coefficient of friction (Ayers, 1987). These soil shear strength parameters were then incorporated in another model to predict cone penetrometer resistance and related soil moisture profiles (Ayers and Bowen, 1987). Results of bulk density predictions relative to actual values are shown in figure 2.112. The differences found in bulk density prediction could be reduced significantly by using corrections based on measured rather than predicted values of moisture content.

Stickiness

The stickiness of soil to implement, track, and tire surfaces has a major impact on the interaction continuity of these units with the soil during field operations. Stickiness is measured in two modes. One is the tearing of a flat surface away from the soil surface in a direction perpendicular to the plane of the soil surface (fig 2.113). The second mode is one in which the soil is forced to slide along a material surface, and the force required to cause this sliding (per unit soil-surface contact) is measured (fig 2.114). This second

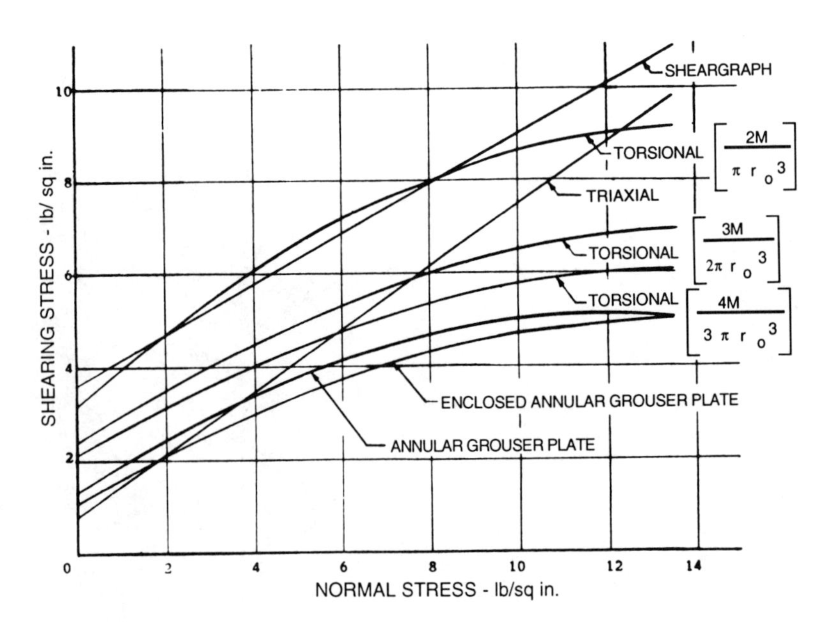

Figure 2.109–Comparison of Sheargraph results with those obtained with other methods of shear-strength measurement (from Bailey and Weber, 1965).

mode, soil-hard surface sliding resistance per unit area at zero normal load, is the property generally termed adhesion in analyses of the resistance of soil flow over hard surfaces. In both modes, some initial normal stress may be applied between the material surface and the soil but removed prior to loading.

Figure 2.115 shows that for a clay soil having 49.3% of its particles smaller than 0.01 mm in diameter, when circular steel or Teflon plates were pressed onto the soil surface to achieve sinkage levels of 10 to 40 mm (with increasing normal stresses associated with increases in sinkage) (Mil'tsev, 1966) increased normal stresses resulted in increased sticking stresses (stresses required to tear the plate normally from the soil surface) at all moisture contents. For both materials, maximum sticking stresses occurred at an intermediate moisture content. For this soil and a second soil with 58% of the particles smaller than 0.01 mm in diameter, the sticking stresses for an assortment of materials are presented in table 2.11 for each soil at its respective moisture content associated with maximum stickiness. In these cases, the plates were initially pressed against the soil surface with a normal stress of 20 kPa for 3 min.

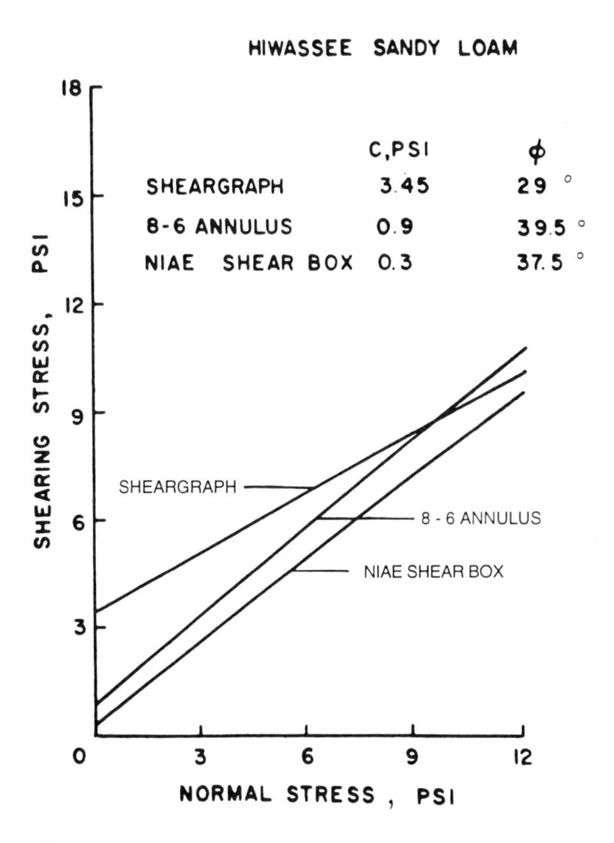

Figure 2.110–Comparison of Sheargraph results with those obtained with other methods of shear strength measurement (from Dunlap et al, 1966).

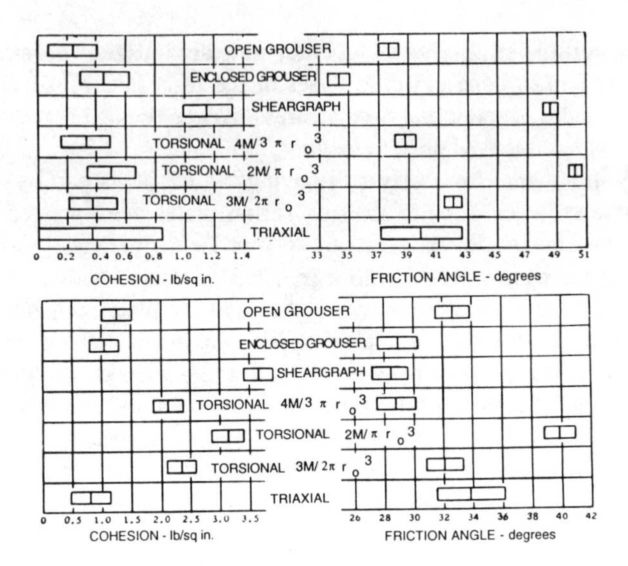

Figure 2.111–Ranges of values of cohesion and angle of internal friction measured with different methods, including the Sheargraph, using artificial soils. Upper panel is for fire clay plus 10% SAE 5W oil. Lower panel is for 50% sand and 50% ball clay plus 17% SAE 140 gear lubricant (from Bailey and Weber, 1965).

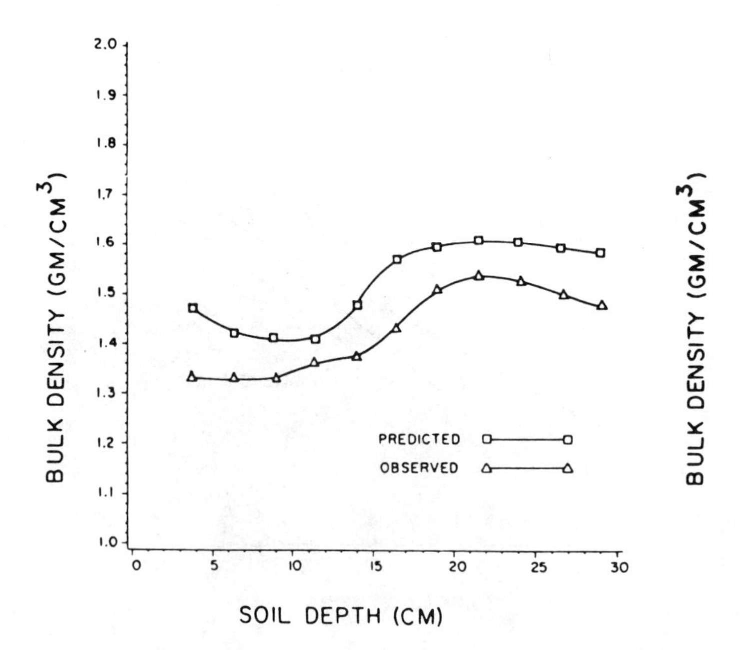

Figure 2.112–Predictions of dry bulk density obtained using Sheargraph readings and cone index, as compared with observed values (from Ayers and Bowen, 1987).

In another series of tests (Nikolaeva and Bakhtin, 1975), in which the steel disk had an area of 25 cm^2 or the sliding soil sample had a contact area of 25 cm^2 and a height of 3 cm, the initial contact stress of 20 kPa for 3 min was again used.

For a soil having 60% of its particles smaller than 0.01 mm in diameter, the data for stickiness in "perpendicular tearing" and "parallel shear" on a steel plate are given in fig 2.116. Again, maximum values were found at intermediate moisture contents.

An investigation of the effects of initial contact stress, contact time, and material temperatures (Zadneprovski, 1975), found also that the sticking stress (normal tearing) tended to increase linearly when the contact stress increased, but that the ratio of stickiness increase to contact stress increase varied from soil to soil (fig 2.117). It was also found that increases in contact time resulted in greater stickiness stresses, but, that little increase in stickiness occurred for contact times greater than 50 s. Increased temperatures tended to reduce stickiness markedly, and stickiness tended to be lower when cold soil was placed in contact with a heated surface than when the reverse was true (fig 2.118).

In a study of the effects of soil texture on cohesion, friction, and stickiness (Domzal, 1970), in general, higher stickiness stresses were found with soils having higher clay contents but that increased levels of organic matter tended to result in reduced stickiness tendencies (table 2.12).

When pure clays were examined, it was found that stickiness stresses were much higher than for typical field soils. The moisture content at which maximum stickiness

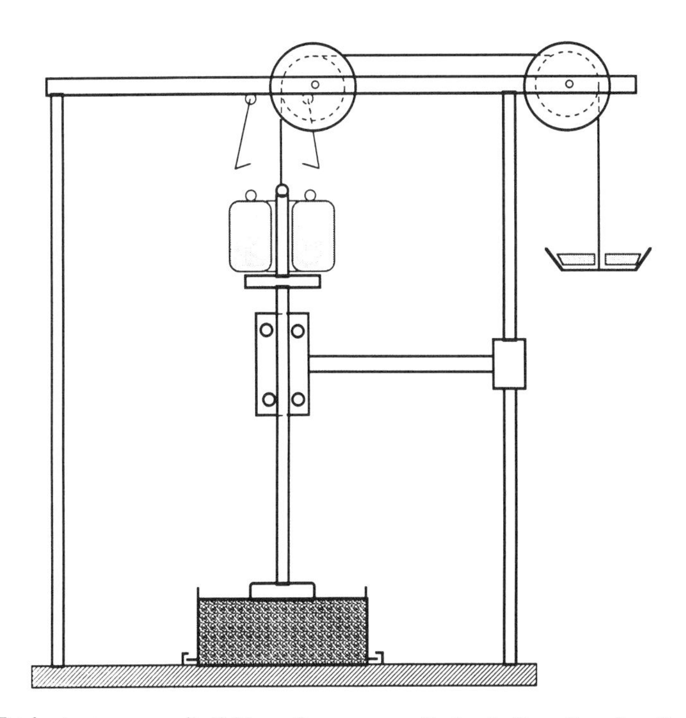

Figure 2.113–Device to measure soil stickiness forces perpendicular to the soil surface (from Nikolaeva and Bakhtin, 1975).

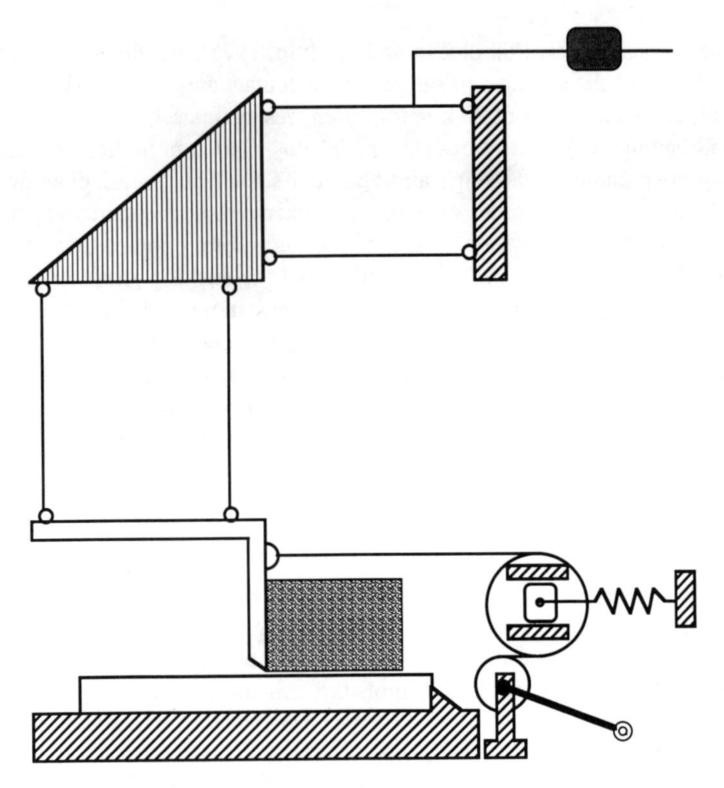

Figure 2.114–Device to measure soil stickiness forces parallel to the soil surface (from Nikolaeva and Bakhtin, 1975).

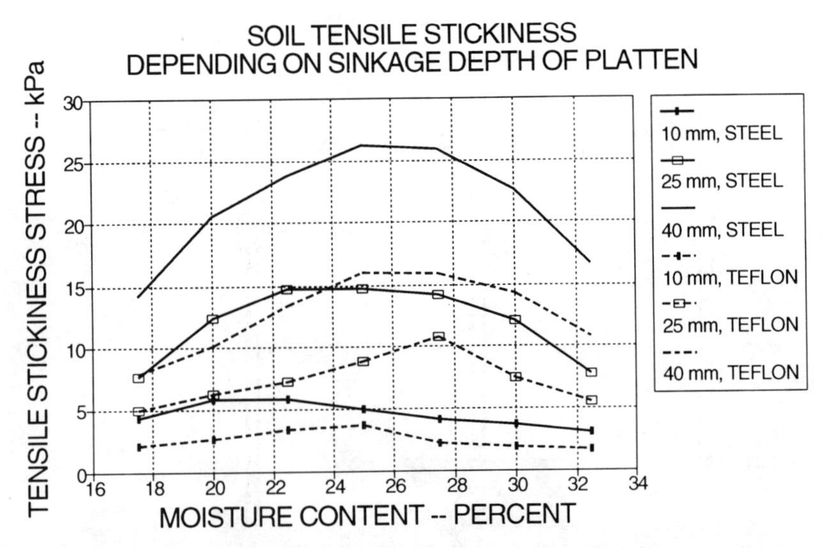

Figure 2.115–Effect of increasing depth of sinkage (and increasing normal stress) on the stickiness stress at various moisture contents (horizontal axis) for a soil with 49.3% of particles smaller than 0.01 mm for steel and Teflon plattens (from Mil' tsev, 1966).

Table 2.11. Maximum perpendicular stickiness stress* (from Mil'tsev, 1974)

Material†	Stress (Pa) Soil 1 (49.3% < 0.01 mm)	Soil 2 (58% < 0.01 mm)
Steel L 65	1800	2270
Teflon 4	865	824
High-density polyethylene	1620	1910
Low -density polyethylene	1840	2060
Perspex	1640	2010
Capron	1900	2210
Co-polymer SNP	1600	2170
Plastic and fiberglass SKM-I	1770	2360
Silicate lacquer K-I	1700	2360
Copper ME	1610	1970
Stainless steel IX 13	1610	2200
Verneer with paraffin	1560	1790
Enamel T-I	1570	1890

* Trade names are used solely to provide specific information. Mention of a trade name does not constitute an endorsement of the product to the exclusion of other products not mentioned.

† Initial normal contact stress was 20 kPa for 3 min.

occurred decreased as the level of initial contact stress increased (fig 2.119). The type and concentration of salts in the pore solution also affected stickiness stresses (fig 2.120). When liquids other than water were used as the pore fluid, stickiness stresses decreased

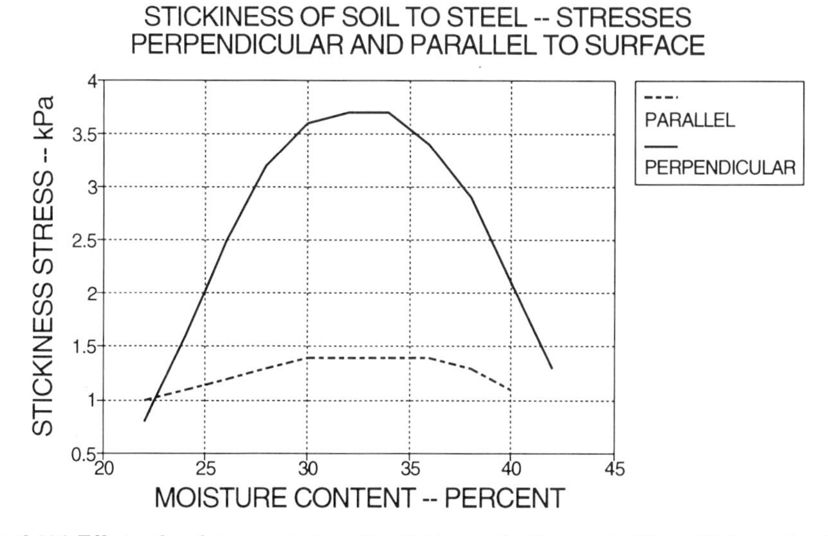

Figure 2.116–Effects of moisture content on the stickiness of soil on steel. The solid line is for forces perpendicular to the soil surface, while the dashed line is for forces parallel to the soil surface (from Nikolaeva and Bakhtin, 1975).

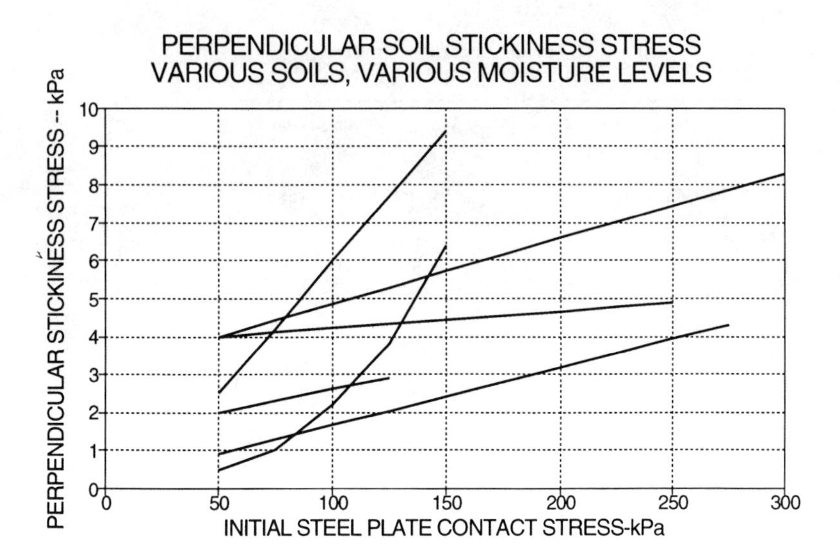

Figure 2.117–The stickiness forces perpendicular to the soil surface per unit area relative to the stresses with which steel plates were pressed against the soil surface for various soils at various moisture contents (from Zadneprovskii, 1975).

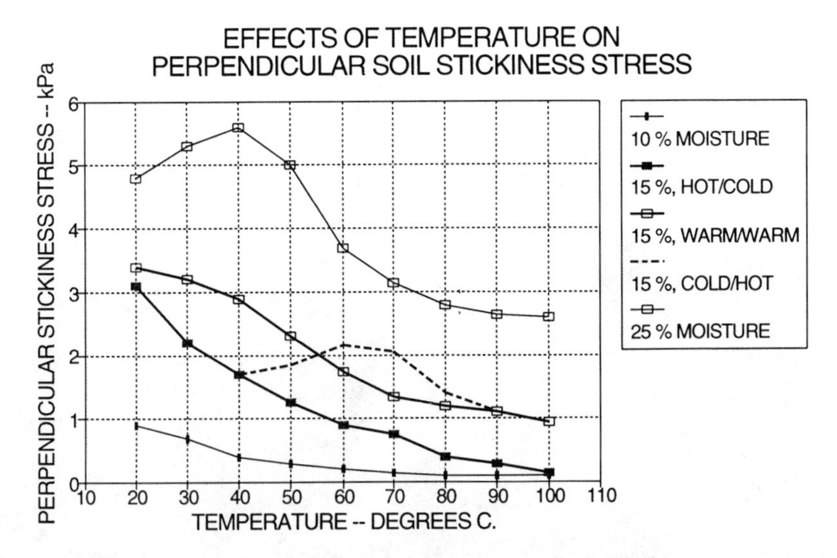

Figure 2.118–Effects of temperature on the stickiness forces per unit area perpendicular to the soil surface at various moisture contents. At 15% moisture, the lower solid line is for contact of cold soil with a heated steel surface, the intermediate solid line is for both materials of the same temperature, and the dashed line is for warm soil on a cold steel surface (from Zadneprovskii, 1975).

Table 2.12. Maximum stickiness stress (from Domzal, 1970)

Soil Parameters		Moisture Content	Maximum
% < 0.002 mm	% Organic Carbon	at Max. Stress	Stickiness (Pa)
1	1.00	26.5%	500
3	1.51	29%	1000
6	1.11	29%	900
16	1.35	32%	1500
22	1.15	28%	900
34	0.69	33%	2700
52	3.24	38%	2000

greatly (fig 2.121). Different types of clays exhibited widely different stickiness characteristics.

Shatter Resistance

In order to develop a basis for evaluating the efficiency of tillage operations, Gill and McCreery (1960) measured the degree of shattering (reported as the change in mean weight-diameter of soil clods and subclods) that resulted when a given mass of soil was dropped a varying number of times from a given height onto a flat, rigid surface. The input to such a process could be quantified in terms of energy per unit mass or energy per unit volume of soil. The quantity, energy per unit volume, is equivalent to the parameter stating draft resistance of a plow per unit furrow cross-sectional area:

$$\frac{\text{draft}}{\text{area}} \times \frac{\text{distance traveled}}{\text{distance traveled}} = \frac{\text{energy}}{\text{volume}} \tag{2.77}$$

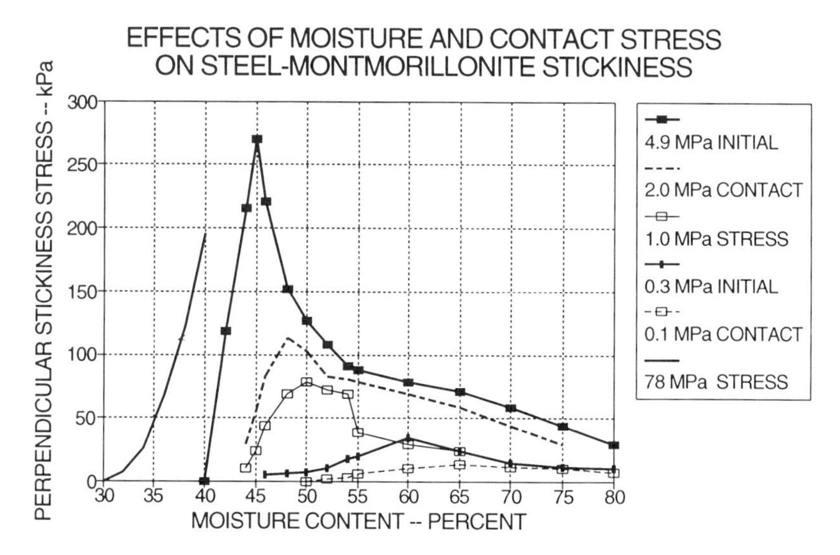

Figure 2.119–Effects of moisture content and contact pressure on the stickiness forces per unit area of steel perpendicular to a montmorillonite clay surface (from Kalachev, 1974).

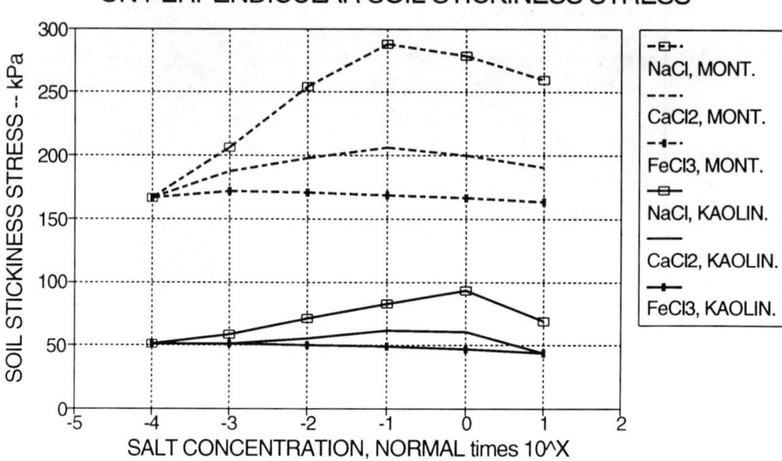

Figure 2.120–The effects of salts in the pore water on stickiness of clay (vertical scale) to a steel surface. The horizontal scale is the concentration (fraction of Normal) of the various salts for kaolinite (lower panel) and montmorillonite (upper panel) (from Kalachev, 1974).

Consequently, energy per unit volume from drop-shatter tests can be related to energy per unit volume required to operate a tillage implement, when the mean-weight-diameter of the soil clods produced is equivalent. There may be differences, however, due to the more highly dynamic nature of drop-shatter tests.

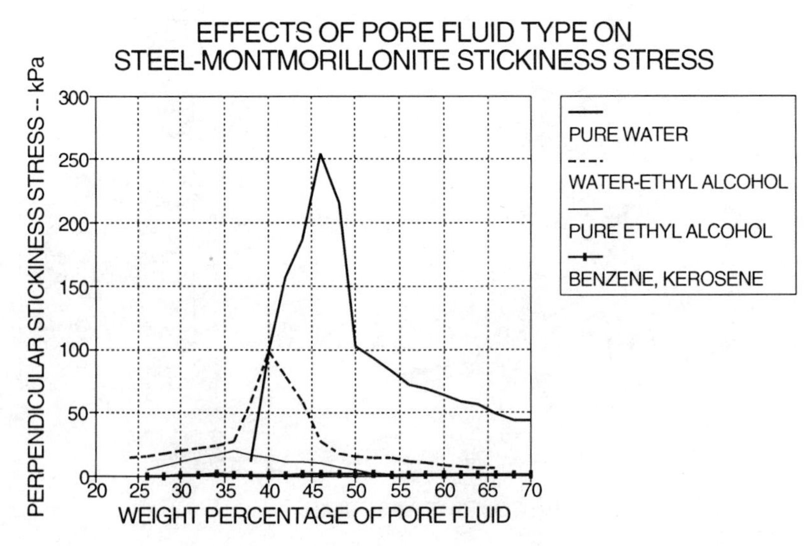

Figure 2.121–Effects of the chemical nature of the pore fluid on the stickiness of montmorillonite clay (vertical scale) to a steel surface (from Kalachev, 1974).

If the energy per unit volume to shatter soil is plotted vs surface per unit volume of the soil clods produced, the slope of the plot represents the energy per unit new surface produced by the soil breaking operation (fig 2.122) (Vomocil and Chancellor, 1969). The surface per unit volume for both spheres and cubes can be computed from:

$$\text{surface / volume} = \frac{6}{\text{dimension}} \tag{2.78}$$

in which dimension = diameter of a sphere or edge-length of a cube.

For a silty clay loam soil, the slope for such a plot, or the average amount of energy per unit new surface generated for the drop-shatter method was 880 N/m over the range of clod sizes from 25 to 200 mm with a standard deviation of 195 N/m (Gill and McCreery, 1960). For another soil, the mean value for the drop-shatter process, was 252 N/m over a range of clod sizes from 17 to 83 mm with a standard deviation of 57 N/m (Gill and VandenBerg, 1968).

Wire loop cutting, a very efficient laboratory-scale soil-breaking process (but not a drop-shatter process), requires values of energy per unit new surface produced for a silty clay soil ranging from 49 N/m at low moisture content and high confining stress to 1.2 N/m at a high moisture content and no confining stress (Vomocil and Chancellor, 1969). A survey of other literature on this topic indicated values of this parameter (involving various processes) ranging from 6.4 to 482 N/m (Vomocil and Chancellor, 1969).

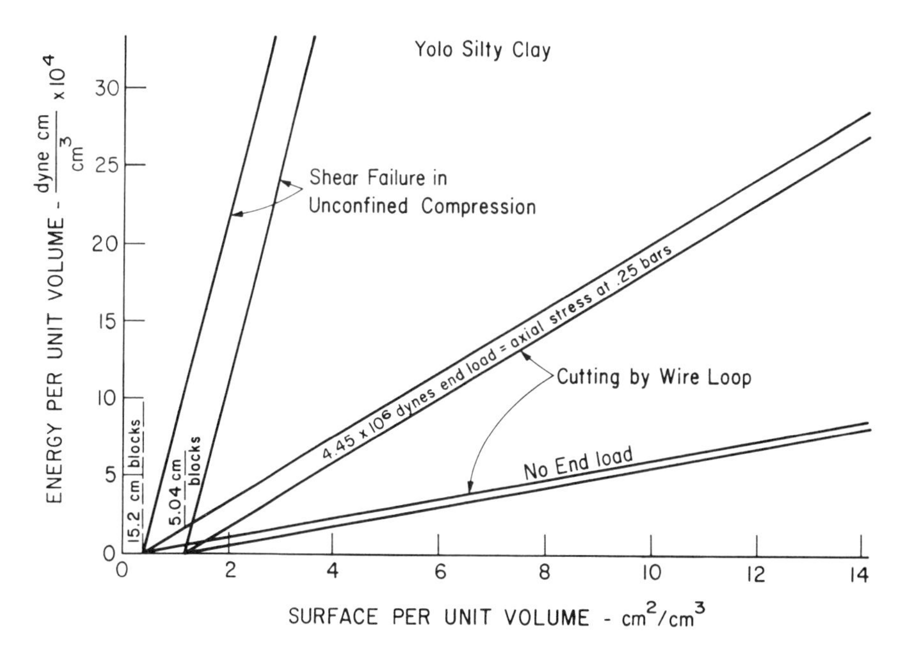

Figure 2.122–Relations between energy per unit volume for soil failure and surface per unit volume (surface/volume = dimension/6) for three different soil failure processes. The slope of the lines (energy per unit surface) is the inverse of a parameter of efficiency in producing breakup of a soil monolith (from Vomocil and Chancellor, 1969).

An investigation in which the drop-shatter method was compared with a process involving the striking of soil with blades attached to a charpy-type impact-testing machine (blade velocity = 4.2 m/s) (Bateman et al, 1965), indicated that at high soil density and low moisture content, the two processes required approximately equivalent amounts of energy per unit soil volume to produce the same mean-weight diameter of clods (81 N/m). At low density and high moisture content, the impact tester required only from 25 to 45% of the energy per unit volume as did the drop-shatter process to produce clods of a given mean-weight-diameter (150 N/m). When the same soils were cut at very low speed (0.35 m/min), the energy per unit volume required to produce clods of a given mean-weight-diameter averaged approximately 18% of that for the impact process. For the impact process, the energy per unit volume to produce clods of a given mean-weight-diameter increased rapidly with increases in density, while for the drop-shatter process, this energy requirement parameter appeared to decrease with increases in density (Bateman et al, 1965).

For clods of a clay loam soil that were crushed by applying the force of a single round tine against the clod (Sitkei, 1985), values of energy per unit of new surface formed ranged from 1000 N/m at 4% moisture, down to 350 N/m at 20% moisture. In this same study, these clay loam clods were imbedded in sand and the matrix was tilled with tines of 25 to 30 mm diameter at various velocities. The critical tine velocity for clod breakage is shown in figure 2.123. Test results confirmed the analytical derivation that the relations among tool velocity (V), time of impact (Δt), clod weight (G), diameter (d), and failure energy per unit of new fracture surface (γ) would be of the logarithmic nature shown in figure 2.124. Also included in this study was a statistical analysis of the factors involved in determining the probability that a clod of a given size would be impacted by

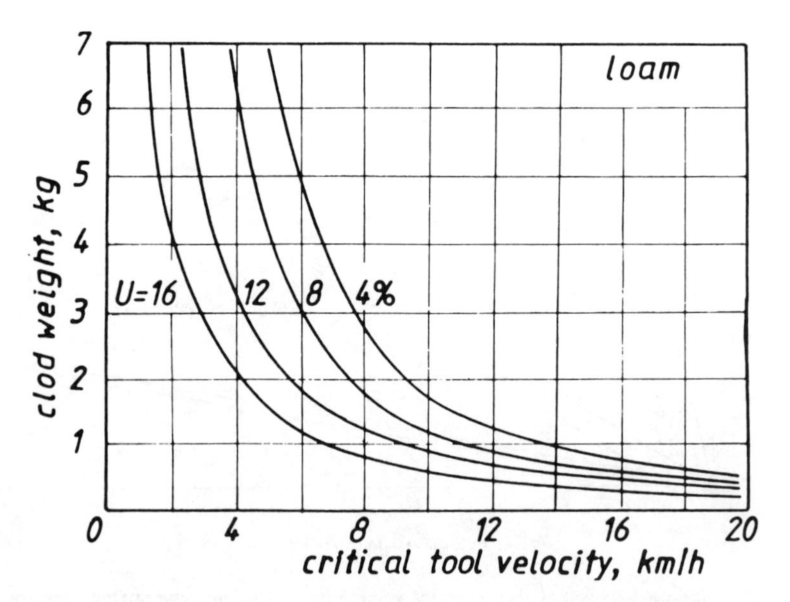

Figure 2.123–Clay loam clods of various sizes embedded in sand tended to be broken by tines moving through the matrix only when tool velocity was above the critical velocity. U is the moisture content (from Sitkei, 1985).

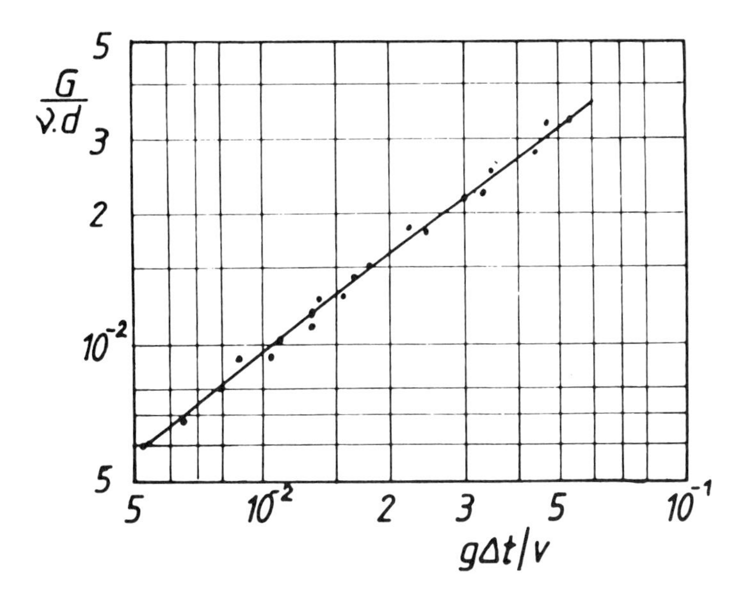

Figure 2.124–Relations between energy required per unit new surface formed, γ, and the parameters of the tine-clod collision process. G = clod weight, d = clod diameter, Δt = duration of tine-clod impact, v = tine velocity, and g = acceleration of gravity (from Sitkei, 1985).

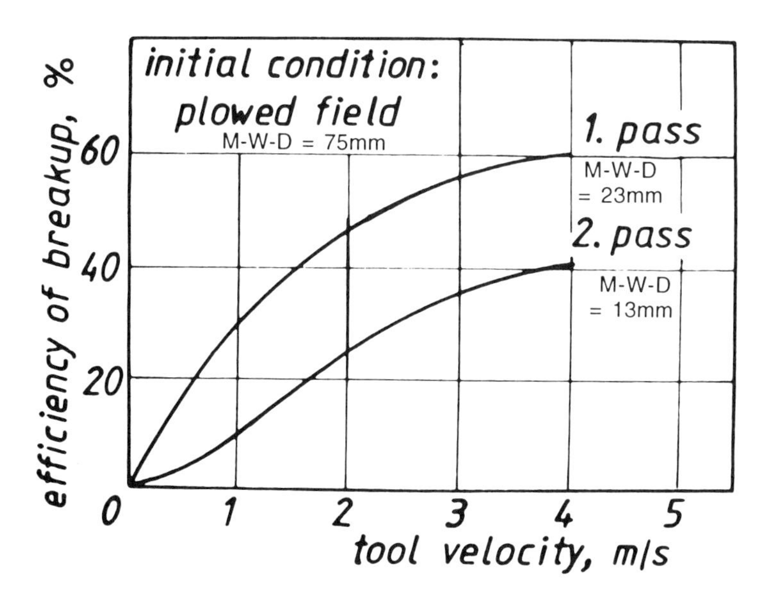

Figure 2.125–Efficiency of clod breakup as a function of tool velocity for a plowed field (from Sitkei, 1985).

Table 2.13. Cutting resistance data

Cutting cylindrical soil samples with a thin sharp blade (from Kitani et al, 1985)

Soil Texture	Moisture Content (%)	Cutting resistance (N/m = N•m/m²)*
Loam	69.1	253
Clay Loam	21.0	247
Silty Clay	32.7	91

Cutting soil ridge with a thin wire (from Gill and VandenBerg, 1968)

Soil Texture	Moisture Content (%)	Dry bulk Density (g/cm³)	Cutting resistance (N/m)*	
			Quasi static	14.4 km/h
Clay	23.8	1.18	131	699
Sandy Loam	11.4	1.56	175	569
Sand	6.5	1.65	96	314

Cutting a soil ridge with a thin wire (from Stefannelli, 1962)

Soil type	Moisture Content (%)	Vertical Stress (kPa)	Dry bulk Density (g/cm²)	Cutting resistance (N/m)*
River Sand	13.6	1.52	1.50	426-461
Farm Soil	9.8	1.22	1.20	167
Farm Soil	11.9	1.41	1.20	176-196
Farm Soil	12.7	1.05	1.00	284-324

Cutting a soil ridge with a 0.25-mm-diameter wire (from Bigsby and Bockhop, 1964)

Soil Texture	Moisture Content (%)	Cutting Resistance (N/m)*
Silty Clay	15	54
Silty Clay	30	197

Cutting a soil cylinder with an encircling wire loop (from 0.66 mm diameter) (from Vomocil and Chancellor, 1969)

Soil Texture	Moisture Content (%)	Cutting Resistance (N/m)*
Silty Clay	12	14.7
Silty Clay	20	1.2
Silty Clay	24	1.2

* Values given are the energy in Joules (N•m) to produce 1 m² of new cut surface considering both sides of the cut as new surface.Thus, the values given are just half of the horizontal force on a wire to cut a soil ridge of a given width.

tines of given width and spacing in a sufficiently direct manner to cause breakage (tine center line within 0.5 times clod radius of the clod center line). The efficiency of clod breakup as a function of tine velocity and mean clod size are shown in figure 2.125.

Cutting Resistance

Most analyses of the soil cutting action of tools or tool elements have been focused on the forces involved in displacing the soil relative to the tool. It is acknowledged, however, that in addition to the force and energy required for these processes, there is an additional force input per unit blade length, or energy input per unit new surface generated, required to produce the rupture in the soil monolith (Soehne, 1956). This rupture energy per unit surface or "cutting resistance" represents an independent property of the soil. It is difficult to measure this property without involving the forces or energies involved in operating the rupture process as distinguished from those going directly to the process. One of these measurement processes is the drop-shatter process mentioned in the preceding section. It is clear that the drop-shatter procedure is not a particularly efficient process. Other tests have been developed to obtain some estimate of cutting resistance. One of these is the use of sharp, thin blade to cut a soil sample (Kitani et al, 1985). Another test method is to stretch a thin wire between two standards and force the wire horizontally through a ridge of soil (Gill and VandenBerg, 1968; Bigsby and Bockhop, 1964; Stefanelli, 1968). A third method is to encircle a cylindrical monolithic soil sample with a thin wire and to apply tension to the wire causing it to cut into the sample (Vomocil and Chancellor, 1969). Data from these tests are given in table 2.13.

In some cases (Stefannelli, 1962; Gill and VandenBerg, 1968), the draft resistance of the wire was nearly the same as that of a shallowly inclined flat-plate tool. When the wire diameter was decreased from 0.25 to 0.15 mm, draft force was reduced somewhat (Bigsby and Bockhop, 1964).

The values shown in table 2.13 generally correspond to the lower portions of the ranges found for the drop-shatter process and are noticeably lower than those for the clod-crushing process reported in the preceding section.

The very low values for the encircling wire loop cutting process indicate that if the rupture process can be implemented with high efficiency, then this energy-consuming aspect of the soil-tool interaction can be a source for overall efficiency improvement.

Soil Mass Dynamic Load-Deformation Properties Definable within a System of Mechanics (Stress-Strain)

Agricultural soils may be described in a physical sense in the same way as other materials for which a comprehensive system of mechanics has been developed to characterize the load-deformation properties of the material. The properties of agricultural soils are not as regular, unchanging, and standardizable as those of steel, for example, but the system of mechanics for describing the dimensional changes under load may be used for agricultural soils as well as for steel, provided the same properties of the materials are used. For agricultural soils, this usually requires that each analysis be confined to a very limited range of conditions over which the properties of the soil do not change appreciably.

Poisson's Ratio

When a prism of material is loaded in one dimension, there is usually a deformation in the same dimension. The ratio of this deformation to the original length in the specified dimension is the strain:

$$\varepsilon = \frac{l - l_0}{l_0} \tag{2.79}$$

in which ε = strain, l_0 = original length of the prism without load, and l = length of prism after load is applied.

With most materials, when the prism is loaded in only one dimension, strain occurs not only in that dimension, but also in the two dimensions perpendicular to that in which the load is applied. If the applied load is compressive and the resulting strain in the direction of the load is also compressive (dimension gets shorter), then it might be expected that in the two dimensions perpendicular to that in which the load is applied, that the prism would undergo expansive strain (dimension becomes greater). The instantaneous ratio between the incremental strain in any one nonloaded dimension to the negative value of the incremental strain in the loaded dimension is called *Poisson's ratio*, μ:

$$\text{Poisson's ratio} = \mu = \frac{\Delta\varepsilon_3}{-\Delta\varepsilon_1} \tag{2.80}$$

in which $\Delta\varepsilon_3$ = incremental expansive strain in dimension perpendicular to that in which load is applied, and $\Delta\varepsilon_1$ = incremental expansive strain in the dimension in which load is applied.

Poisson's number is defined as the inverse of Poisson's ratio (Kézdi, 1974).

$$\text{Poisson's number} = m = \frac{1}{\mu} = \frac{-\Delta\varepsilon_1}{\Delta\varepsilon_3} \tag{2.81}$$

The ratio of the applied load to the cross-sectional area of the prism, the plane of which is perpendicular to the direction of load application, is the stress, σ_1.

$$\text{stress} = \sigma_1 = \frac{\text{load (N)}}{\text{area}\left(\text{m}^2\right)} \tag{2.82}$$

To describe the strain response of a material in the direction of load application, the elastic modulus, E, (fig 2.126) (Lambe and Whitman, 1969) is defined as:

$$E = \frac{\sigma_1}{\varepsilon_1} \tag{2.83}$$

for the case of uniaxial loading.

The maximum shear stress, τ_{max}, is defined as:

$$\text{maximum shear stress} = \tau_{max} = \frac{\sigma_1 - \sigma_3}{2} \tag{2.84}$$

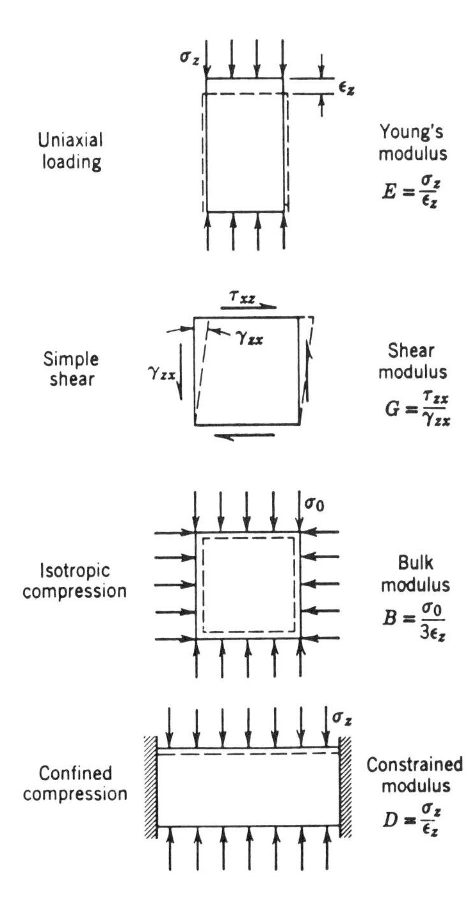

Figure 2.126–Description of the elastic (Young's) modulus, E, shear modulus, G, bulk modulus, B, and constrained modulus, D. Reprinted with permission of John Wiley & Sons, Inc. from *Soil Mechanics*, Lambe, T. W. and R. V. Whitman, copyright © 1969.

in which σ_1 = stress in the direction of the main load, and σ_3 = stress in the direction perpendicular to that of the main load application and for which the stress is the smallest of the three mutually perpendicular stresses.

Maximum shear strain, γ_{max}, may be described by:

$$\gamma_{max} = \varepsilon_1 - \varepsilon_3 \tag{2.85}$$

The relationship between shear stress and shear strain may be described (Lambe and Whitman, 1969) by the shear modulus, G:

$$G = \frac{\tau_{max}}{\gamma_{max}} = \frac{E}{2(1 + \mu)} \tag{2.86}$$

The volumetric strain of the prism may be described by:

$$\Delta V/V = \varepsilon_1 + \varepsilon_2 + \varepsilon_3 \tag{2.87}$$

in which ΔV = incremental volume change, and V = original volume of prism.

For the special case in which $\sigma_1 = \sigma_2 = \sigma_3 = \sigma_0$ (hydrostatic compression), the ratio between stress level and volume strain is described (Lambe and Whitman, 1969) by the bulk modulus, B:

$$\text{bulk modulus} = B = \frac{\sigma_0}{\Delta V / V} = \frac{E}{3(1 - 2\mu)} \tag{2.88}$$

For soils, the constrained modulus, D, is often determined because it is necessary to confine a granular or amorphous material during test. The constrained modulus is measured as shown in figure 2.126, using axial compression in a cylinder of fixed diameter. For a test such as this, assuming linear elastic behavior and zero friction with the walls at the cylinder:

$$\sigma_x = \sigma_y = \frac{\mu}{1 - \mu} \sigma_z \tag{2.89}$$

$$D = \frac{E(1 - \mu)}{(1 + \mu)(1 - 2\mu)} \tag{2.90}$$

For an elastic material (Lambe and Whitman, 1969):

$$\varepsilon_x = \frac{1}{E}\left[\sigma_x - \mu(\sigma_y + \sigma_z)\right] \tag{2.91}$$

$$\varepsilon_y = \frac{1}{E}\left[\sigma_y - \mu\left(\sigma_z + \sigma_x\right)\right] \tag{2.92}$$

$$\varepsilon_z = \frac{1}{E}\left[\sigma_z - \mu\left(\sigma_x + \sigma_y\right)\right] \tag{2.93}$$

For the case of uniaxial loading (in the σ_z direction) of a cylinder of elastic test material:

$$\frac{\Delta V}{V} = \varepsilon_x + \varepsilon_y + \varepsilon_z = (1 - 2\mu)\frac{\sigma_z}{E} \tag{2.94}$$

$$\tau_{max} = \frac{\sigma_z}{2} \text{ and } \gamma_{max} = \frac{\sigma_z}{2G} \tag{2.95}$$

For confined compression of an elastic material:

$$\frac{\Delta V}{V} = \varepsilon_x + \varepsilon_y + \varepsilon_z = \frac{(1 + \mu)(1 - 2\mu)\,\sigma_z}{E(1 - \mu)} \tag{2.96}$$

$$\tau_{max} = \frac{\sigma_z}{2}\frac{(1 - 2\mu)}{(1 - \mu)} \text{ and} \tag{2.97}$$

$$\gamma_{max} = \frac{\sigma_z}{2G}\frac{(1 - 2\mu)}{(1 - \mu)} \tag{2.98}$$

The value of Poisson's ratio is, therefore, an important element in relating the stresses, strains, and various basic properties of a material that are to be used in a system of mechanics that describes the results of various load-deformation processes.

The above relationships (Lambe and Whitman, 1969) are for elastic, isotropic, and homogeneous materials and may not always be applicable to agricultural soils.

Values of Poisson's ratio, μ, have been tabulated for common materials in table 2.14 (Head, 1986; Lambe and Whitman, 1969).

The most usual ways in which Poisson's ratio for soils is measured are by using either a triaxial test or an unconfined compression test in which both the length and the diameter of the cylindrical soil sample (either monolithic or remolded) are measured throughout the test. Because soils can have an amorphous or granular form, the way in which the sample material is held together in the form of a prism becomes a major determinant of the values of Poisson's ratio that are obtained from a given test. In the case of the unconfined compression test, the diameter of the sample is usually measured physically (fig 2.127) (Kézdi, 1974). For the triaxial test, either a special diameter

Table 2.14. Values of Poisson's ratio and elastic modulus*

Material	Poisson's Ratio	Elastic Modulus (N/m^2) × 10^9
Steel	0.28 – 0.29	200
Aluminum	0.34 – 0.36	55 – 76
Limestone	0.27 – 0.30	87 – 107
Granite	0.23 – 0.27	73 – 86
Quartzite	0.12 – 0.15	82 – 96
Slate	0.15 – 0.20	35
Ice	0.36	7.1
Normally consolidated clays	0.49	0.005 to 0.030
Consolidated Kaolin	0.10	0.003 to 0.005
London Clay	0.40	0.060 to 0.150
Sand	0.48	loose 0.020 to 0.180 dense

* Data from Head (1986) and Lambe and Whitman (1969)

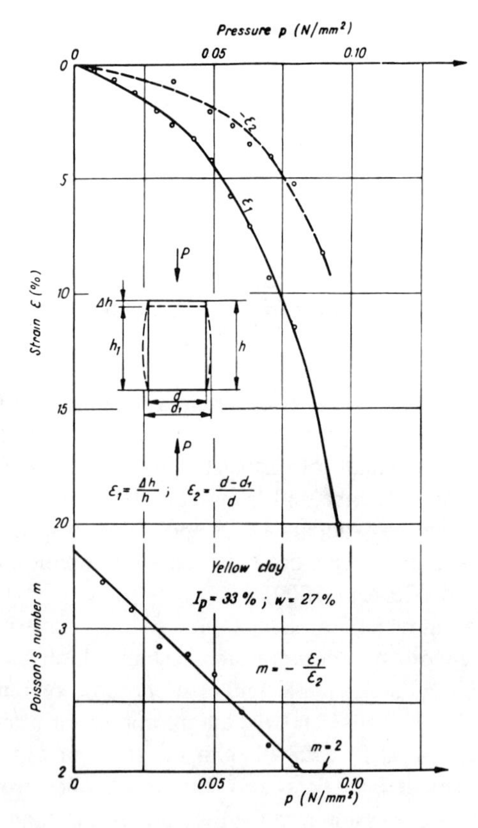

Figure 2.127–Determination of Poisson's number, m, (m = 1/Poisson's ratio) by measuring the axial and lateral deformation of a compressed cylindrical soil sample (from Kézdi, 1974).

measuring apparatus is used inside the test cell (Gill and VandenBerg, 1968) or the incremental volume strain ($\Delta V/V$) is measured (in terms of the volume of air or water expressed from the drained sample), along with the incremental axial strain ($\Delta\varepsilon_1$), and the incremental radial strain ($\Delta\varepsilon_3$) is determined from:

$$\Delta\varepsilon_3 = \left(\frac{\Delta V}{V} - \Delta\varepsilon_1\right) / 2, \text{ or, } \frac{\Delta V}{V} = \Delta\varepsilon_1 + 2\Delta\varepsilon_3 \tag{2.99}$$

in which compressive strains and volume reductions are considered to be negative.

Volume strain, which depends on a sizable number of factors, is thus an important element in the determination of Poisson's ratio for agricultural soils.

Equation 2.99 may be reformed as:

$$-\frac{\Delta\varepsilon_3}{\Delta\varepsilon_1} = 1/2 - \frac{\Delta V / V}{2\Delta\varepsilon_1} = \mu \tag{2.100}$$

Thus, when a material is of essentially constant volume irrespective of the stresses applied, as is the case for rubber or a saturated agricultural soil, the Poisson's ratio is 0.5. For materials like cork, for which compression in one dimension has little effect on other dimensions, Poisson's ratio is near zero. Overconsolidated sands, when subjected to compressive strain in one direction, may shear and dilate causing the volume actually to increase, in which case the Poisson's ratio can be greater than 0.5 (Lambe and Whitman, 1969; Chancellor and Korayem, 1965). Some soil mechanics texts (Kézdi, 1974; Spangler, 1960; Lambe and Whitman, 1969) state that Poisson's ratio for soils is extremely variable between the values of 0.0 and 0.5 and that an actual value will probably lie closer to 0.5 than to 0.0. Assumed values are usually in the 0.3 to 0.5 range, depending on whether or not it is expected that volume strain will take place.

Data from which the Poisson's ratio may be determined for agricultural soils have been presented for several different regimes of stresses for maintaining the sample prism as a structural element:

1. A monolithic cylindrical soil sample is loaded axially in an unconfined compression test with no lateral constraints (Chancellor et al, 1969).
2. A monolithic cylindrical soil sample is loaded axially with a fixed level of lateral stress applied (Chancellor et al, 1969).
3. A remolded cylindrical sample of length equal to diameter is loaded axially while the friction of the end supports on the sample prevent catastrophic sample failure and allow the sample to undergo extensive shear strain (Chancellor and Korayem, 1965).
4. A remolded cylindrical sample of length approximately twice diameter is loaded axially while radial expansion is constrained by a spring-loaded band with various spring constants (Kitani and Persson, 1967).
5. A cylindrical sample of length approximately twice the diameter is subjected simultaneously to increasing levels of both σ_1 and σ_3, while these two stresses are

constrained to ratios of 1.5, 2.0, 2.5, and 3.0, all of which were below the ratio necessary to cause shear failure (Grisso et al, 1987a, b).

In the latter two cases (4 and 5) external stresses were applied in the $\sigma_2 = \sigma_3$ direction either passively or actively. Under these circumstances $(-\Delta\varepsilon_3 / \Delta\varepsilon_1)$ is not truly the Poisson's ratio, and the effects of these lateral stresses must be considered in determining the actual Poisson's ratio.

For the case of a remolded sample in unconfined compression, figure 2.127 illustrates the case of clay soil in which the Poisson's number ($m = 1/\mu$) decreases linearly with increases in applied axial stress from the value of 3.6 at zero stress ($\mu = 0.28$) to 2.0 ($\mu = 0.5$) at a high stress level, and then remains at 2.0 ($\mu = 0.5$ = constant volume) as stresses increase further. When a monolithic sample with an intrinsic stress of 40 kPa was tested in unconfined compression (fig 2.128), lateral expansion was very small at low axial strain levels and increased more than proportionally with axial strain as failure was approached. Nevertheless, even at failure, the constant volume state had not been reached, and the Poisson's ratio, μ, was only 0.27, as shown in table 2.15.

When cylindrical monolithic samples were tested in confined compression with a confining stress of 73.7 kPa (Chancellor et al, 1969), much higher levels of axial strain were tolerated before failure occurred. Under these circumstances, the constant volume

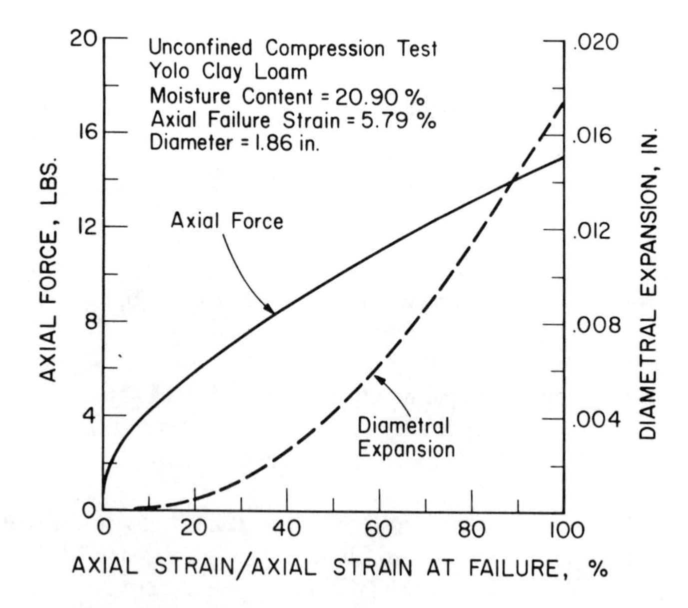

Figure 2.128–Typical response of a long, monolithic, cylindrical soil sample during axial compression. Note that diametral expansion is very low at the beginning of the process (low Poisson's ratio) and then becomes much greater toward the end of the process (from Chancellor et al, 1969).

Table 2.15. Poisson's ratio determinations during axial compression of a long monolithic cylindrical soil sample*

Axial Strain Range (%) Failure Strain	$-\varepsilon_1$	$-\Delta\varepsilon_1$	ε_3	$\Delta\varepsilon_3$	Poisson's Ratio (μ)
0 – 20	0.0116	0.0116	0.00027	0.00027	0.023
20 – 40	0.0232	0.0116	0.00134	0.00112	0.096
40 – 60	0.0347	0.0116	0.00328	0.00189	0.163
60 – 80	0.0463	0.0116	0.00618	0.0029	0.250
80 – 100	0.0579	0.0116	0.0093	0.00312	0.269

* See figure 2.128.

conditions (at which μ = 0.5) were encountered at axial strain levels well below the axial strain associated with sample failure (figs 2.129 and 2.130).

When the lateral expansion of short cylindrical samples of two very different soils over a broad range of moisture and porosity conditions were constrained by frictional forces on the end of the sample (the samples were squashed), Poisson's ratios were initially near zero at low strain levels and then increased to nearly 0.5 at shear strain levels (γ_{max}) of 0.5 or 0.6 (figs 2.131 and 2.132) (Chancellor and Korayem, 1965) (Note:

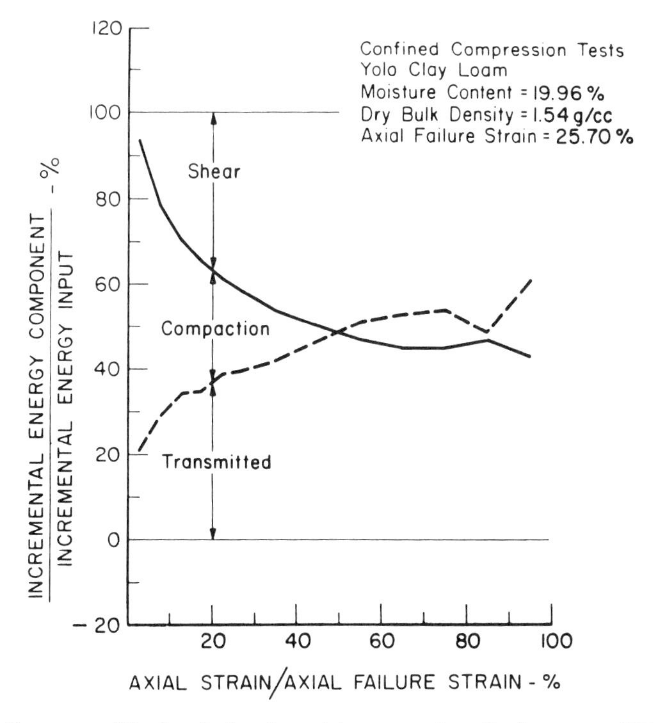

Figure 2.129–Energy partitioning during the axial compression of a long, monolithic cylindrical soil sample. Note that compaction or volumetric strain ends (Poisson's ratio becomes 0.5) midway through the process (from Chancellor et al, 1969).

in figs 2.131, 2.132 and 2.133, the term $\Delta\gamma_{max} / \Delta\varepsilon_1$ is equal to $\mu + 1$). The relationships among the stress levels acting and Poisson's ratio for these tests are illustrated in figure 2.133).

For the case of tests in a triaxial cell in which $\sigma_2 = \sigma_3$ and this cell pressure value was held constant while σ_1 was increased, Chi et al (1993b) has proposed the following relationship to represent Poisson's ratio, μ:

$$\mu = a + b \frac{\sigma_1 - \sigma_3}{(\sigma_1 - \sigma_3)_f} \tag{2.101}$$

in which $(\sigma_1 - \sigma_3)_f$ = deviatoric stress at failure, and a and b are experimentally determined coefficients.

Chi et al (1993a) used a modified form of this relationship:

$$\mu = a + b \frac{(1 - \sin\phi)\,(\sigma_1 - \sigma_3)}{2\,c\,\cos\phi + 2\,\sigma_3\,\sin\phi} \tag{2.102}$$

in which c = cohesion, ϕ = angle of internal friction, and

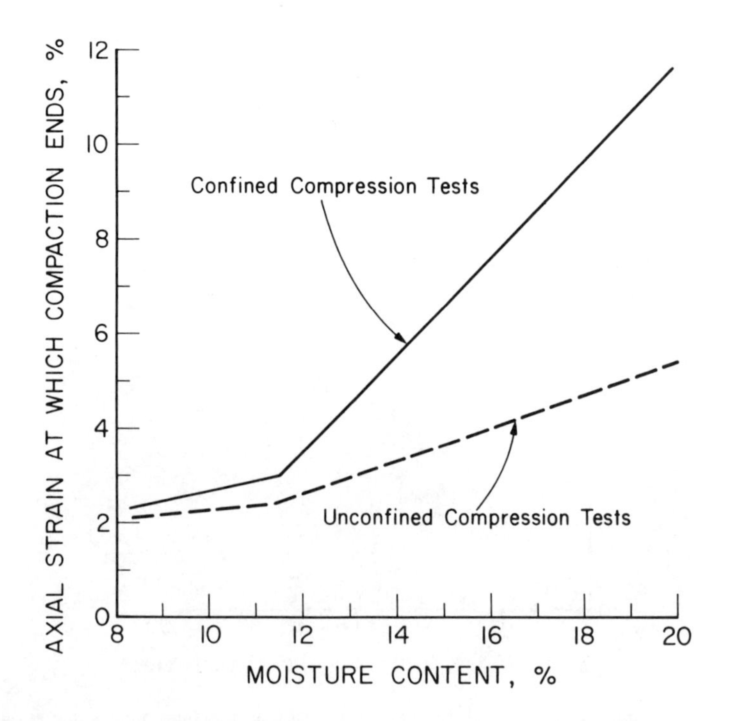

Figure 2.130–Strain levels of long, monolithic, cylindrical soil samples at which volumetric strain ceases (Poisson's ratio becomes 0.5) (from Chancellor et al, 1969).

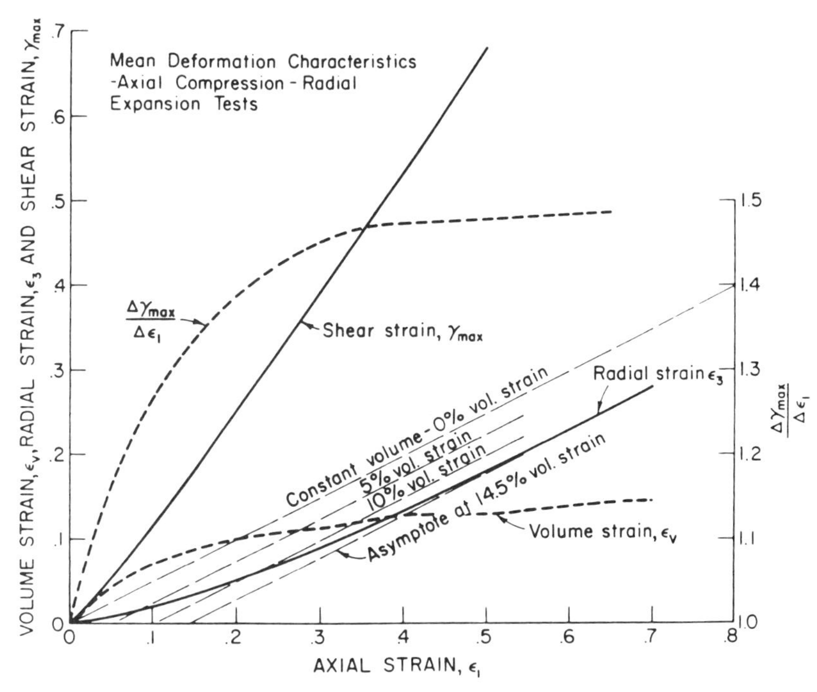

Figure 2.131–Composite deformation characteristics of several short, wide, cylindrical soil samples loaded axially. Poisson's ratio is equal to ($\Delta\gamma_{max}/\Delta\varepsilon_1$) - 1.0 (from Chancellor and Korayem, 1965).

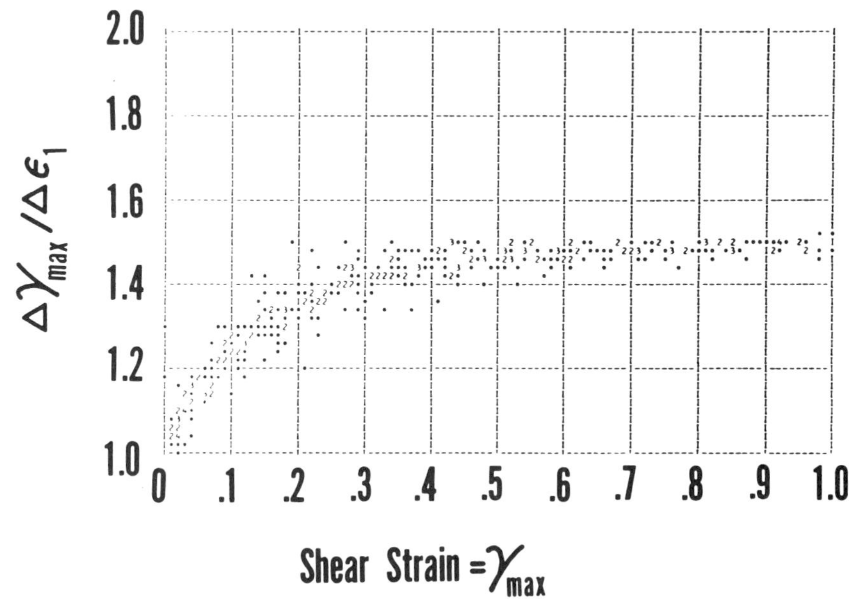

Figure 2.132–Relations between Poisson's ratio and shear strain for several short, wide, cylindrical soil samples loaded axially. Poisson's ratio is equal to ($\Delta\gamma_{max}/\Delta\varepsilon_1$) – 1.0 (from Chancellor and Korayem, 1965).

$$\frac{2\ c\ \cos\phi + 2\ \sigma_3\ \sin\phi}{1 - \sin\phi}$$

is the deviatoric stress at failure as predicted by the Mohr-Coulomb failure criterion (see table 2.20).

The term *Poisson's ratio* is used to refer to the ratio $-\Delta\varepsilon_3/\Delta\varepsilon_1$, which is the relationship usually specified for cases of strictly uniaxial loading of cases in which lateral stresses do not change. However, in cases in which applied stresses occur in the $\sigma_2 = \sigma_3$ direction as well as in the σ_1 direction, the ratio, $-\Delta\varepsilon_3/\Delta\varepsilon_1$, is not truly the Poisson's ratio. Materials such as steel have some sort of internal bonding stresses which tend to maintain the shape of the prism when it is loaded uniaxially. Monolithic (and even preconsolidated, remolded) soil samples are similar in this respect to steel. When the stresses to maintain prism shape are caused, even in part, by externally active or passive phenomena, the definition of Poisson's ratio is not a property of the material alone.

When lateral stresses are subject to change for cylindrical samples having $\sigma_2 = \sigma_3$:

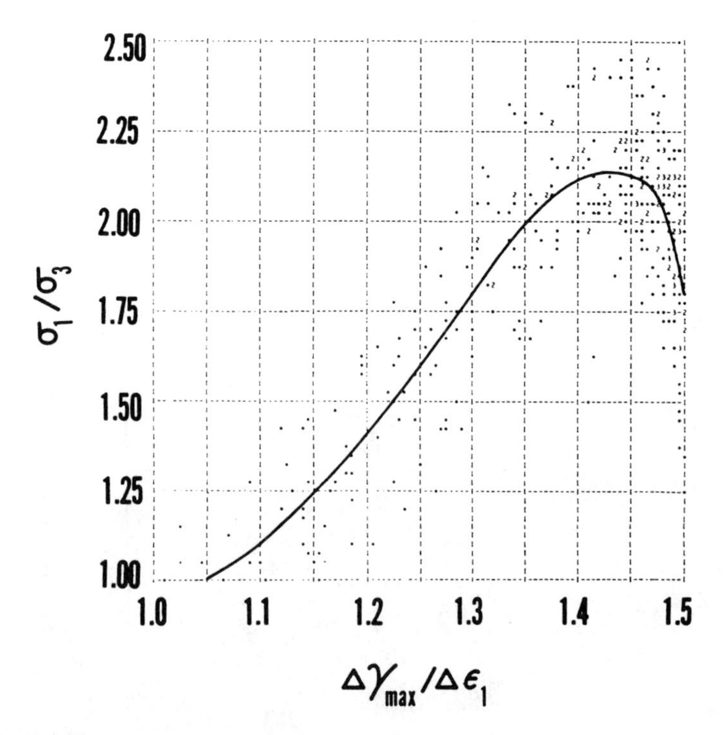

Figure 2.133–Relations between Poisson's ratio and the principal stress ratio for several short, wide, cylindrical soil samples loaded axially. Poisson's ratio is equal to ($\Delta\gamma_{max}/\Delta\varepsilon_1$) – 1.0 (from Chancellor and Korayem, 1965).

$$d\varepsilon_1 = \frac{d\sigma_1}{E} - \frac{2\mu}{E}(d\sigma_3)$$

$$d\varepsilon_3 = \frac{d\sigma_3}{E} - \frac{\mu}{E}(d\sigma_1 + d\sigma_3)$$

$$= (1-\mu)\frac{d\sigma_3}{E} - \frac{\mu d\sigma_1}{E}$$

Thus,

$$-\frac{d\varepsilon_3}{d\varepsilon_1} = \frac{\mu d\sigma_1 - (1-\mu)\, d\sigma_3}{d\sigma_1 - 2\,\mu\, d\sigma_3} \tag{2.103}$$

If $d\sigma_3 = 0$, then $-\dfrac{d\varepsilon_3}{d\varepsilon_1} = \mu$

When the lateral expansion of the sample was constrained by a spring-loaded sample casing to which various springs with different spring constants were applied, axial and lateral stresses and strains could be measured. In this case it was found that, irrespective of the spring constants used, the resulting ratio of σ_3/σ_1 was approximately a constant (0.141) at all stress levels. This permitted true values of Poisson's ratio to be computed. Results are shown in figure 2.134 and in table 2.16 (Kitani and Persson, 1967) for the highest and lowest spring constants used. Data in table 2.16 indicate the differences

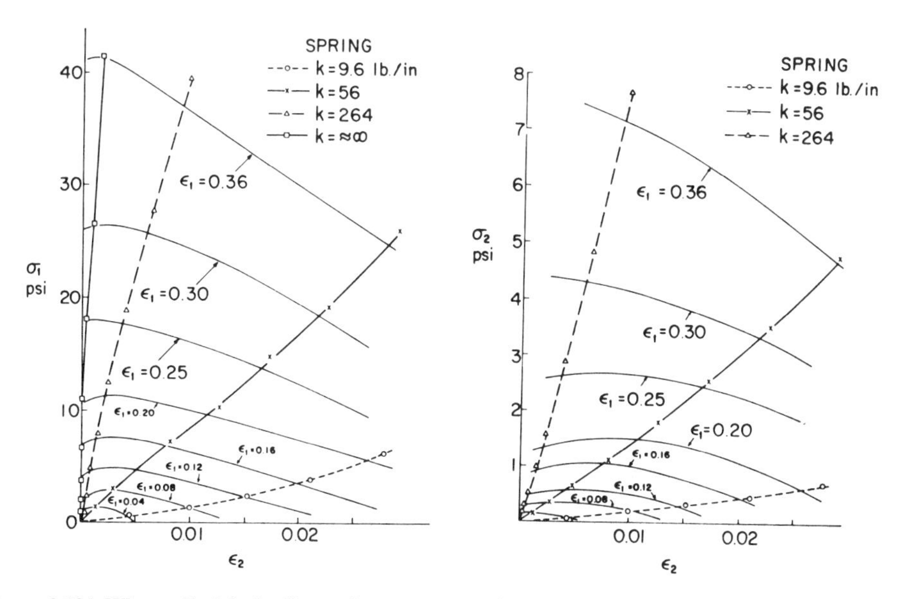

Figure 2.134–When cylindrical soil samples were constrained laterally by an expandable spring-loaded enclosing cylinder, the ratios of $\varepsilon_2/\varepsilon_1$ were very low (from Kitani and Persson, 1967).

Table 2.16. Poisson's ratio for soils constrained by a spring-loaded encasement*

ε_1	$\Delta\sigma_3/\Delta\sigma_1$	$-\Delta\varepsilon_3/\Delta\varepsilon_1$	μ	
0.06	10.26	0.0058	0.094	
0.10	13.47	0.0060	0.075	
0.14	12.56	0.0075	0.085	
0.18	15.72	0.0133	0.084	Highest spring
0.225	2.67	0.0242	0.286	constant
0.275	4.84	0.0335	0.197	(46,220 N/m)
0.33	3.87	0.0414	0.234	
0.06	8.60	0.1065	0.1952	
0.10	13.50	0.1123	0.1738	Lowest spring
0.14	7.11	0.1073	0.2120	Constant
0.18	7.72	0.1248	0.2240	(1681 N/m)

* Data from Kitani and Persson (1967).

between values of Poisson's ratio and $-\Delta\varepsilon_3/\Delta\varepsilon_1$, for this case in which values of $\sigma_2 = \sigma_3$ were increased by a passive stability mechanism.

The results in table 2.16 bear some similarity to those in figure 2.133, although in the case represented in the table, the constant volume state was never reached.

For a case in which long, cylindrical, remolded, samples of four different soils were subjected to principal stress ratios (σ_1/σ_3) of 1.5, 2.0, 2.5, and for one soil 3.0, as stress levels were simultaneously increased, the values of ε_1 and ε_3 were recorded (Grisso et al, 1987a, b). These stress ratios were below the level required to produce shear failure of the samples. In this case of an active stability mechanism, it was possible to input compressive energy to the soil in the $\sigma_2 = \sigma_3$ direction, as well as in the σ_1 direction. Under these circumstances, when the ratio of octahedral shear stress, τ_{oct}, to octahedral normal stress, σ_{oct}, was in the range of 0.49 to 0.59 (stress ratio, σ_1/σ_3, for this particular case in the range of 2.6 to 2.8), the soils displayed a value of $-\Delta\varepsilon_3/\Delta\varepsilon_1 = 0.0$, ie, there was no radial strain. Application of equation 2.104 to this zero-radial-strain condition resulted in a range of Poisson's ratio values from 0.263 to 0.277. The octahedral shearing stress is described as:

$$\tau_{oct} = \frac{1}{3}\left[(\sigma_1 - \sigma_3)^2 + (\sigma_2 - \sigma_3)^2 + (\sigma_3 - \sigma_1)^2\right]^{\frac{1}{2}} \qquad (2.104)$$

and for this triaxial cell case, τ_{oct}, can be represented by:

$$\tau_{oct} = \frac{1}{3}\left[2(\sigma_1 - \sigma_3)^2\right]^{\frac{1}{2}} \qquad (2.105)$$

The octahedral normal stress, σ_{oct}, is described by:

$$\sigma_{oct} = \frac{1}{3}(\sigma_1 + \sigma_2 + \sigma_3) \quad (2.106)$$

and for this triaxial cell case can be represented by:

$$\sigma_{oct} = \frac{1}{3}(\sigma_1 + 2\sigma_3) \quad (2.107)$$

For the orthotropic case in which σ_2 and σ_3 differ from each other an elastic model of stress-strain relationships can be described by:

$$\varepsilon_1 = \frac{1}{E_1}\sigma_1 - \frac{\mu_{12}}{E_1}\sigma_2 - \frac{\mu_{13}}{E_1}\sigma_3$$
$$\varepsilon_2 = -\frac{\mu_{12}}{E_1}\sigma_1 + \frac{1}{E_2}\sigma_2 - \frac{\mu_{23}}{E_2}\sigma_3$$
$$\varepsilon_3 = -\frac{\mu_{13}}{E_1}\sigma_1 - \frac{\mu_{23}}{E_2}\sigma_2 + \frac{1}{E_3}\sigma_3$$

in which

E_1 = Young's modulus in the major principal direction
E_2 = Young's modulus in the intermediate principal direction
E_3 = Young's modulus in the minor principal direction
μ_{12} = Poisson's ratio for strain in the intermediate principal direction due to the major principal stress
μ_{13} = Poisson's ratio for strain in the minor principal direction due to the major principal stress
μ_{23} = Poisson's ratio for strain in the minor principal direction due to the intermediate principal stress.

Khan (1993) conducted tests on samples of Lloyd clay at 17% moisture. A triaxial cell apparatus was used which permitted σ_2 to be varied independently of σ_3. For one test at a constant level of mean principal stress, σ_{oct} = 137.9 kPa, and $\Delta\sigma_2 = 0.5\ \Delta\sigma_3$ (the minor and intermediate principal stresses were reduced as the major principal stress was increased to keep the mean principal stress constant), the results are presented in table 2.17.

It can be seen from table 2.17 that values of μ_{12} and μ_{13} remained near 0.5 while the value of μ_{23} increased from about 0.15 to 0.55 as the shear stress, τ_{oct}, increased. Values of μ_{13} above 0.5 indicate dilation of the sample at high shear stress levels.

Modulus of Elasticity

The modulus of elasticity, E, is the ratio of stress to strain in one dimension for a prism of material undergoing uniaxial loading (see fig 2.126). It is also known as

Young's modulus. For elastic materials, such as steel, at stresses below the proportional limit, a single value of modulus of elasticity would apply over a broad range of stress levels and for tension as well as compression. For soils, however, the modulus of elasticity changes throughout the range of stresses and changes as moisture content, porosity, and other factors change.

In order to adapt the system of mechanics used for materials like steel to problems involving agricultural soil, approximations of values for the elastic modulus are made and applied over limited ranges of stress conditions. In one case, where an attempt was made to use the finite element method to predict the effect of surface loads on subsurface stresses and volume changes (Raper and Erbach, 1988a, b), a sensitivity analysis was made as to the effect of the use of various values of E and of Poisson's ratio on the resultant stresses and volume strains. When Poisson's ratio was varied from 0.13 to 0.38, there was an approximate 10% change in subsurface vertical stress values. When Poisson's ratio was kept constant and the modulus of elasticity varied from 506 to 5236 kPa, vertical stress differences ranged from –1.6 to +2.9%. However, when the effects on subsurface volume strain were considered, the changes in Poisson's ratio caused differences of about 47%, while changes in the elastic modulus resulted in changes averaging 90.5%. Thus, the use of appropriate values of the elastic modulus was of particular importance.

Because of the high degree of variability of the elastic modulus for agricultural soils, a number of methods of obtaining values from the results of compressive or tensile tests of soil material samples have been used (see fig 2.135) (Head, 1986). Among these are the tangent modulus, either at the origin or at another specified point, and the secant modulus between various specified points on the stress-strain curve. Another method for obtaining a value of the elastic modulus, which represents the more truly "elastic" properties of the soil sample, is to load it partly to failure, unload the sample, and then

Table 2.17. Elastic parameters for Lloyd clay (17% moisture) (from Khan, 1993)*

τ_{oct} kPa	E_1 kPa	E_2 kPa	E_3 kPa	μ_{12}	μ_{13}	μ_{23}
29.78	6871	11835	11104	0.471	0.501	0.154
37.23	6636	10900	10326	0.471	0.495	0.180
44.68	6313	9459	8611	0.458	0.505	0.262
52.13	6071	8481	7590	0.451	0.507	0.314
59.57	5857	7816	6991	0.449	0.506	0.338
67.02	5583	7203	6433	0.449	0.507	0.347
74.47	5211	6488	5693	0.448	0.515	0.356
81.92	4786	5666	4720	0.442	0.539	0.385
89.37	4352	4855	3770	0.435	0.576	0.425
93.08	3926	4101	2955	0.427	0.624	0.469
96.81	3509	3432	2315	0.423	0.676	0.508
100.53	3150	2915	1882	0.422	0.718	0.533
104.25	2858	2519	1566	0.422	0.757	0.554

* $\Delta\sigma_2 = 0.5\ \Delta\sigma_3$, $\sigma_{oct} = 137.9$ kPa

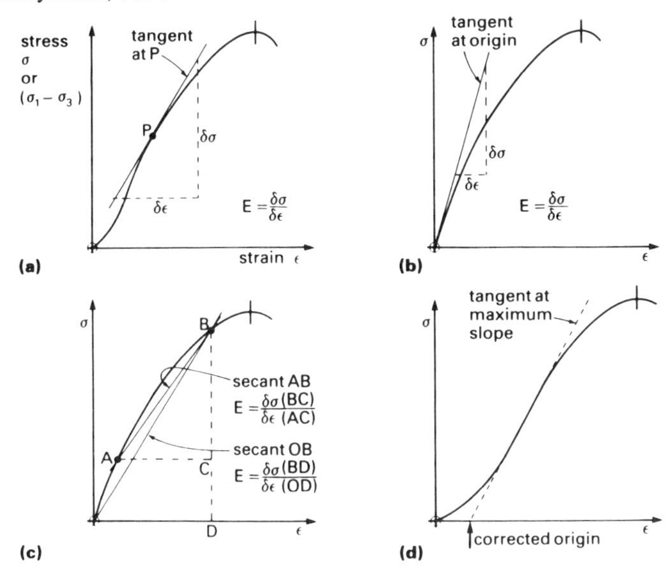

Figure 2.135–Illustration of the secant modulus and several forms of the tangent modulus as estimates of the elastic (Young's) modulus for soil samples loaded in a triaxial cell (from Head, 1986).

reload the sample, measuring the tangent to the slope of the reloading stress-strain curve and using this value of the tangent as the elastic modulus (fig 2.136) (Head, 1986).

The rate of propagation of compressive-tensile waves through a rod of material is theoretically:

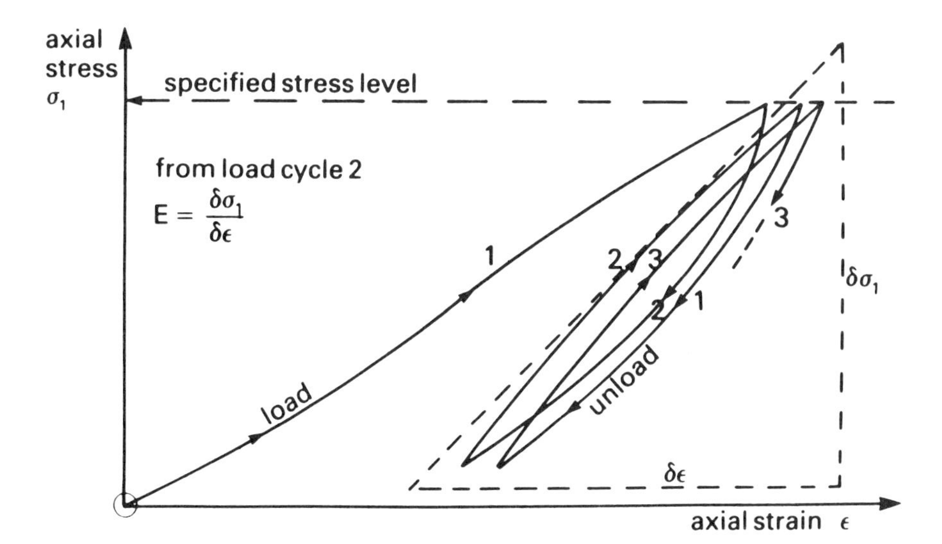

Figure 2.136–Illustration of the determination of Young's modulus using repeated loading cycles of cylindrical soil samples in a triaxial cell (from Head, 1986).

$$\text{wave velocity} = C_L = \sqrt{\frac{E}{\rho}} \tag{2.108}$$

in which ρ = the mass density of the material, C_L = wave propagation velocity, and E = modulus of elasticity.

By placing accelerometers on both ends of a long, cylindrical soil sample and vibrating one end at sonic frequencies, it is possible to determine the velocity of wave transmission. If amplitudes of vibration are kept very small, the above relationship may be employed to determine the elastic modulus (Rickman, 1971) (fig 2.137). However, if larger amplitudes are to be involved, the damping characteristics of the material need to be taken into account. Various relationships for accounting for the damping characteristics (Kocher and Summers, 1988) include use of the complex modulus, in which:

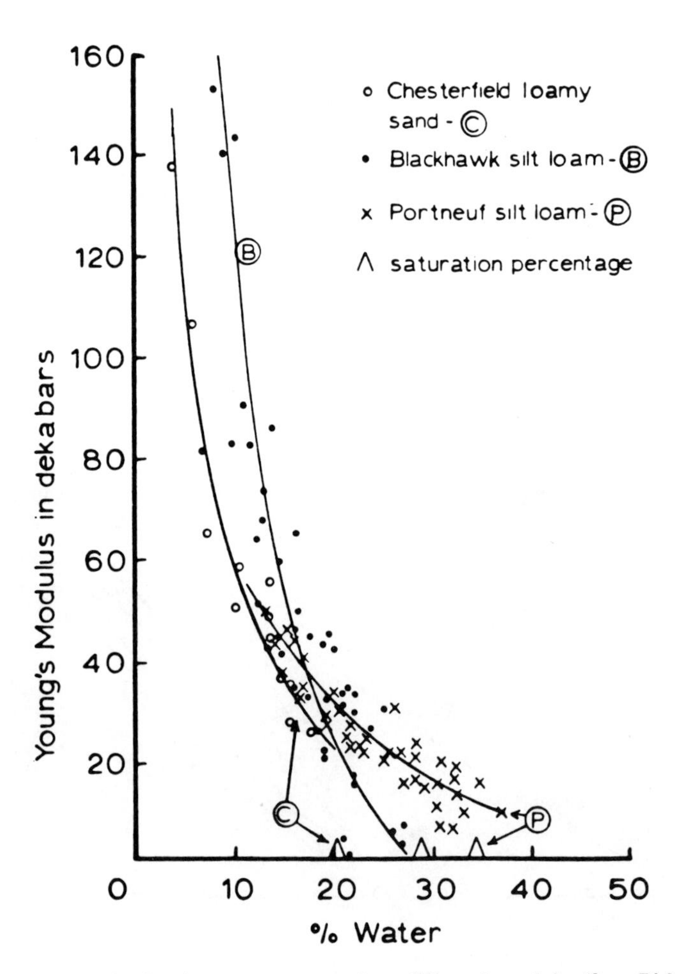

Figure 2.137–Effects of soil moisture content on values of Young's modulus (from Rickman, 1971).

$$E = \frac{\sigma}{\varepsilon\left(1 + j\,\delta\right)} \tag{2.109}$$

in which $j = \sqrt{-1}$, and δ = phase shift energy adsorption function, as well as various viscous or viscoelastic models including a second-order viscoelastic model:

$$\sigma = E\,\varepsilon + \alpha\,\frac{\partial \varepsilon}{\partial t} + \xi\,\frac{2^2 \varepsilon}{\partial\, t^2} \tag{2.110}$$

in which α and ξ are values relating to the viscous properties of the material (see table 2.3).

It was found that the second-order viscoelastic model gave the best representation of soil characteristics for the range of frequencies from 200 to 2000 Hz (Kocher and Summers, 1988).

Very general values of the elastic modulus are tabulated in the foregoing discussion on Poisson's ratio. In figure 2.137, values ranging from 160 MPa down to nearly zero were found for three soils as moisture content increased.

When an Ottawa sand was subjected to wave velocity tests for a long cylindrical section with varying levels of confining stress applied, a near-linear relationship on a log-log plot was found between values of elastic modulus and those of confining stress. Data ranged from an elastic modulus of 109 MPa at 17 kPa confining stress to 517 MPa at 324 kPa confining stress. These values for E were from 2 to 3 times higher than values measured directly in triaxial tests (Lambe and Whitman, 1969).

However, it should be noted that in wave velocity tests, strain levels are very small. The effect of strain level on the modulus of elasticity of sand is illustrated in figure 2.138 (Lambe and Whitman, 1969). General values for the elastic modulus of sands are tabulated in table 2.18.

Figure 2.139 (Lambe and Whitman, 1969) illustrates modulus of elasticity values for non-agricultural clay soils relative to the triaxial cell pressure at which the tests were conducted. As shear stresses approach those at failure, modulus values decrease. The effect of overconsolidation of the materials is shown to have limited effect. It was also found that as the remolded material aged from 3 to 60 days, the modulus of elasticity values approximately doubled. As strain rates increased from failure in 10 min to failure in 1 s, or in another case, from 1% strain in 500 min to 1% strain in 1 min, the modulus values increased from 1.5 to 2 times (Lambe and Whitman, 1969). Furthermore, modulus of elasticity values obtained with repeated loadings were 1.4 to 1.5 times those at the initial loading (Lambe and Whitman, 1969).

In a series of tests with monolithic samples of an agricultural clay loam soil (Vomocil and Chancellor, 1969), the secant modulus of elasticity between unstressed conditions and those at failure were measured for tensile, unconfined compression and confined compression loading over a range of moisture contents. The results are given in figure 2.140 to 2.142. It can be seen that the modulus of elasticity in tension was several times that in compression. Modulus of elasticity values were little affected by the

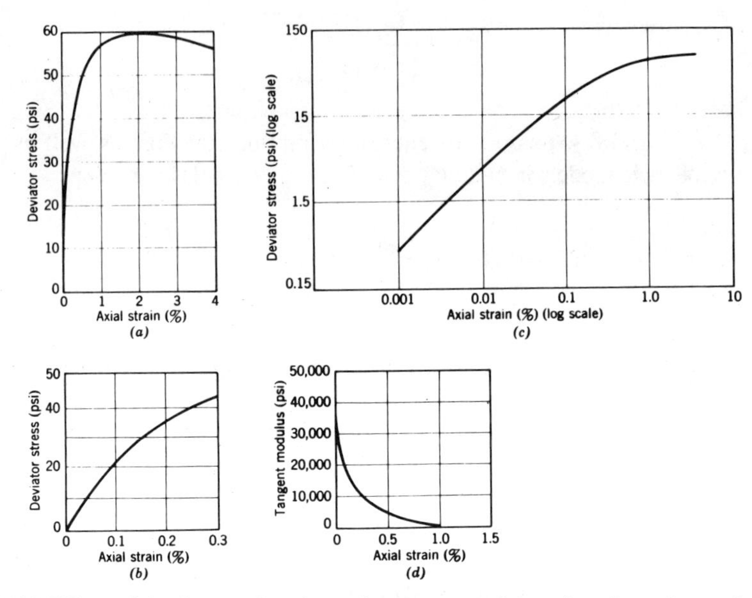

Figure 2.138–Effects of the degree of strain on the tangent modulus of a subangular sand sample in a triaxial cell. Reprinted with permission of John Wiley & Sons, Inc. from Chen, 1948, as reproduced in *Soil Mechanics*, Lambe, T. W. and R. V. Whitman, copyright © 1969.

application of a confining stress of 74 kPa. The characteristic stress-strain curves for these three modes of loading are shown in figure 2.143 to 2.145. The characteristic curve for unconfined compression shows initially a low modulus value which then increases as axial strain increases. It is believed that this is due to the action of the intrinsic stress in these monolithic samples. After the axial strain was sufficient to allow the externally applied stress to reduce the effects of tensile bonds in the axial direction, the externally applied stress was then no longer assisted by the intrinsic stress.

Monolithic samples of the same soil were subjected to repeated loading in both the confined (95 kPa) and unconfined modes (fig 2.146) (Aref et al, 1975), the values for the tangent modulus of elasticity upon repeated loading (see fig 2.136) are shown in table 2.19 (Aref, 1973).

Table 2.18. Elastic secant modulus to 1/2 maximum deviator stress for sands at 99 kPa confining stress (Lambe and Whitman, 1969)

		E (MPa)	
Description	Loading	Loose	Dense
Angular, breakable particles	Initial	13.8	34.5
Hard, rounded particles	Initial	55.1	103.4
Fine crushed quartz, angular	Repeated	117.2	206.8
Medium Ottawa sand, rounded	Repeated	206.8	358.5

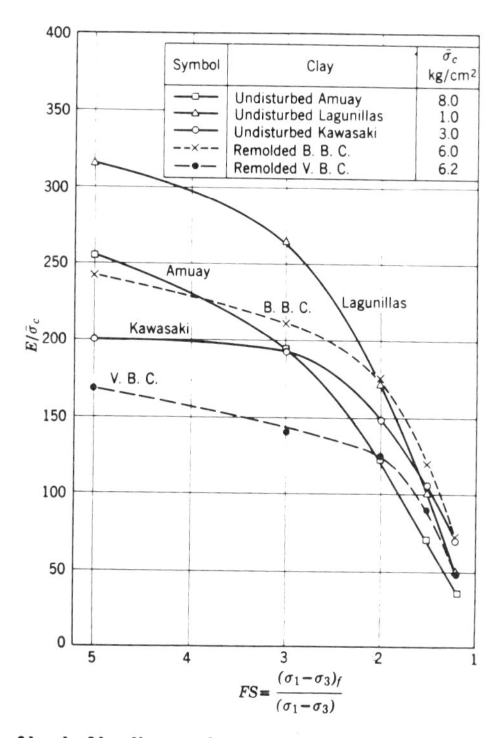

Figure 2.139–The effect of level of loading on the secant modulus (normalized according to the effective intergranular confining pressure, $\bar{\sigma}_c$) for several undrained clay samples. FS is the factor of safety or the ratio of the shear stress at failure to that at which the modulus was determined. Reprinted with permission of John Wiley & Sons, Inc. from Ladd, 1064 as reproduced in *Soil Mechanics*, Lambe, T. W. and R. V. Whitman, copyright © 1969.

It can be seen that these tangent modulus values upon repeated loading were considerably higher than the secant modulus values for this same soil shown in figures 2.141 and 2.142.

The combined effects of confinement and shear stress level relative to the failure shear stress were considered by Duncan (1980) in formulating a general expression for the elastic modulus of soil samples tested in a triaxial cell with $\sigma_2 = \sigma_3 =$ constant.

$$E_t = \left[1 - \frac{R_f(1 - \sin\phi)\,(\sigma_1 - \sigma_3)}{2c\cos\phi + 2\sigma_3 \sin\phi}\right]^2 \cdot K \cdot P_a \left(\frac{\sigma_3}{P_a}\right)^n \qquad (2.111)$$

in which E_t = tangent elastic modulus in the axial direction, and R_f = failure ratio =

$$\frac{(\sigma_1 - \sigma_3)\,\text{failure}}{(\sigma_1 - \sigma_3)\,\text{ultimate}}$$

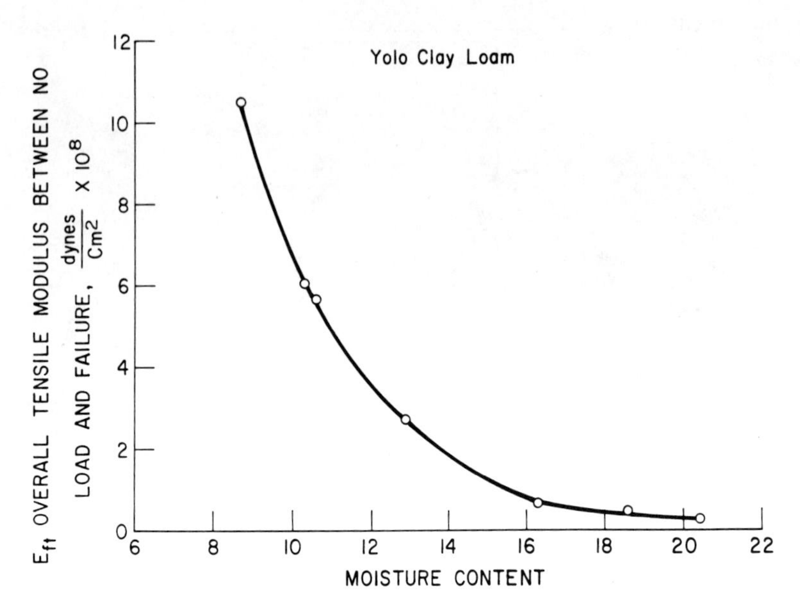

Figure 2.140–Overall (secant) modulus at various moisture contents for a Yolo clay loam soil loaded in tension (from Vomocil and Chancellor, 1969).

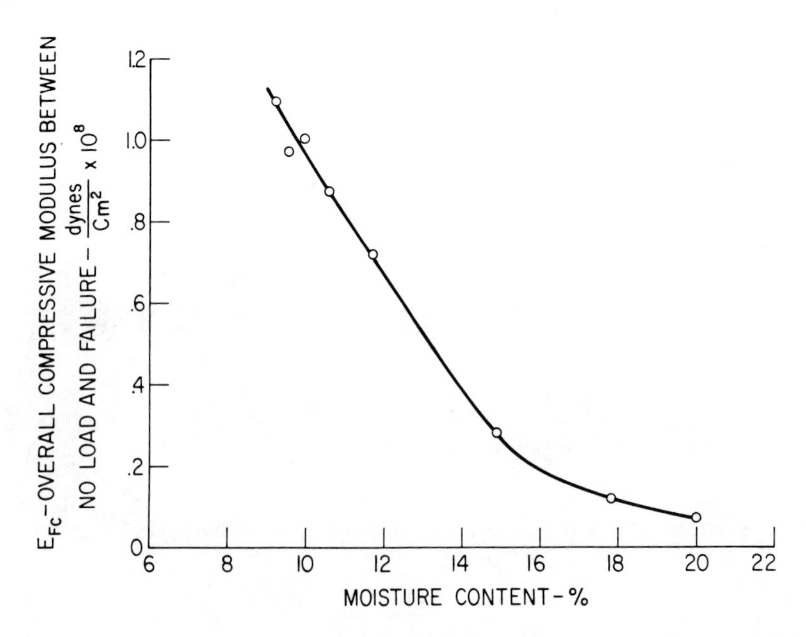

Figure 2.141–Overall (secant) modulus at various moisture contents for a Yolo clay loam soil loaded in unconfined compression (from Vomocil and Chancellor, 1969).

to represent the case in which shear stress peaks and then decreases to a fixed level as shear strain continues, P_a = atmospheric pressure (in the same units as σ_3), and K and n are dimensionless constants. It should be noted that according to the Mohr-Coulomb failure criterion.

$$\frac{(2c\cos\phi + 2\sigma_3\sin\phi)}{(1-\sin\phi)} = (\sigma_1 - \sigma_3)$$

Data for two soils at two depths were determined by Chi et al (1993a) and appear in table 2.20.

For analysis of soil properties the case in which σ_2 and σ_3 are not equal, Khan (1993) has made use of an elastic model that allows for anisotropic properties to be determined. In the case in which tests were made in a triaxial cell which allowed σ_2 to be controlled independently of σ_3 such that $\Delta\sigma_2 = 0.5\ \Delta\sigma_3$ as σ_2 and σ_3 were reduced as σ_1 was increased to maintain a constant level of mean normal stress, σ_{oct}, values of the modulus of elasticity decreased as shear strain increased (see table 2.17). These decreases were greater in the σ_2 and σ_3 directions than in the σ_1 direction, although all three values of the modulus of elasticity were of the same general magnitude. Khan (1993) also found that the modulus of elasticity values decreased more or less in proportion to the magnitude of the mean normal stress, σ_{oct}, and that the rates of modulus of elasticity decrease with shear strain increases were approximately the same irrespective of the level of the mean normal stress (fig 2.147).

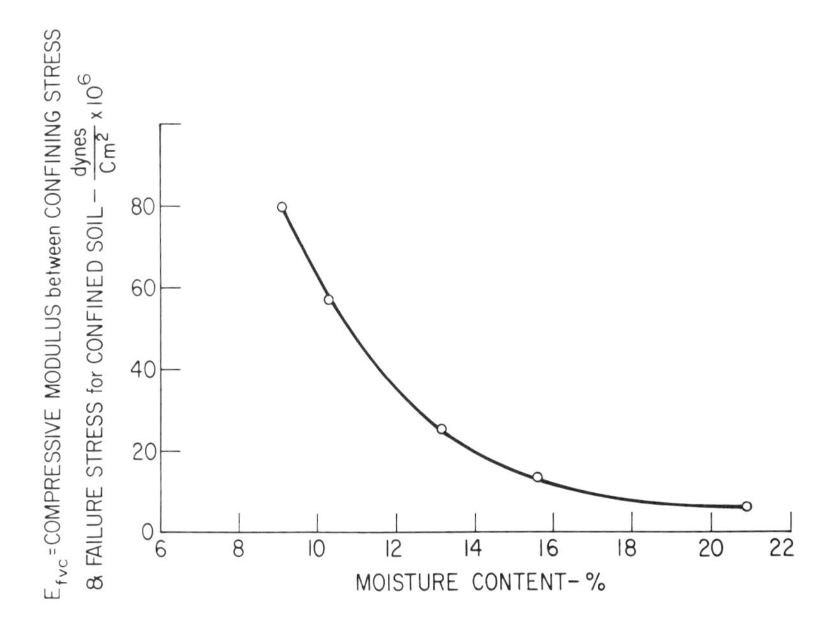

Figure 2.142–Overall (secant) modulus at various moisture contents for a Yolo clay loam soil loaded in confined compression (from Vomocil and Chancellor, 1969).

Table 2.19. Effect of moisture content and confinement on elastic modulus*

Moisture Content (%)	Confinement	Modulus of Elasticity (MPa)
2.25	Unconfined	87.1
6.9	Confined	47.5
7.8	Unconfined	58.5
10.0	Unconfined	34.2
11.75	Confined	19.1
20.9	Unconfined	1.87
22.8	Confined	3.07

* Data from Aref (1973), confining stress = 95 kPa.

Bulk Modulus Under Hydrostatic Stress

The bulk modulus, B, is defined as:

$$B = \Delta\sigma_0 / (\Delta V / V) \tag{2.112}$$

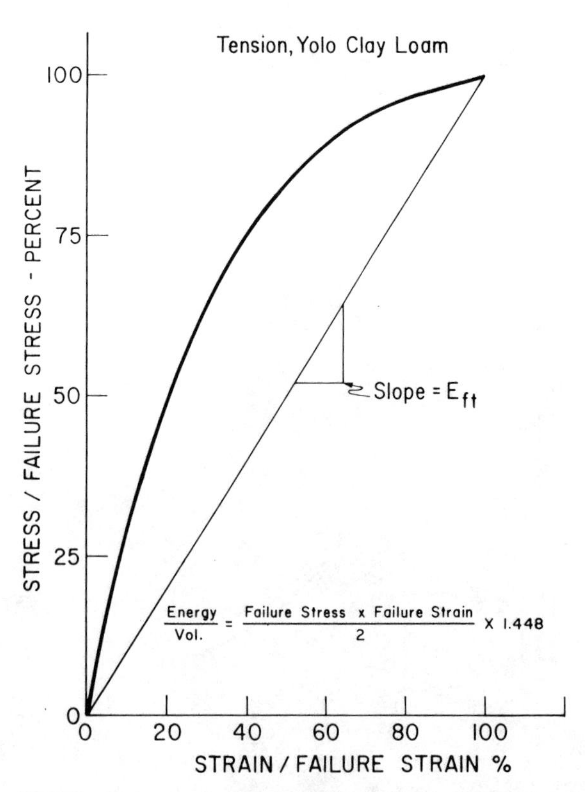

Figure 2.143–Normalized composite stress-strain curve for Yolo clay loam at several moisture contents, loaded in tension. Slope, E_{ft}, is proportional to the modulus of elasticity (from Vomocil and Chancellor, 1969).

in which B = bulk modulus, and $\Delta\sigma_0$ = incremental increase in hydrostatic stress. Hydrostatic stress = $\sigma_1 = \sigma_2 = \sigma_3$, and $\Delta V/V$ = resulting incremental volumetric strain = $\varepsilon_1 + \varepsilon_2 + \varepsilon_3$.

Large volumetric strains can occur during isotropic compression as the result of the collapse of arrays of particles. Each such collapse causes rolling and sliding between particles, and as a result, tangential forces occur at the contact points between particles. However, such tangential forces average out to zero over a surface passing through many contact points. Thus, the shear stress on any plane is zero even though large shear forces exist at individual contacts (Lambe and Whitman, 1969).

The most usual method of determining the bulk modulus is to use a triaxial cell in which either the sample diameter can be physically monitored or the change in volume can be measured by determining the volume of pore fluid (air or water) expressed. Another method of bulk modulus measurement is to calibrate the pressure-volume characteristics of a closed container, followed by remeasuring the pressure-volume characteristics of the container after a weighed sample of soil sealed in a balloon has been placed in the container (Upadhyaya et al, 1982). In this case, the compressibility

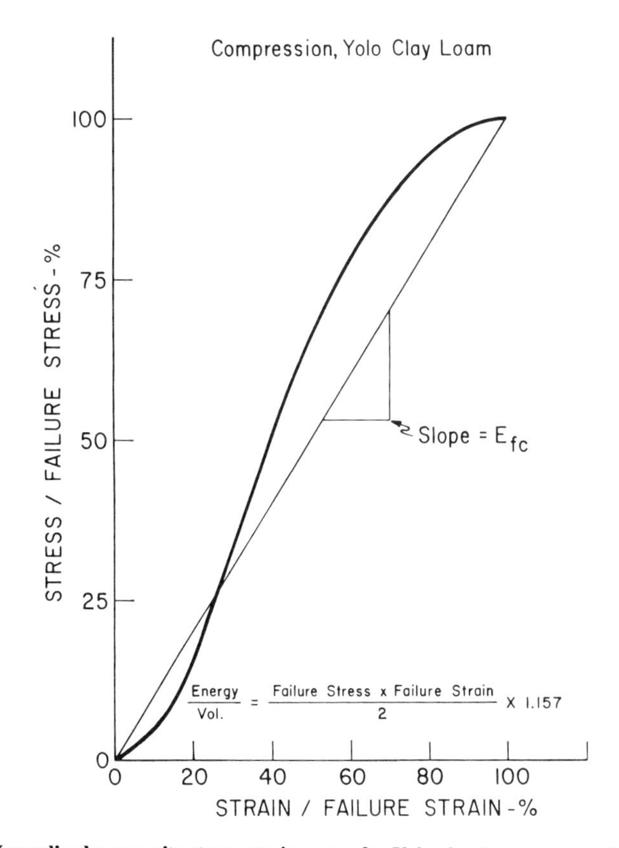

Figure 2.144–Normalized composite stress-strain curve for Yolo clay loam at several moisture contents, loaded in unconfined compression. Note the reverse in curvature during the early stages of loading (from Vomocil and Chancellor, 1969).

characteristics of the air in the sample must be taken into account when evaluating the results.

The bulk modulus is computed as the absolute value of the reciprocal of the slope of the volume strain vs pressure curve (fig 2.148). Generally, the bulk modulus of agricultural soils during the initial compression increases with the pressure level applied to the soil. The bulk modulus of water also increases approximately linearly with increases in the applied stress (Eshbach, 1952). Upon repeated loading the bulk modulus of soil is not only noticeably higher than upon initial loading, but also its value does not change much with pressure level (Johnson et al, 1984) (fig 2.148). When, upon repeated loading, the stress exceeds that to which the soil has been previously subjected, the bulk modulus reverts to its initial-compression characteristics (see fig 2.148).

Two general models have been used to describe the characteristics of a soil with regard to the relation between the hydrostatic applied stress, σ_h, and the change in sample volume, ΔV. The first of these (Bailey et al, 1984) is:

$$\frac{V}{W} = m \log \sigma_h + b \tag{2.113}$$

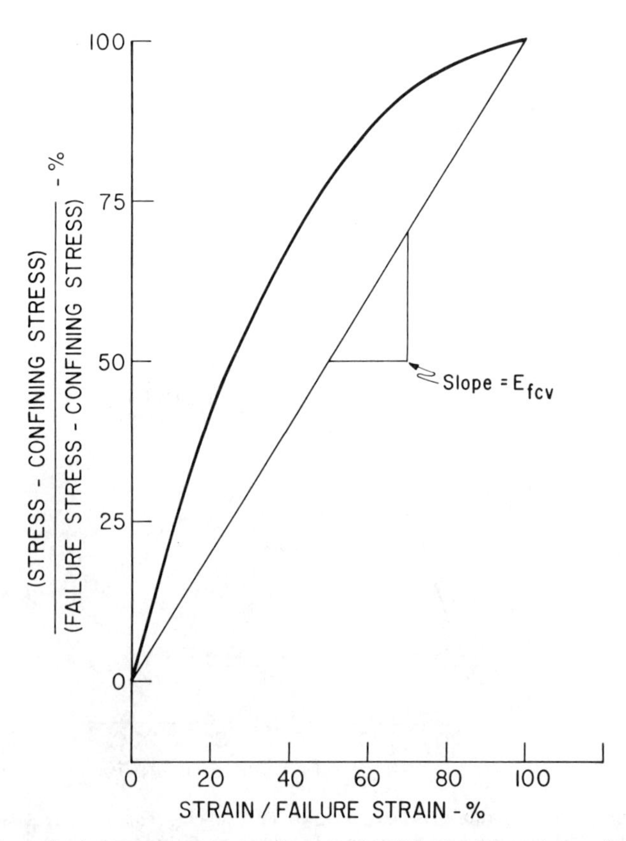

Figure 2.145–Normalized composite stress-strain curve for Yolo clay loam at several moisture contents, loaded in confined compression (from Vomocil and Chancellor, 1969).

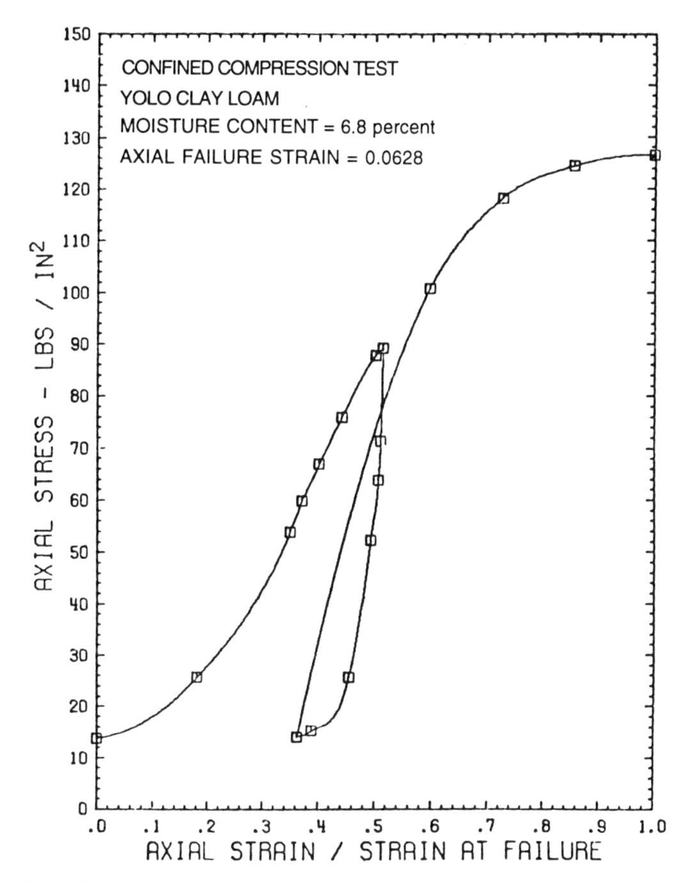

Figure 2.146–Stress-strain characteristics (slope is proportional to the elastic modulus) for a soil sample subjected to compression, release, and then compression to failure (from Aref, 1973).

Table 2.20. Elastic parameters for two soils at two depths (from Chi et al, 1993a)

	Mawcook Gravel-Sandy Loam		Ste-Rosalie Clay	
Parameter	<25 cm depth	>25 cm depth	<25 cm depth	>25 cm depth
K	84.3	105.58	10.35	73.39
n*	0.0	0.0	1.71	0.0
Poisson's ratio (before failure)†				
a	0.1216	0.0804	0.1712	0.1264
b	0.3336	0.3415	0.2991	0.3437
(after failure)	0.4725	0.4728	0.4771	0.4553
Moisture content (%)	24.6	16.1	22.2	20.2
Dry bulk density (Mg/m^3)	1.12	1.28	1.31	1.52
Cohesion (c) (kPa)	25.66	29.00	56.28	61.70
Internal friction angle (ϕ)	31.4°	32.2°	18.4°	29.6°
Failure ratio (Rf)	0.7566	0.7652	0.7886	0.7611

* In this case n is the exponent for $[(\sigma_3 + P_a)/P_a]$ rather than for (σ_3/P_a) as shown in equation 2.112.

† See equation 2.102 for the interpretation of constants, a and b.

in which V = sample volume, W = sample dry weight, m = compressibility coefficient, σ_h = hydrostatic stress, and b = value of V/W at σ_h = 1.0.

The bulk modulus, B, can then be described (Wood, 1990) as:

$$B = \frac{(V/W) \cdot \sigma_h}{(-m)}$$

and is thus indicated to increase as the hydrostatic stress level increases.

Another representation of this phenomenon (Desai and Siriwardane, 1984) is:

$$B = B_o + \alpha \cdot \sigma_h$$

in which B_o = an initial value of the bulk modulus, and α = a material parameter coefficient.

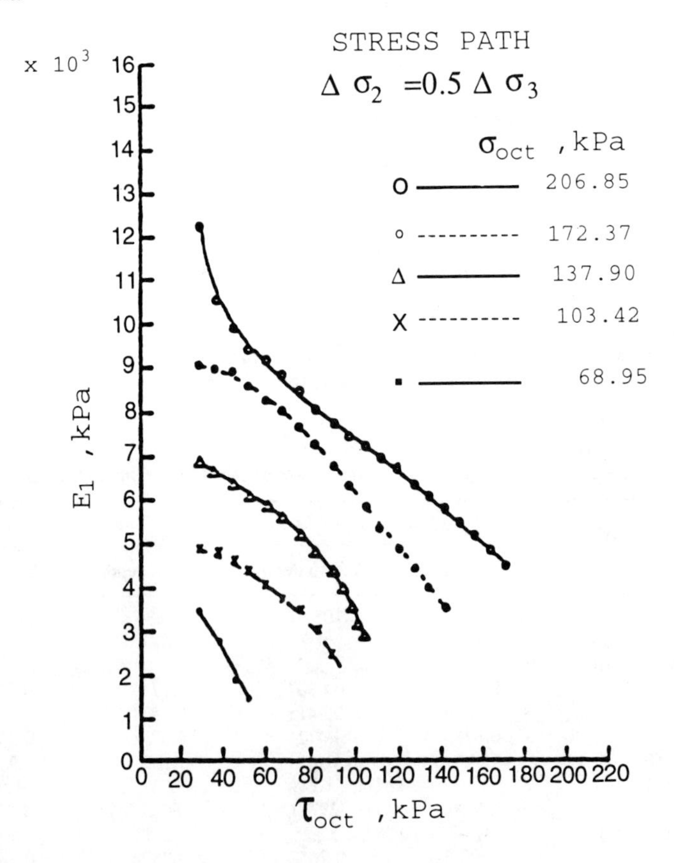

Figure 2.147–Effects of shear stress, τ_{oct}, and mean normal stress, σ_{oct}, on the modulus of elasticity in the axial direction for samples of Lloyd clay at 17% moisture loaded in shear at constant σ_{oct} values by simultaneously increasing σ_1 and decreasing σ_2 and σ_3 independently (from Khan, 1993).

The second model (Johnson et al, 1984; Bailey et al, 1984; Grisso et al, 1984; Bailey et al, 1986) is of the form:

$$\ln\left(1 + \frac{\Delta V}{V_0}\right) = \ln\left(\rho_o / \rho\right) = \left(A + B \cdot \sigma_h\right) \cdot \left(1 - e^{-C\sigma_h}\right) \qquad (2.114)$$

in which ρ_o = initial dry bulk density of the soil, ρ = resultant dry bulk density of the soil, e = logarithmic base, 2.718, V_0 = initial sample volume, and A, B, and C = constants.

Both models seem to fit most experimental data well, with the exception that the first model is not defined at $\sigma_h = 0.0$, and that the actual soil bulk density may not match that which the model would predict at very low stress levels (Bailey et al, 1984; Bailey et al, 1986).

The second model was found capable of representing both the initial hydrostatic compression of soil as well as repeated loading of soil after release of initial compression stresses (Johnson et al, 1984). The coefficients A, B, and C were different for the two cases, however.

It was found that the stress level above which the coefficients for the virgin loading curve should be used, and below which the coefficients for the reloading curve should be used, could be estimated by simultaneously solving for the stress level and associated density which would satisfy the two equations formulated from the one equation form and the two sets of coefficients. A single set of coefficients could not be used to represent both the reloading conditions and the conditions of compression at stresses

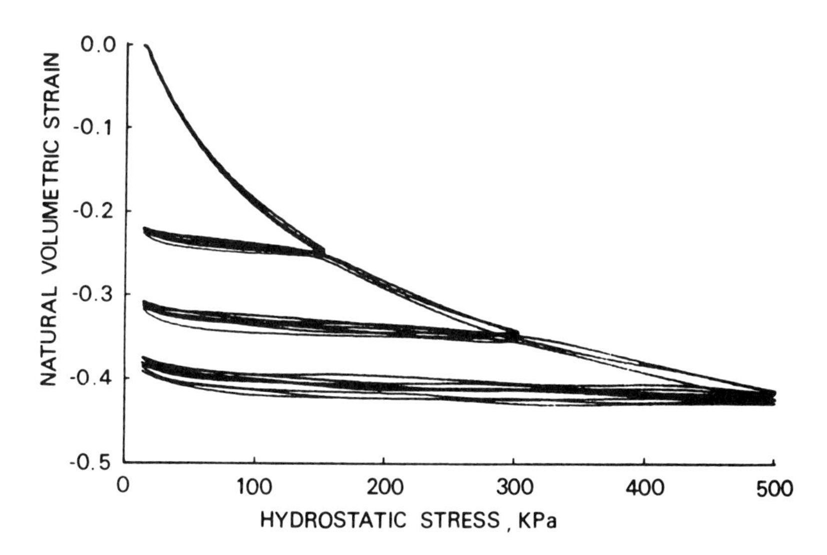

Figure 2.148–Hydrostatic stress – volumetric strain characteristics of a soil under repeated loading. The inverse of the slope is the bulk modulus (from Johnson et al, 1984).

above the stress and density level at which compression had previously stopped and stresses relaxed prior to reloading.

Coefficients A, B, and C, as well as values for ρ_0, at $\sigma_h = 0$, were found to vary with moisture content for a given soil (Bailey et al, 1986). For the first mentioned model, the value of m tends for most soils to not change much with moisture content, while the value of B tends to change in a regular way with moisture content (fig 2.149) (Chancellor, 1977).

When drained tests are conducted rapidly with all soils or even comparatively slowly with clay soils, a point is approached at high moisture contents at which all the pores are nearly completely filled with water. Under these circumstances, compressibility declines markedly (fig 2.149) and the above models do not apply without major changes in the parameters. For these near-saturated conditions (Lambe and Whitman, 1969):

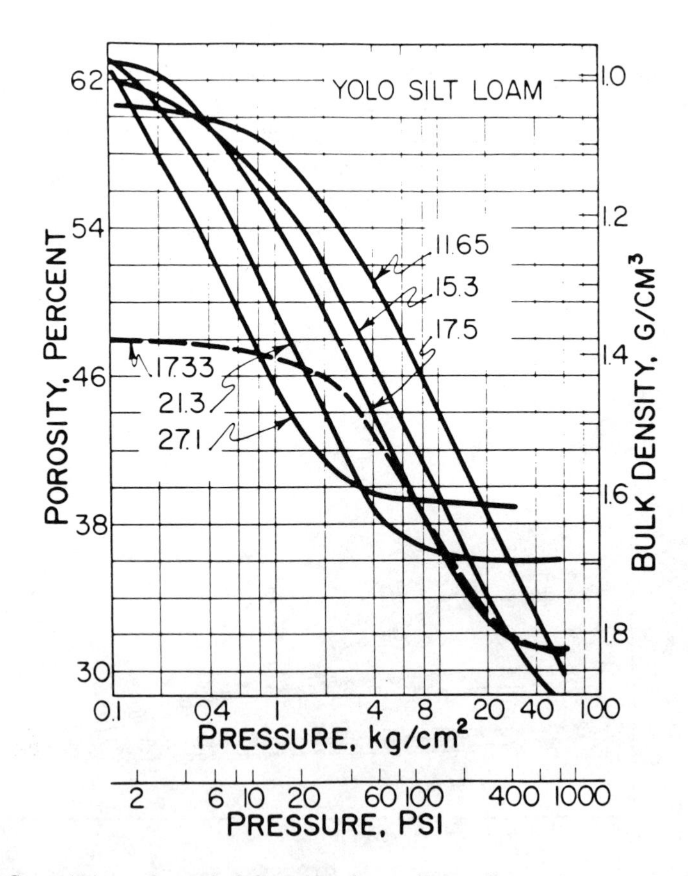

Figure 2.149–Stress (σ_1) vs volumetric deformation for a soil at various moisture contents and densities. Semi-log form of the graph in combination with the linear characteristics shown, indicate increased bulk modulus values at higher stresses. Dashed line represents a preconsolidated sample (from Chancellor, 1977).

$$\frac{1}{B} = \frac{\Delta V}{V} \times \frac{1}{\Delta \bar{\sigma}}, \text{ and}$$

$$\Delta \bar{\sigma} = \Delta \sigma - \Delta u, \text{ and}$$

$$\Delta u = \frac{\Delta \sigma}{1 + \eta \left(B_{soil} / B_{water}\right)}$$

in which B = bulk modulus, $\Delta \bar{\sigma}$ = effective all-around compressive stress increment, Δu = neutral stress (water pressure) increment, and η = porosity of soil.

In the above discussion pertaining to hydrostatic stress application, the volume changes can be related either to the major principle stress, σ_1, or to the mean stress:

$$\sigma_m = \frac{\sigma_1 + \sigma_2 + \sigma_3}{3}$$

since $\sigma_m = \sigma_1$, under hydrostatic conditions. In the absence of the constraint that σ_1, σ_2, and σ_3 need be equal, shear stresses, and shear strains, may prevail. In such cases, the relationship between volume strain and σ_m may be different than was the case when $\sigma_1 = \sigma_2 = \sigma_3$ (fig 2.150) (Grisso et al, 1987a), indicating a stress-path dependency due to plasticity when shear stresses and strains exist.

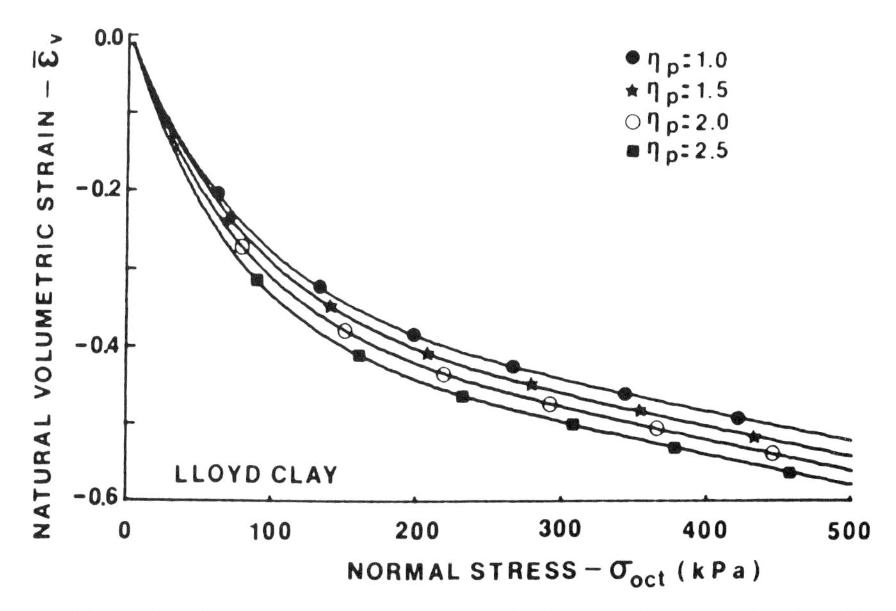

Figure 2.150–Octahedral normal stress vs volumetric strain characteristics for a soil loaded with four different ratios, η_p, of σ_1 to σ_3 (from Grisso et al, 1987a).

When the bulk modulus is interpreted in terms of the change in the voids ratio, e, instead of in terms of total soil sample volume, the bulk modulus is termed *modulus of compression*, E_v, (Kézdi, 1974):

$$E_v = B \frac{1}{1 - \eta}$$

in which η = soil porosity.

Values for E_v (Kézdi, 1974) are given in table 2.21. For water at 20° F (atmospheric pressure), B = 2200 MPa

Shear Stress – Shear Strain Parameters

The main shear stress – shear strain parameter is the shear modulus, G, illustrated in figure 2.126.

$$G = \frac{\Delta\tau}{\Delta\gamma} \tag{2.115}$$

in which $\tau = \tau_{max}$ = shear stress = $(\sigma_1 - \sigma_3)/2$, and $\gamma = \gamma_{max}$ = shear strain = $\varepsilon_1 - \varepsilon_3$. ($\sigma_1$ and ε_1 are the major principal stress and strain, while σ_3 and ε_3 are the minor principal stress and strain).

The shear modulus is not a commonly used parameter in soil dynamics considerations relative to tillage and traction operations because most soil failure in shear (or maximum shear stress mobilization) takes place between 3 and 30% shear strain, while the level of shear stress associated with failure is dependent on not only soil properties, but also the level of applied and confining stresses or the stress-strain compliance characteristics of matrix of which the soil element under examination is a part. However, the shear

Table 2.21. Modulus of compression, E_v, for various soils MPa (From Kézdi, 1974)

Soil Type	Soil Condition			
Granular Soil		Loose	Medium	Dense
Sandy gravel		30 – 80	80 – 100	100 – 200
Sand		10 – 30	30 – 50	50 – 80
Non-plastic silt		8 – 12	12 – 20	20 – 30
Cohesive Soils	Soft	Plastic	Hard	Very Hard
Silty sand	5 – 8	10 – 15	15 – 20	20 – 40
Silt	3 – 6	6 – 10	10 – 15	15 – 30
Lean clay	2 – 5	5 – 8	8 – 12	12 – 20
Fat clay	1.5 – 4	4 – 7	7 – 12	12 – 30
Organic silt		0.5 – 5		
Organic clay		0.5 – 4		
Peat		0.1 – 2		

modulus is the main soil parameter involved in a model for predicting cone penetrometer resistance (Rohani and Baladi, 1981).

The shear modulus is directly linked to the shear wave velocity when a shear wave passes through a body of soil (Lambe and Whitman, 1969):

$$\text{shear wave propagation velocity} = C_s = \sqrt{G / \rho} \tag{2.116}$$

in which ρ = mass density of the soil.

Data for an Ottawa sand at a dry bulk density of ρ = 1704 kg/m^3 indicate a straight-line relationship on a log-log plot between C_s and the confining stress, σ_c, between the confining stress levels given table 2.22.

A general equation given for sand with angular particles and several clays (Lambe and Whitman, 1969) is:

$$G = 1230 \frac{(2.973 - e)^2}{1 + e} (\sigma_c)^{0.5} \tag{2.117}$$

in which e is the void ratio (in decimal form), σ_c is the confining stress in lb/in.^{2}, and G is the shear modulus in lb/in.^{2}

This equation was compared with experimental results obtained in vibratory (180 to 200 Hz) torsion tests with some remolded and some essentially undisturbed samples of three agricultural soils at confining stresses in the 0.0 to 3 kPa range (Womac et al, 1988). It was found that the above equation generally under-predicted the shear modulus. In these tests, it was found that the shear modulus tended to decrease slightly with increases in the strain amplitude (range = 4 to 12 × 10^{-5} cm/cm), and that the remolded samples had shear modulus values averaging about 80% of those of comparable minimally disturbed samples (table 2.23).

The shear modulus is an important soil parameter in the sonic exploration of soils. The magnitudes of shear strain that take place in the wave propagation phenomena, however, are very small compared to those in most tillage and traction operations.

Shear modulus determinations under these circumstances are most frequently made using a triaxial or unconfined compression test. A typical set of test results is shown in

Table 2.22. Wave velocity and stress-strain moduli for confined Ottawa sand*

			σ_c = 19 kPa	σ_c = 310 kPa
Cs	=	shear wave velocity (m/s)	131	366
G	=	shear modulus (MPa)	29	228
C_L	=	compressive wave velocity (m/s)	253	579
E	=	elastic modulus (MPa)	109	517

* Data from Lambe and Whitman (1969), σ_c = confining stress.

Table 2.23. Shear modulus values determined from shear wave velocity*

Soil Type	Moisture Content (%)	G (MPa) Remolded	G (MPa) Minimally Disturbed
Clay loam	16.4	13.5	23.1
Fine sandy loam	16.7	28.8	34.2
Silt loam	17.0	18.4	22.1

* Data from Womac et al (1988).

figure 2.151 (Bailey and Weber, 1965). In such a case, the incremental shear strain, $\Delta\gamma$, is equal to:

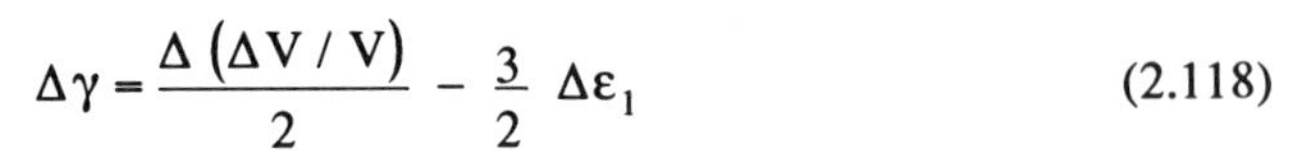

$$\Delta\gamma = \frac{\Delta\left(\Delta V / V\right)}{2} - \frac{3}{2}\,\Delta\varepsilon_1 \qquad (2.118)$$

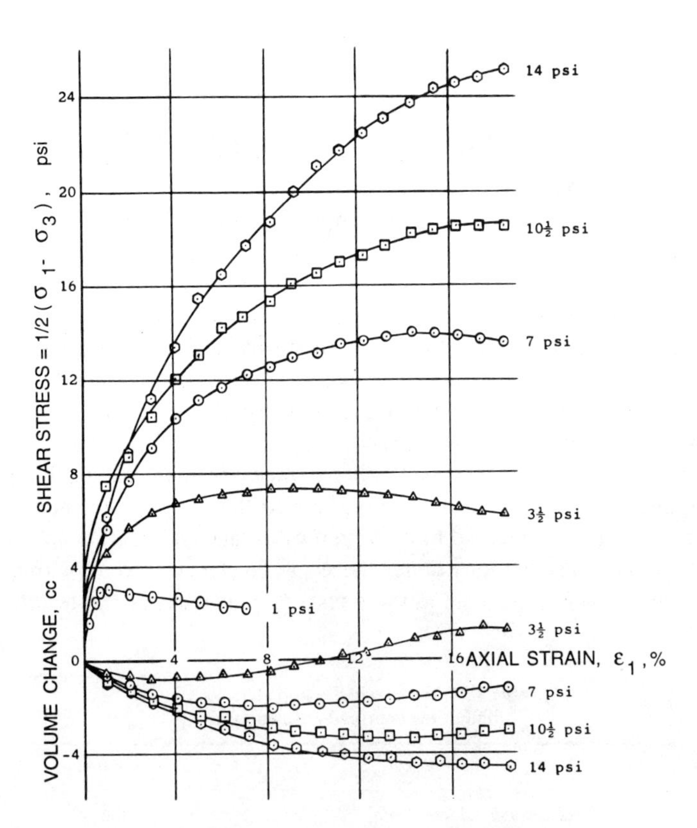

Figure 2.151–Typical stress-strain results as obtained with an artificial soil (fire clay plus 10% SAE 5W oil) in a triaxial cell using various cell pressures. Because volume strain was unusually small for these 120-cm³ samples, shear strain can be approximated as 1.5 times the axial strain, ε_1, (see equation 2.118), and the slope of the curves may be interpreted as representing 67% of the shear modulus, G (from Bailey and Weber, 1965).

in which $\Delta\gamma$ = incremental shear strain, $\Delta\,\varepsilon_1$ = increment in axial strain (ε_1 negative for compression), and $\Delta(\Delta V/V)$ = increment in volume strain ($\Delta V/V$ negative for compression)

Since $G = \Delta\tau / \Delta\gamma$, it can be seen from figure 2.151 that at the beginning of a triaxial test G will be high, while toward the end of the test, G will be very low and may even be negative. Figure 2.151 also illustrates the effect that the confining stress can have on the value of G at various levels of shear strain.

For the case of triaxial tests of saturated clay soils, the incremental shear strain, $\Delta\gamma$, can be represented by:

$$\Delta\gamma = -\,1.5\;\Delta\varepsilon_1 \qquad (2.119)$$

(which assumes a Poisson's ratio of $\mu = 0.5$).

Undrained triaxial tests of a saturated clay soil (Hassan, 1968) indicated that the reciprocal of the shear modulus increased linearly with axial strain, ε_1, throughout the tests and that the slope of this increase of $1/G$ with increases in ε_1 as the tests progressed to failure, decreased in a regular way with increases in the confining stress (fig 2.152).

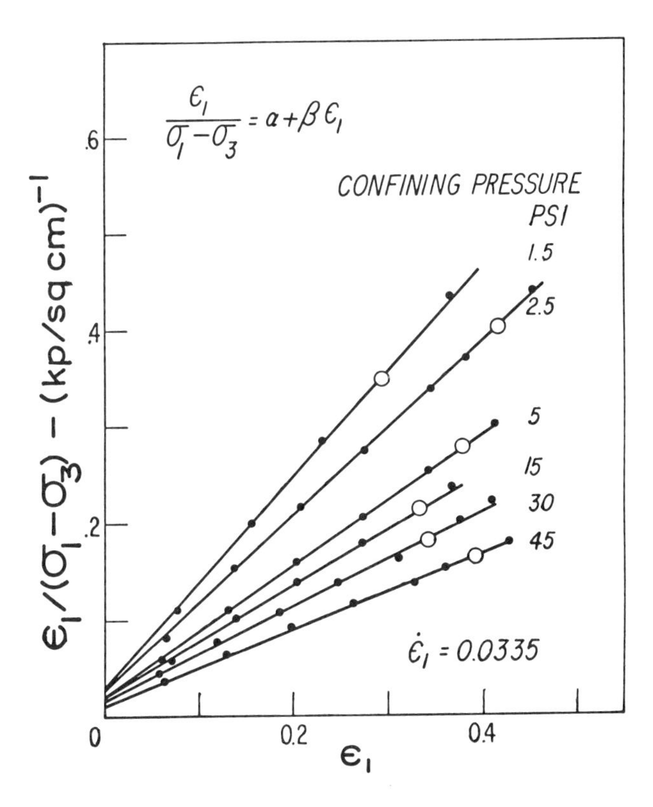

Figure 2.152–Effect of confining stress levels on the shear modulus, G, of a saturated clay soil. The ordinate, $\varepsilon_1/(\sigma_1-\sigma_3)$, is equal to 1/3G (from Hassan, 1968).

In another series of tests in which the lateral stress on a cylindrical sample was increased in direct proportion to lateral expansion with varying proportions between σ_3 and ε_3, the relationships between τ and γ tended to change in a nonlinear manner (fig 2.153), with the ratio $\Delta\tau/\Delta\gamma = G$ increasing as the test, using axial compression, progressed. When σ_3 increased rapidly with ε_3, higher values of G were reached than when σ_3 increased only slowly with ε_3.

When a test was used that employed cylindrical samples with length equal to diameter to prevent catastrophic failure, it was found that shear stress-shear strain data from two very different soils over a broad range of moisture contents and porosities could be generalized by dividing the shear stress by the value of (σ_{normal})* $\times \tan\phi$, an approximation of the shear stress at failure, (Chancellor and Korayem, 1965) (fig 2.154). The resulting general shear strength characteristic was one in which a near-constant value of $G/(\sigma_n \tan\phi) = 6$ prevailed up to a shear strain level, γ, of approximately 0.2. Above γ levels of 0.5, $G/(\sigma_n \tan\phi)$ assumed a value of approximately –0.7.

In a series of triaxial tests in which σ_3 was controlled to be a fixed proportion of σ_1 at all times (a proportion, however, which was sufficiently large to prevent catastrophic sample failure), the shear modulus was very low at the beginning of the test and increased as the test progressed to higher levels of stress and strain. However, when the same final stress state was obtained by first applying a hydrostatic stress, and then, while

* σ_{normal} is the normal stress that prevailed at any time on the plane of shear failure which made an angle of $45° + \phi/2$ with the plane of the base of the cylindrical samples.

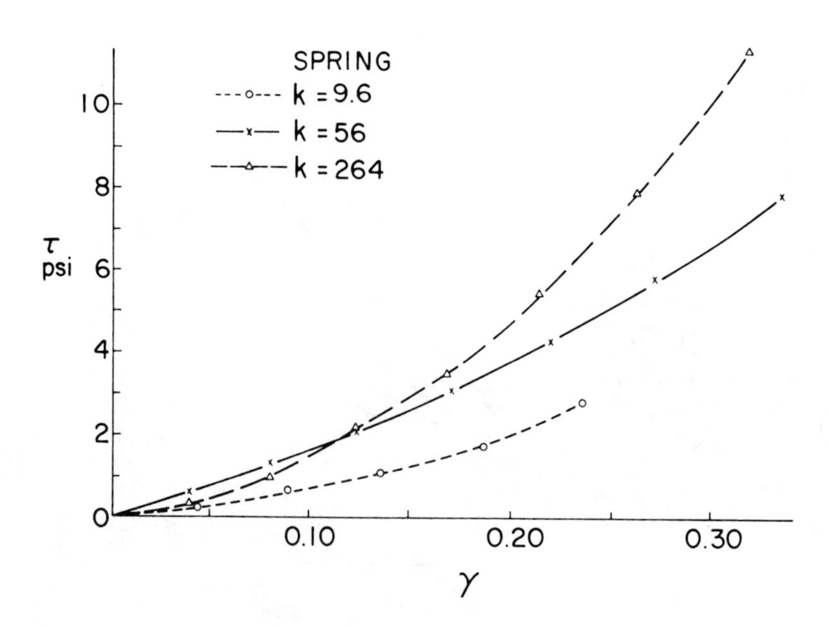

Figure 2.153–Shear stress, τ, vs shear strain, γ, for samples loaded axially while a spring-loaded enclosure with varying spring constants acted laterally. The shear modulus, G, is the slope of the curves shown (from Kitani and Persson, 1967).

the sum of $\sigma_1 + \sigma_2 + \sigma_3$ (the mean normal stress times 3) was kept constant, the shear stress was increased, the shear modulus started out at very high levels and decreased as the test progressed (fig 2.155).

The changes in apparent shear modulus for soils are thus affected by the stress sequence involved in the loading. If the deformation process was separated into plastic and elastic components, it is possible that the shear modulus pertaining to the strictly elastic properties (although the modulus might be variable) would not vary with changes in the stress application sequence or stress path.

One general form for representing the shear modulus (Desai and Siriwardane, 1984) is:

$$G = G_o + W\sqrt{\gamma} \tag{2.120}$$

in which G_o = an initial value of shear modulus, γ = two-dimensional shear strain, and W = a material property coefficient.

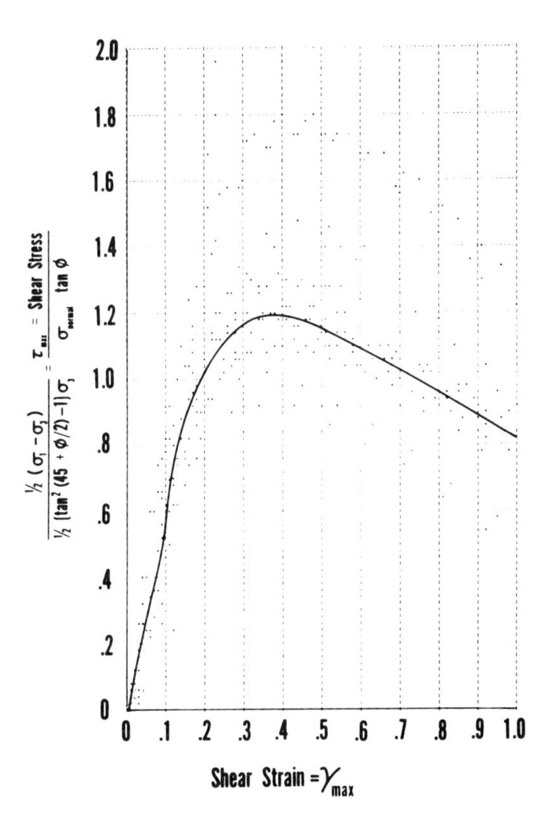

Figure 2.154–Generalized shear stress (normalized) vs shear strain characteristics for a broad range of moisture and density conditions of two very different soils. The slope of the curve is proportional to the shear modulus, G, (from Chancellor and Korayem, 1965).

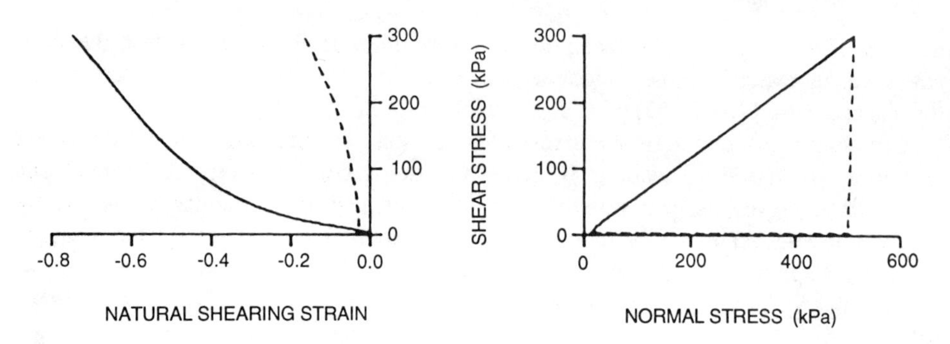

Figure 2.155–Shear stress vs shear strain curves for a soil loaded to the same stress state with a fixed ratio between σ_1 and σ_3 (solid lines) and a sequence of normal stress increase followed by shear stress increase (dotted lines) (from Grisso et al, 1987a).

A relationship between the shear modulus, the bulk modulus and Poisson's ratio was given by Wood (1990) as:

$$G = \frac{3(1 - 2\mu)B}{2(1 + \mu)} \tag{2.121}$$

in which μ = Poisson's ratio, and B = bulk modulus. Because of the tendencies of the bulk modulus to increase with increased levels of mean principal stress and of Poisson's ratio to increase with increases in shear strain, it could be anticipated that the shear modulus, G, would increase with increases in the mean principal stress and decrease with increases in shear strain.

Although shear displacements can be measured under field conditions in which tillage and traction operations take place, it is very difficult to measure shear strain because this requires some information on the relations between horizontal displacements and depth values associated with those displacements. An attempt was made to overcome this problem by using the value of normal pressure, P, on a field soil surface as an indicator of the depth to which the horizontal displacement effect would extend (Taylor and VandenBerg, 1966). Thus, the ratio J/P of horizontal surface displacement, J, to normal pressure, P, could be used instead of shear strain in relating shear stresses to strainlike displacements. It was also found that J/P was linearly related on a log-log plot to tan ϕ (the coefficient of internal friction), a relationship that could be described by an intercept, T, and a slope, n (fig 2.156) (Taylor and VandenBerg, 1966). Data obtained using this method of analysis in combination with annular grouser tests of soil shear strength (fig 2.157) (Taylor and VandenBerg, 1966) indicated that at a given level of normal pressure, P, the ratio of $\tau/(J/P)$ (a parameter similar to the shear modulus, G) decreased as J/P increased.

At a given level of shear stress, τ, the ratio of $\tau/(J/P)$ increased as the normal stress, P, increased.

Although the ratio $\tau/(J/P)$ may be a parameter related to the shear modulus, it is not the shear modulus G, as used in the foregoing discussion.

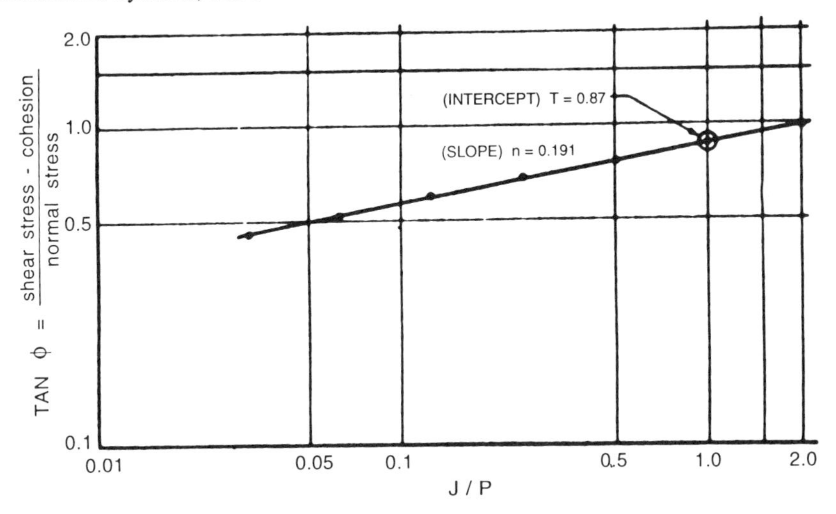

Figure 2.156–The value of J/P (ratio of surface shear displacement to normal stress) for an annular grouser plate on the soil surface was found to be logarithmically related to the shear stress (normalized by the normal stress) (from Taylor and VandenBerg, 1966).

Volumetric Strain with Stress State Changes

There has long been an interest in relating volumetric strain to changes in the state of stress associated with a volume element of soil. Such a relationship would allow the prediction of volume strain from knowledge of only the loads applied and the mechanism of load distribution as stress. A system of this type (Soehne, 1958) was used to predict the extent of soil compaction under wheels of farm equipment. Used in this system was the following relationship between stress and porosity:

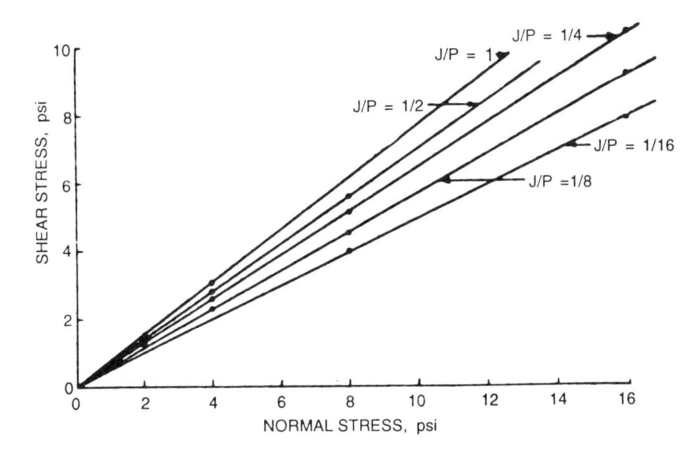

Figure 2.157–For an annular grouser plate on the soil surface, at a given level of normal stress, P, the ratio of shear stress to J/P (a parameter similar to shear strain)–the ratio being similar to the shear modulus, G – decreased as J/P increased (from Taylor and VandenBerg, 1966).

$$\eta = -A \ln \sigma_1 + C \quad (2.122)$$

in which η = porosity of the soil, σ_1 = major principal stress, and C and A = constants depending on soil characteristics.

However, it has been shown (Soehne, 1958; Chancellor, 1977) that a given soil at a given moisture content may have one value of C when loaded axially in a cylinder ($\varepsilon_3 = \varepsilon_2 = 0$, and $\gamma_{max} = \varepsilon_1 = \Delta V/V$) and another value when loaded in such a way that the soil can undergo a significant extent of shearing strain (see fig 2.158). Furthermore, even when major differences in shearing strain are not involved, it has been shown that there are some minor, but regular, differences in the porosity values that result for different values of the ratio σ_3/σ_1 during the compression process (fig 2.159) (Koolen and Vaandrager, 1984). The equation given above does not consider any effects of the value of σ_3.

Because it is difficult to predict shear strain from information on stress state, considerable work has been done to find ways to relate changes in stress state alone to

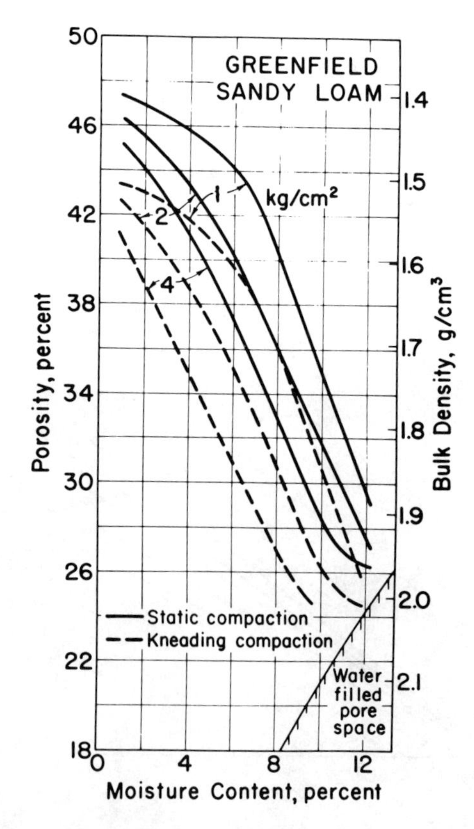

Figure 2.158–The porosity vs major principal stress characteristics of this coarse sandy loam soil are shown to be different when kneading compaction was used, as opposed to when static compaction in a cylinder was used (from Chancellor, 1977).

volumetric strain. Two general approaches have been used. The first of these approaches is to use σ_1 as the only stress parameter while incorporating a sizeable amount of information from the details of the compression process. The other approach is to use the mean normal stress, σ_m:

$$\sigma_m = \left(\frac{\sigma_1 + \sigma_2 + \sigma_3}{3}\right)$$

and the maximum shear stress:

$$\tau_{max} = \left(\frac{\sigma_1 - \sigma_3}{2}\right)$$

as the two stress parameters involved.

One of the three additional approaches using only σ_1 is that employed in conventional soils engineering practice (Jumikis, 1962) in which the void ratio, e, is predicted by:

$$e = -\frac{1}{C_v} \ln (p_i + p) + C \tag{2.123}$$

in which the various factors involved can be related by fig 2.160 (Jumikis, 1962).

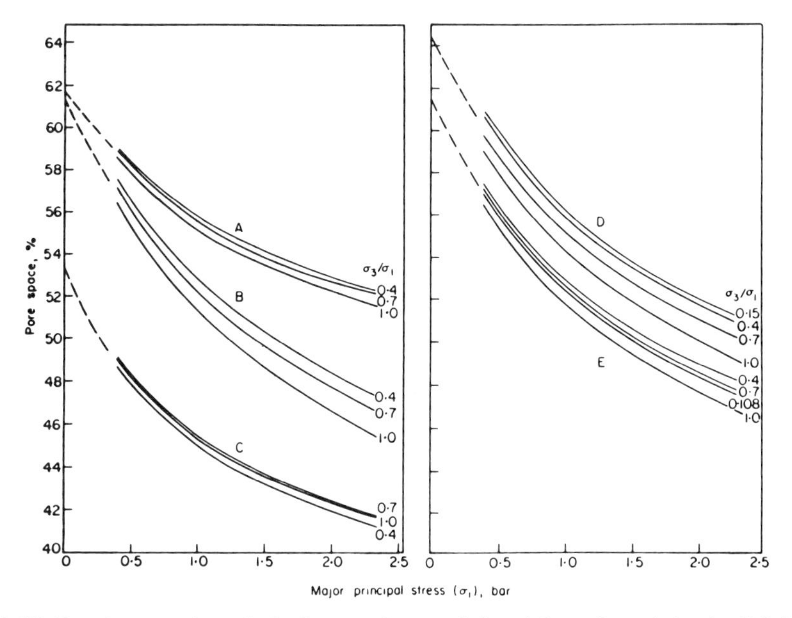

Figure 2.159–Porosity vs major principal stress characteristics of five soils varied only slightly due to changes in the value of σ_3/σ_1 applied during the compression process (from Koolen and Vaandrager, 1984).

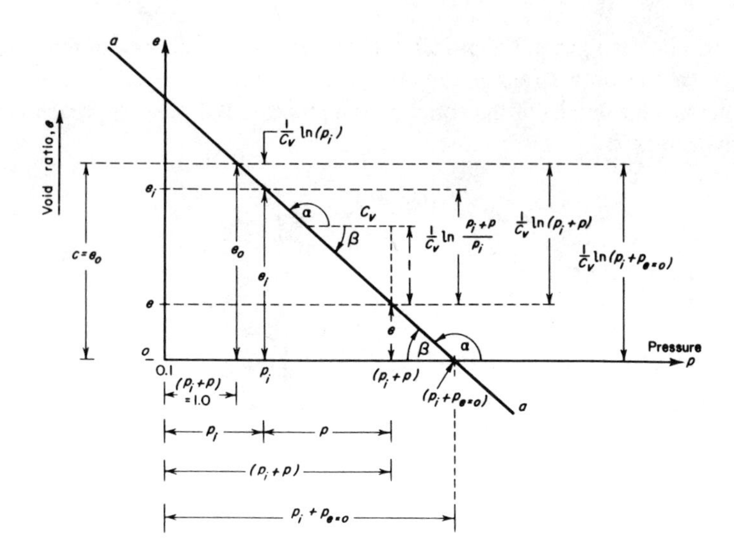

Figure 2.160–Semi-logarithmic relationship of pressure, p, vs void ratio, e, for compression of an unsaturated soil (from Jumikis, 1962).

$-1/C_v$ = a constant = tan α = tan (180° – β) = the slope of the line a-a (C_v is called the coefficient of compressibility)

p_i = a value of σ_1 which corresponds to an actual data point on the line a-a in figure 2.160 for which the void ratio is equal to that at the beginning of the compression process

C = the value of void ratio at the point on line a-a for which the value of σ_1 = $(p_i + p) = 1.0$ irrespective of the units of stress used (the point σ_1, C must lie on the line a-a, but it need not be an actual point from the data of the soil compression process)

Another approach which includes the effects of moisture content on the stress-volume strain relationship (Amir et al, 1976) and is limited to soils for which the volumetric water content lies within the range of 40 to 90% of the pore volume.

In this case:

$$\eta = A_n - B_n \ln\left(P_r + P\right) - C_n \ln \theta \qquad (2.124)$$

in which η = porosity; A_n, B_n, and C_n are constants; θ = the volumetric moisture content of the soil; P = the applied value of σ_1; and P_r = the residual pressure (for the initial compression of an originally loose soil, $P_r = 0$, while for a preconsolidated soil P_r is the pressure, which when added to the applied pressure, P, allows the above equation to predict the residual value of soil porosity).

An illustration of the sort of porosity prediction obtained from the above equation appears in figure 2.161.

A porosity-stress model similar to that immediately above, in that it also includes the effects of moisture content, (Gupta et al, 1985a) is as follows:

$$\rho = [\rho_k + \Delta_T (S_1 - S_k)] + C \log (\sigma_a / \sigma_k) \qquad (2.125)$$

in which

ρ = soil dry bulk density
ρ_k = dry bulk density of the soil at a known value of $\sigma_1 = \sigma_k$
Δ_T = slope of the dry bulk density vs degree of water saturation curve at $\sigma_1 = \sigma_k$
S_k = degree of saturation corresponding to ρ_k and σ_k
S_1 = the desired degree of saturation
σ_k = value of σ_1 at a point at which ρ_k is known
σ_a = applied normal stress, σ_1

In experiments with cylindrical soil samples that were loaded axially and allowed to expand radially against a spring-loaded encasement (using various spring constants) (Kitani and Persson, 1967), it was found that the ratio of radial stress to axial stress (σ_3/σ_1) was a constant irrespective of the value of the spring constant used in providing radial expansion resistance. Volume strain data from tests at all values of spring constant could be accurately represented by the equation:

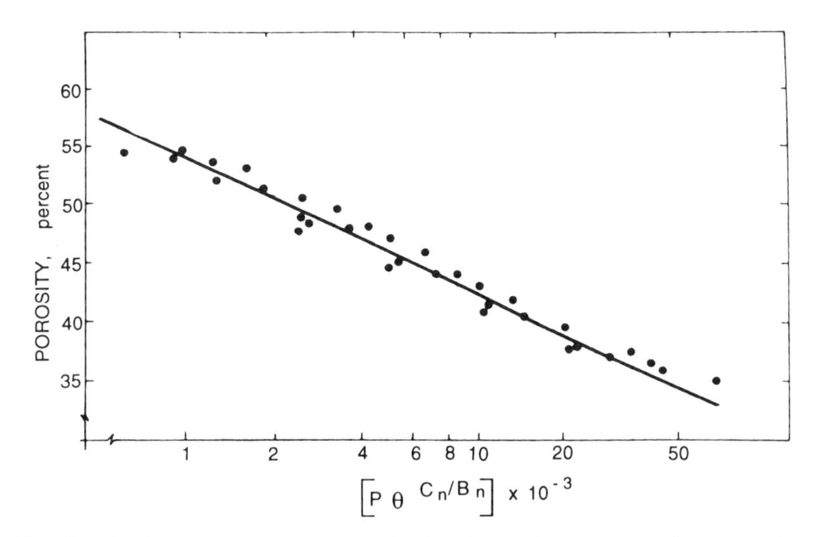

Figure 2.161–Relationship between porosity and the logarithm of the product of pressure, P, and the volumetric moisture content to the Cn/Bn power, $\theta^{Cn/Bn}$, to correspond to the equation: porosity = A_n - $B_n$lnP - $C_n$lnθ (see equation 1.124) in which A_n, B_n, and C_n are constants. Data points plotted are from Soehne (1958) and the constants for the relationship illustrated are: A_n = 89.3, B_n = 5 and C_n = 10.5 when P is given in lbs/in.^{2} while porosity and θ are in percent (from Amir et al, 1976).

$$\sigma_m = \frac{1 + 2\,\sigma_3 / \sigma_1}{3}\, a\left[-\ln\left(1 + \frac{\Delta V}{V_0}\right)\right]^n \tag{2.126}$$

in which σ_m = the mean normal stress

$$= \frac{\sigma_1 + \sigma_2 + \sigma_3}{3}$$

a and n = constants, V_0 = initial sample volume, and $\Delta V/V_0$ = volume strain.

If both sides of the above equation were to be multiplied by σ_1 and then divided by:

$$\sigma_m = \frac{\sigma_1 + \sigma_2 + \sigma_3}{3}$$

(in this case $\sigma_2 = \sigma_3$), it would have the form:

$$\sigma_1 = a\left[-\ln\left(1 + \frac{\Delta V}{V_0}\right)\right]^n \tag{2.127}$$

or, in another form:

$$\sigma_1 = a\left(-\ln \frac{\rho}{\rho_o}\right)^n \tag{2.128}$$

in which ρ = dry bulk density, and ρ_0 = initial dry bulk density. This equation differs only in form from the original equation:

$$\eta = -A \ln \sigma_1 + C \tag{2.129}$$

A fourth system for using σ_1 as the only stress parameter employs a hyperbolic rather than a logarithmic equation form to represent the stress vs. volume strain relationship (Koolen and Kuipers, 1983) (see fig 2.162):

$$\frac{e - e_f}{e_i - e_f} = \frac{1}{1 + \frac{\sigma_1}{c}} \tag{2.130}$$

in which, as σ_1 goes to ∞ the hyperbola approaches the asymptote, $e = e_f$, a theoretical final void ratio.

c = the stress that gives a value of e which lies halfway between e_i and e_f.

Values for e_i, e_f and C can be derived from experimentally obtained compression data as follows:

1. Determine n (for instance 20) experimental points (p_j, e_j), $j = 1, \ldots n$, which should be evenly spread along the experimental curve.
2. From these, select one experimental point (p_k, e_k) that approximately lies in the middle of the expected hyperbolic part of the experimental curve.
3. Compute $B_j = (p_j - p_k) / (e_k - e_j)$ for each experimental point (p_j, e_j).
4. Plot B_j against p_j. A great part of this curve is a straight line, $B_j = mp_j + q$.
5. Determine m and q graphically.

It can be shown that $c = (q/m)$ and $e_i - e_f = (1 + (p_k/c))/m$. These expressions, together with the equation obtained by substituting the numerical values of p_k and e_k into the hyperbolic equation, are sufficient to compute e_i, e_f, and c. This method of curve-fitting should be preferred over statistical methods because it gives a result that is not affected by the deviating initial and final parts of the experimental curves.

One of the reasons many researchers have found the use of the stress σ_1 well correlated with volume strain is that under conditions in which the soil volume element is compressed in only one of the three orthogonal directions, the sum of the energy inputs required for the volume strain and for the required shear strain associated with the

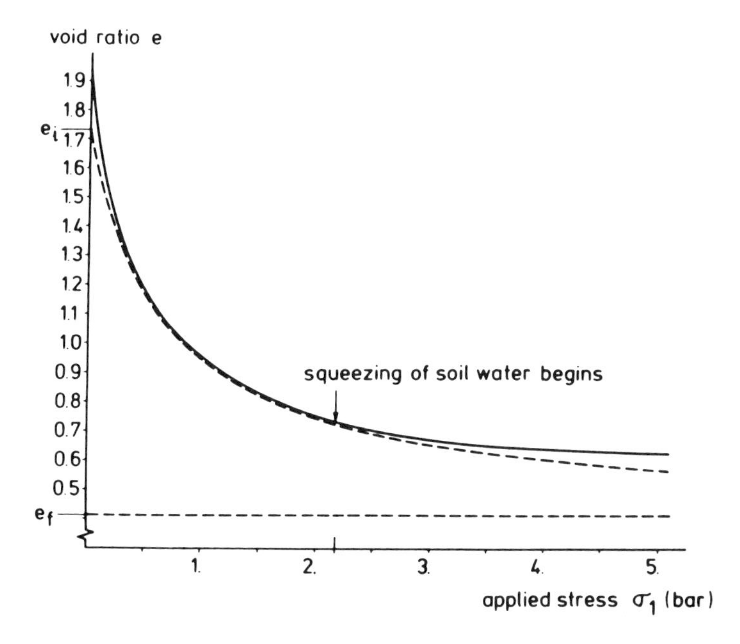

Figure 2.162–Diagram for applying a hyperbolic form to the stress, σ_1, vs void ratio, e, relationship for a soil (from Koolen and Kuipers, 1983).

absence of compression in the σ_2 and σ_3 directions, per unit of volume strain that takes place, is equal to the major principal stress, σ_1 (Chancellor and Korayem, 1965). This does not, however, apply to cases in which compression also takes place simultaneously in the σ_2 and/or σ_3 directions. Furthermore, this identity between σ_1 and the minimum energy required per unit volume strain does not affect whether a particular soil will have σ_1 vs volume-strain characteristics which will follow a logarithmic form, a hyperbolic form, or any other form.

The energy required for volume strain alone, per unit of volume strain, is equal to $\sigma_m = (\sigma_1 + \sigma_2 + \sigma_3) / 3$. Investigators attempting to introduce two aspects about the stress state (instead of only one, σ_1) into the predictive relationship between volume strain and stress state have usually used σ_m (the mean normal stress) and a shear stress parameter such as $\tau_{max} = (\sigma_1 - \sigma_3) / 2$.

One of the first of these (Roscoe et al, 1958) proposed for saturated clay soils a three-dimensional plot (at shear failure conditions) of $2 \bullet \tau_{max}$, σ_m' (effective mean normal stress) and moisture content (fig 2.163). In the case of these saturated soils moisture

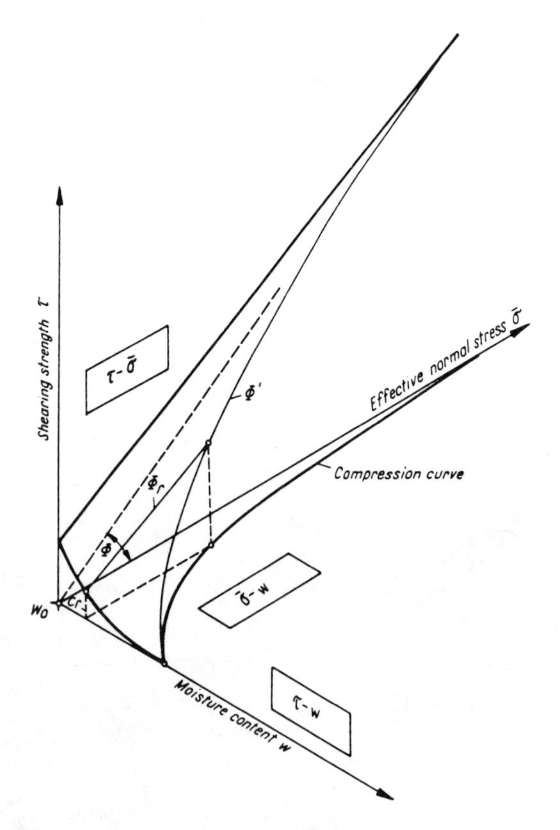

Figure 2.163–Diagram of the three-factor relationship among effective normal stress, $\overline{\sigma}$, shear strength, τ, and moisture content of a saturated clay soil. Under these circumstances, moisture content is directly proportional to porosity (from Kézdi, 1974).

content is a measure of soil porosity. It was assumed that the shearing strength of the soil depended only slightly on the stress path along which stress was applied during shear and, therefore, that the failure surface $2 \cdot \tau_{max} = f(\sigma_m'$ and moisture content) was a unique relationship for a given soil. In figure 2.163, the limiting surface of shear strength is bounded in space by the curved line, Φ', which represents the drained shear strength of a normally loaded material. It was further hypothesized that lying on the three-dimensional surface representing the stress-deformation characteristics of a given soil was a line connecting points at which a sample of the soil in question, at any given confining stress, would reach a porosity (or void ratio) that would not change with further deformation. This line is hypothesized to lie on the failure surface, and so once σ_m' and $2 \cdot \tau_{max}$ reach this line (called the critical state line; see figure 2.164) (McKyes, 1985), deformation could continue without any further change in void ratio, σ_m' or $2 \cdot \tau_{max}$ (Koolen and Kuipers, 1983).

This same approach was applied to unsaturated soils (Bailey and VandenBerg, 1968). Instead of moisture content of a saturated soil, the bulk weight volume [BWV = 1/(dry bulk density)] was used in the three-dimensional plot. In these studies, it was found for unsaturated soils that the conditions of strain at constant volume and those for strain at constant stress did not coordinate into a single three-dimensional diagram as was hypothesized for saturated clay soils. In an effort to derive a model for volume-strain vs stress state relations (Bailey and VandenBerg, 1968), a special stress parameter, ζ, was formulated.*

* It might be noted that in a conventional triaxial test, $\sigma_1 = \sigma_m + 4/3\ \tau_{max}$ (Chancellor and Korayem, 1965).

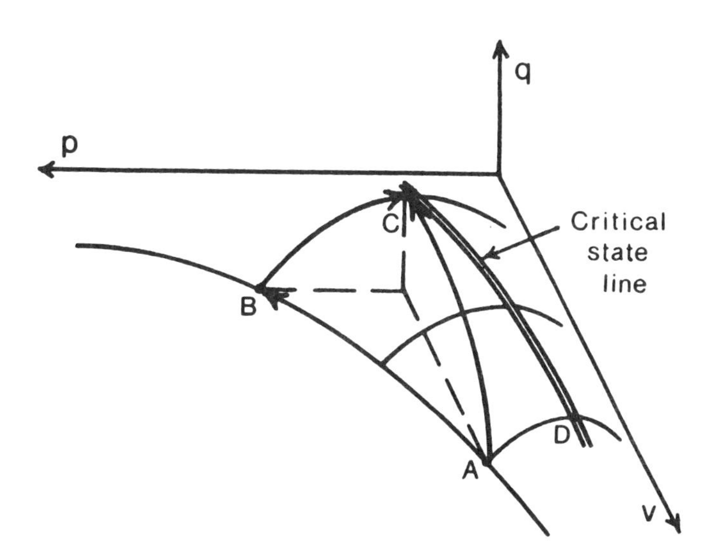

Figure 2.164–Critical state line (of unchanging void ratio) in the three-dimensional relationship among normal stress, p, shear stress, q, and volume of a given mass of soil, v, (from McKyes, 1985).

$$\zeta = \left(\sigma_m^2 + \tau_{max}^2\right)^{0.5} \quad (2.131)$$

This was used in the following model for soil BWV:

$$BWV = m \log \zeta - n\left(\tau_{max} / \sigma_m\right) + b \quad (2.132)$$

in which m, n, and b are constants dependent on soil conditions.

Further work along these lines (Bailey, 1971), using an artificial soil, led to the conclusion that the concept of the generalized shear surface in the three-dimensional stress-BWV yield diagram was not valid. Additional studies with a natural soil (Bailey, 1973) further confirmed this conclusion. The precompaction histories of the soil materials appeared to change their yield and plastic flow characteristics even at stress states sufficiently intense that the effects of the stress levels used for precompaction should have been eclipsed.

A model used to relate volume strain to a hydrostatic ($\sigma_1 = \sigma_2 = \sigma_3$) stress state was expanded to include the case of a continuously increasing shear stress (Grisso et al, 1987a). The model used for the hydrostatic case was:

$$\bar{\varepsilon}_{octH} = \left(A_H + B_H\,\sigma_H\right)\left(1 - e^{-C \cdot \sigma_H}\right) \quad (2.133)$$

in which

$$\bar{\varepsilon}_{oct_H} = \frac{1}{3}\ln\left(1 + \frac{\Delta V}{V_o}\right)$$

σ_H = hydrostatic stress = σ_m
A_H, B_H, and C_H = constants depending on soil conditions
e = natural logarithmic base

For varying shear stress conditions, the model used was:

$$\bar{\varepsilon}_{oct_R} = \beta\left(A_H + B_H\,\sigma_m\right)\left(1 - e^{-C_H \cdot \sigma_m}\right) \quad (2.134)$$

in which $\beta = \varepsilon_{oct_R}$ (varying shear stress) / ε_{oct_H}, (hydrostatic stress), and A_H, B_H, and C_H = constants obtained in the hydrostatic test of the same soil.

In an effort to be able to relate volume strain to stress-state changes, a system was devised to predict the value of β over a broad range of conditions from the value of stress-strain relations at initial conditions and at conditions for which σ_m = 500 kPa = σ_{oct}.

Work along these lines continued with efforts to make the model more conveniently applicable to design situations. A model was developed for cylindrical loading conditions ($\sigma_2 = \sigma_3$) as would apply in a conventional triaxial test (Bailey and Johnson, 1989). It was found that under these conditions, soil dry bulk density tended to be related to the ratio of the octahedral shear stress, τ_{oct}, to the octahedral normal stress, $\sigma_{oct} = \sigma_m$ (fig 2.165). This characteristic was embodied in the following model:

$$\bar{\varepsilon}_v = (A + B\,\sigma_m)\left(1 - e^{(-C\,\sigma_m)}\right) + D\left(\tau_{oct} / \sigma_m\right) \qquad (2.135)$$

in which

$\bar{\varepsilon}_v$ = natural volumetric strain = $\ln\left(1 + \frac{\Delta V}{V_o}\right) = \ln\left(\frac{\rho_0}{\rho}\right)$

A, B, C, and D are constants, values of which depend on the soil condition

$\sigma_m = \sigma_{oct} = \frac{\sigma_1 + \sigma_2 + \sigma_3}{3}$

$\tau_{oct} = \frac{1}{3}\left[(\sigma_1 - \sigma_2)^2 + (\sigma_2 - \sigma_3)^2 + (\sigma_3 - \sigma_1)^2\right]^{0.5}$

for cylindrical conditions, $\tau_{oct} = (\sqrt{2})(1/3)(\sigma_1 - \sigma_3)$

ρ_0 and ρ = initial and final values of dry bulk density

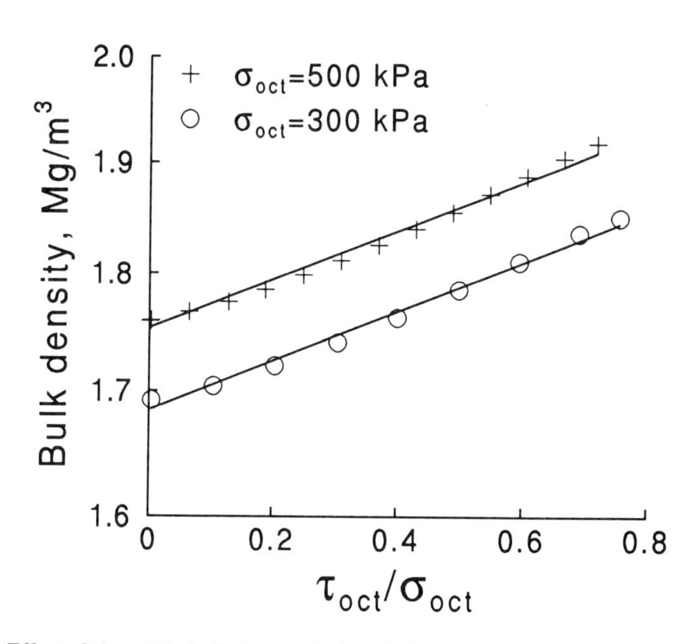

Figure 2.165–Effect of the octahedral stress ratio (octahedral shear stress per unit octahedral normal stress) on the dry bulk density achieved in a compression test of a given soil (Norfolk sandy loam). At a given value of σ_{oct}, an increase in the octahedral stress ratio is accompanied by an increase in the major principal stress, σ_1. The vertical scale is logarithmic (see equation 2.135) (from Bailey and Johnson, 1989).

It was shown (Bailey and Johnson, 1989) that under plastic flow yield conditions that there was a constant linear relationship between σ_m and τ_{oct}.

$$K = (\tau_{oct} \text{ at yield}) / \sigma_m \qquad (2.136)$$

From the basic equation for shear failure:

$$\tau_{max} = \sigma_n \cdot \tan \phi + C \qquad (2.137)$$

it might be anticipated that:

$$\sigma_1 = \sigma_3 \tan^2 (45° + \phi / 2) + 2 C \tan (45° + \phi / 2) \qquad (2.138)$$

which represents the relationship between σ_1 and σ_3 for a soil at the Mohr-Coulomb failure criterion and, therefore, at yield conditions, an approximation of the value K might be computed from the values of C and tan ϕ.

At this stage of development, the model for volumetric strain as a function of stress state involves two stress parameters, σ_m and τ_{oct}, and four constants to describe the soil characteristics, A, B, C, and D. It has not yet been established what relationships prevail among various basic measures of soil physical properties and the values of A, B, C, and D, although it is known that A, B, and C vary with moisture content.

Volumetric Strain, Shear Strain, and Stress State Changes

Experimental evidence (Soehne, 1958; Chancellor, 1977; Raghavan and McKyes, 1977; Greacen, 1960) has indicated that for some soils, when shear strain can accompany compressive stress, soil volumetric strain is greater when the shear strain is present than when it is not — even though the level of compressive stress may be the same in both cases. For some other soils, the presence of shear strain makes little difference in the compressive stress vs volumetric strain relationship (contrast fig 2.166 with fig 2.158).

It is believed that whether or not a soil responds to shearing strain with additional volumetric strain depends on the grain size distribution. Soils with a broad range of grain size distribution (including a sizable proportion of large grains) tend to be those which respond more in this way than do soils with a less broad distribution of grain sizes (contrast the grain-size distribution for the soils in fig 2.158 and 2.166, as shown in fig 2.1).

The effect of particle-size distribution on the response of a soil to kneading compaction was illustrated by Bodman and Constantin (1965) who mixed a coarse sand with silty clay in varying proportions. The results (fig 2.167) show that at certain proportions (about 80% sand in these cases) the void ratio obtained with a given compaction regime was minimized and that the mixture with coarse sand reached lower void ratios than that with fine sand.

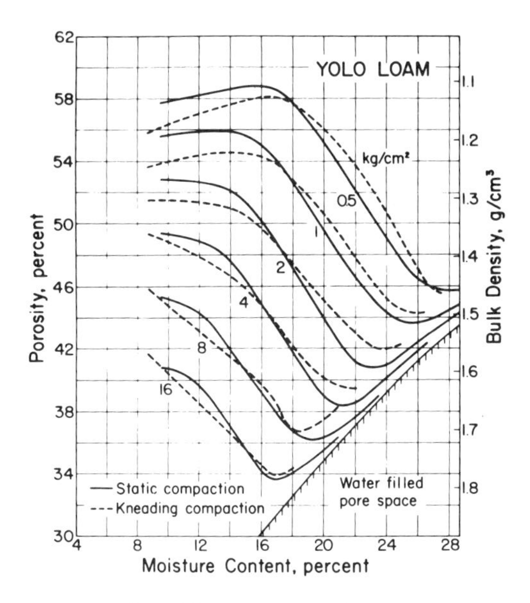

Figure 2.166–For this soil, the porosity achieved by a given level of major principal stress was the same whether the stress was applied by static or kneading compaction (from Chancellor, 1977).

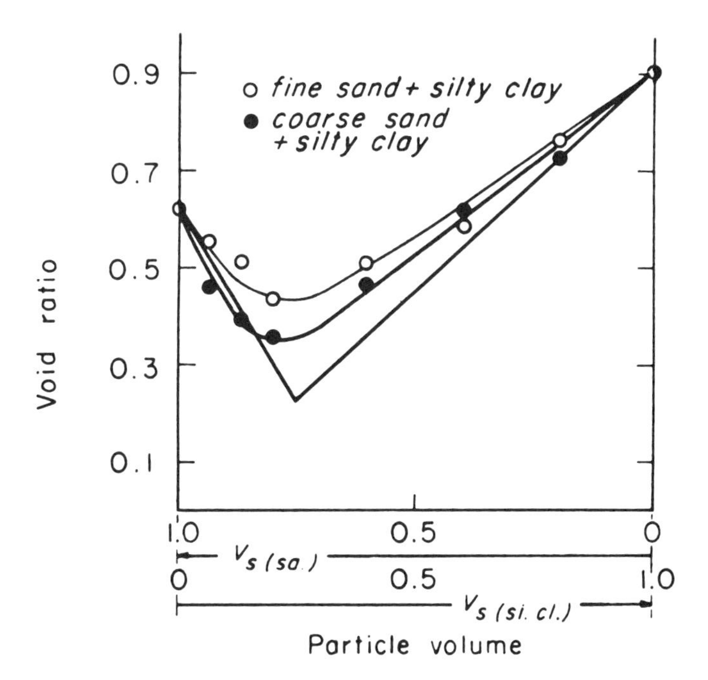

Figure 2.167–When both large particles and small particles exist in the soil in certain proportions, a lower porosity (ordinate) is achievable upon compaction, than when all particles are of approximately the same size (from Bodman and Constantin, 1965).

The tendency for a volumetric strain response to shearing strain was found in one case to be greater for a sandy loam soil than for a sand soil (Raghavan and McKyes, 1977). However, the effect was found to prevail for both sands and clays (fig 2.168) (McKyes, 1985).

Efforts have been made to quantify this effect. One model proposed (VandenBerg, 1966) was of the form:

$$\gamma_d = A + B \ln \sigma_m \left[1 + \ln\left(1 + \gamma_{max}\right)\right] \qquad (2.139)$$

in which γ_d = dry bulk density, σ_m = mean normal stress = $(\sigma_1 + \sigma_2 + \sigma_3) / 3$, γ_{max} = maximum shear strain, and A and B = constants depending on soil characteristics.

This model was validated using upper levels of γ_{max} ranging from 0.3 for a sandy loam to 0.75 for a clay. When the model was applied to other soils for which the upper levels of γ_{max} were above 1.0 (Chancellor and Korayem, 1965), it was found that dry bulk density did not continue increasing indefinitely with shear strain, but rather reached a maximum. The shear strain levels at which the maximum was reached tended to be a function of both the type of loading circumstances prevailing, and, in some cases, the soil condition. For a case in which a cylindrical sample was allowed to expand laterally at stresses that just prevented catastrophic failure (Chancellor and Korayem, 1965), constant values of dry bulk density were reached at γ_{max} values of approximately 0.5 irrespective of soil type, density, or moisture content (see fig 2.132). For monolithic soil samples subjected to both confined and unconfined compression tests (Chancellor et al,

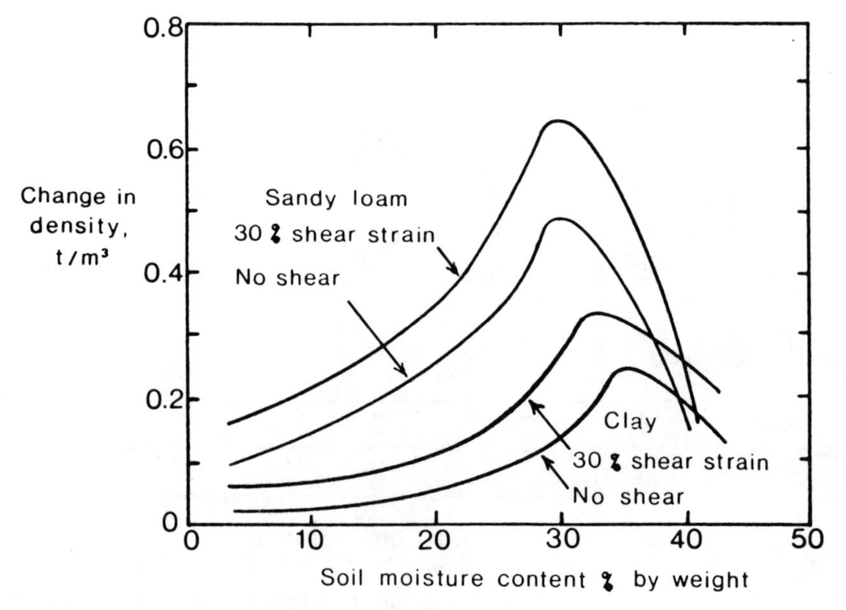

Figure 2.168–When shear strain accompanied the compaction process, greater density increases took place than when shear strain was not present (from McKyes, 1985).

1969), these values of γ_{max} ranged from 0.03 to 0.07 for the unconfined compression tests (see fig 2.130). For both types of tests, this level of strain at which volumetric compression was maximum tended to increase with moisture content. For both types of tests, high moisture content samples reached the point of unchanging volume strain well before strain levels required for failure were achieved. For dryer samples in unconfined compression, failure levels of strain were reached before volume strain had ceased. For dry samples in confined compression, not only did maximum volume strain occur before reaching the shear strain levels associated with failure, but once the point of maximum volume strain had occurred, further shear strain resulted in the expansion of the samples with consequent decrease in dry bulk density.

This dilatation phenomena (fig 2.169) associated with shear strain is commonly found with soils, particularly sands, which are initially very dense. This characteristic has been quantified as "the angle of dilatation"* (fig 2.170) (Koolen and Kuipers, 1983) and values as high as 24 or 30° have been reported.

The effects of moisture content and density of a sand and the changes in density that take place when it is loaded to shear failure are illustrated in figure 2.171 (Kézdi, 1979). It can be seen that densification, dilatation, densification followed by dilatation, or

* The terms "dilation" and "dilatation" are considered to be of equivalent meaning.

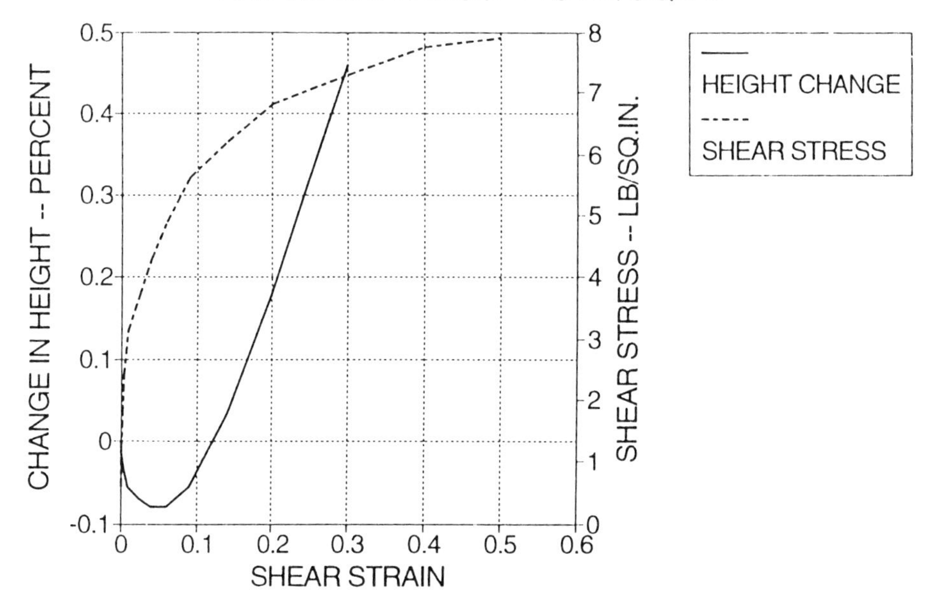

Figure 2.169–Volumetric changes in sand subjected to a fixed level of normal stress, when shear strain took place (from Karol, 1955).

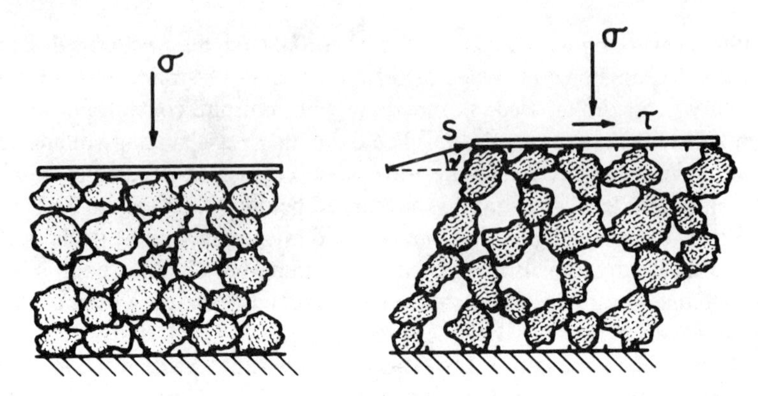

Figure 2.170–Illustration of the expansion that takes place when a dense soil (usually sand) fails in shear. The angle, ν, is termed the angle of dilatation (from Koolen and Kuipers, 1983).

dilatation followed by densification can take place depending on the initial conditions of the soil and the stresses imposed.

In the continuing search to find some generalizable characteristics to represent soil stress-strain behavior, two essentially equivalent approaches have appeared that pertain to cylindrical soil samples as would be tested in a triaxial cell apparatus. Johnson and

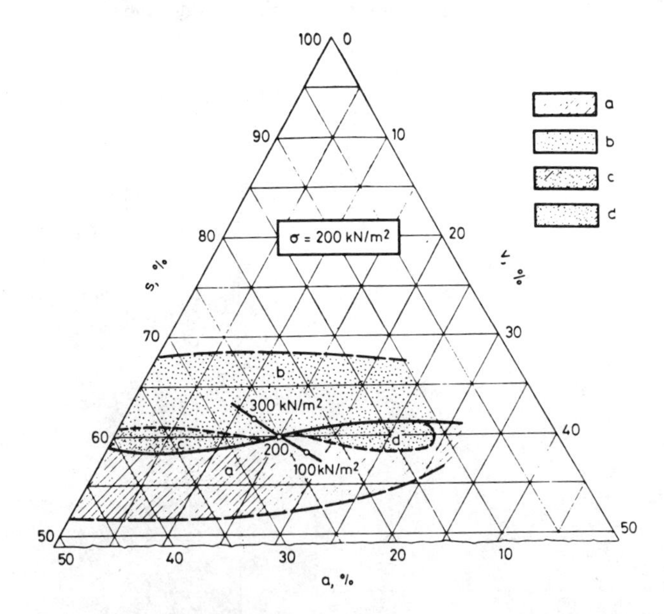

Fields of the changes in the phase composition at shear as a function of the initial state
a loosening
b densification
c first loosening, then densification
d first densification, then loosening
σ applied normal stress

Figure 2.171–The effects of loading a sand soil to shear failure on the porosity, v = volumetric moisture content, a = air-filled porosity, s = solids volume / total volume (from Kézdi, 1979).

Bailey (1990) have proposed that stress-strain characteristics of cylindrical soil samples be represented by a plot of $\gamma_{max} / \bar{\varepsilon}_{v-ss}$ vs τ_{max} / σ_1, in which:

$\bar{\gamma}_{max}$ = $\bar{\varepsilon}_1 - \bar{\varepsilon}_3$, defined as the maximum natural shear strain
$\bar{\varepsilon}_1$ = major natural principal strain
$\bar{\varepsilon}_3$ = minor natural principal strain
$\bar{\varepsilon}_{v-ss}$ = natural volumetric strain occurring after shear stress is initiated
τ_{max} = $(\sigma_1 - \sigma_3)/2$, defined as the maximum shear stress acting on the plane of a cylindrical sample, with the plane oriented at 45° to the direction of major principal stress
σ_1 = major principal stress
σ_3 = minor principal stress

This proposed relationship is essentially composed of the ratio of two general relationships:

$$\bar{\gamma}_{max} = f_1(\tau_{max}) \text{ and } \bar{\varepsilon}_v = f_2(\sigma_1)$$

with the ratio

$$\bar{\gamma}_{max} / \bar{\varepsilon}_{v-ss} = f_3(\tau_{max}, \sigma_1)$$

including the possible form

$$\bar{\gamma}_{max} / \bar{\varepsilon}_{v-ss} = f_4(\tau_{max} / \sigma_1)$$

The ratio τ_{max} / σ_1, or its equivalent, is a parameter that has been generally observed to be linked to various measures of soil deformation.

The quantitative model proposed by Johnson and Bailey (1990) is:

$$\frac{\bar{\gamma}_{max}}{\bar{\varepsilon}_{v-ss}} = \ln\left[1 - \left(\tau_{max} / K' \sigma_1\right)\right] / h \qquad (2.140)$$

or the inverse form:

$$\frac{\tau_{max}}{\sigma_1} = K' \left(1 - \beta_e^{-h \bar{\gamma}_{max} / \bar{\varepsilon}_{v-ss}}\right) \qquad (2.141)$$

in which $\beta = e^{hI}$, and I = the intercept of $\bar{\gamma}_{max} / \bar{\varepsilon}_{v-ss}$ when $\tau_{max}/\sigma_1 = 0$, and K′, β and h are coefficients.

Figure 2.172 illustrates a plot of data for cylindrical samples of one soil loaded with a broad range of varying stress paths and levels, while figure 2.173 illustrates the prediction error relating actual data to the above quantitative model.

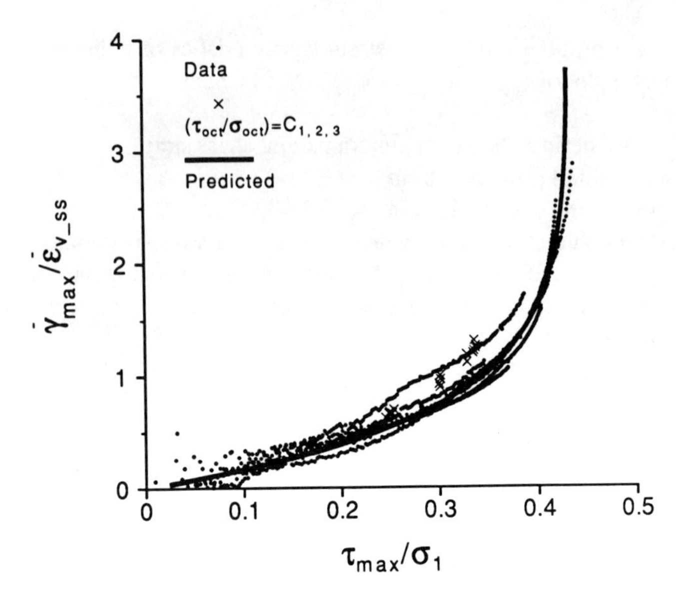

Figure 2.172–Generalized stress-strain relationship for cylindrical samples of Hiwassee Clay loaded at various levels using various stress paths. Average moisture content=18.6%, average dry bulk density=1.077 Mg/m³ (from Johnson and Bailey, 1990).

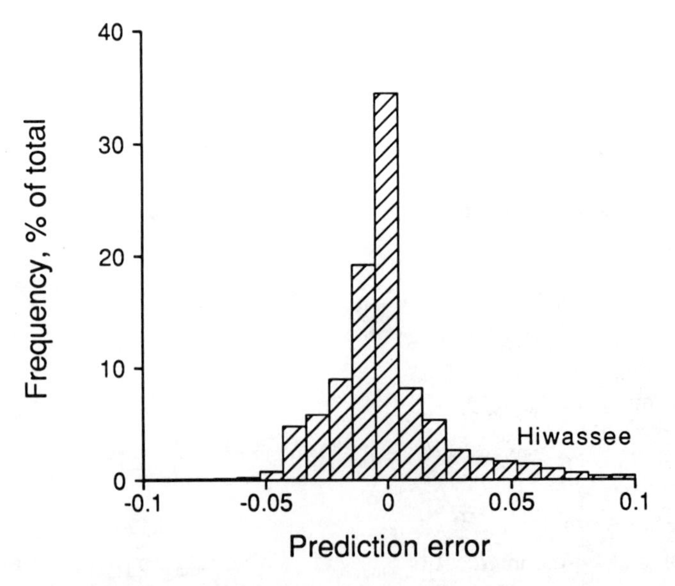

Figure 2.173–Errors between predicted values and data points in figure 2.172. Predictions made using K′ = 0.4335 and h = 1.584 for Hiwassee Clay in the equation:

$$\frac{\tau_{max}}{\sigma_1} = K'\left(1 - \beta e^{-h\,\bar{\gamma}_{max}\,/\,\bar{\varepsilon}_{v\text{-}ss}}\right) \text{ in which } \beta = e^{hI} \text{ and I typically is 0.0}$$

(from Johnson and Bailey, 1990).

An equivalent system is represented by the generalized soil stress-strain response characteristics for cylindrical samples illustrated in figure 2.132 (a plot of $\Delta\ \gamma_{max}/\Delta\ \varepsilon_1$, vs γ_{max}*) and in figure 2.133 (a plot of σ_1/σ_3 vs $\Delta\ \gamma_{max}/\Delta\ \varepsilon_1$) (Chancellor and Korayem, 1965).

For cylindrical stress applications:

$$\frac{\tau_{max}}{\sigma_1} = \frac{\sigma_1 - \sigma_3}{2} \times \frac{1}{\sigma_1} = \frac{1}{2}\left(1 - \frac{\sigma_3}{\sigma_1}\right) \tag{2.142}$$

so, it can be seen that there is an exact relationship between τ_{max}/σ_1 and σ_1/σ_3.

Furthermore, there is a similar direct relationship between $\Delta\gamma_{max}/\Delta\varepsilon$, and $\bar{\gamma}_{max}/\bar{\varepsilon}_{v-ss}$, in that:

$$\Delta\ \varepsilon_v = \Delta\ \varepsilon_1 + 2\ \Delta\ \varepsilon_3 \text{ and } \Delta\ \gamma_{max} = \Delta\ \varepsilon_1 - \Delta\ \varepsilon_3,$$

so that

$$\Delta\ \varepsilon_3 = \Delta\ \varepsilon_1 - \Delta\ \gamma_{max},$$

and substituting

$$\Delta\ \varepsilon_v = \Delta\ \varepsilon_1 + 2\ \Delta\ \varepsilon_1 - 2\ \Delta\ \gamma_{max} = 3\ \Delta\ \varepsilon_1 - 2\ \Delta\ \gamma_{max}$$

$$\frac{\Delta\ \varepsilon_v}{\Delta\ \gamma_{max}} = \frac{3\ \Delta\ \varepsilon_1}{\Delta\ \gamma_{max}} - 2 \tag{2.143}$$

In addition, $\sum \Delta\ \varepsilon_v = \bar{\varepsilon}_v$, and $\sum \Delta\ \gamma_{max} = \bar{\gamma}_{max}$, provided the increment, Δ, is small.

Thus, it is possible to use the generalized data from figure 2.132, to go from a given value of $\gamma_{max} = \bar{\gamma}_{max}$ to a value for $\Delta\ \gamma_{max}/\Delta\ \varepsilon_1$. These $\Delta\ \gamma_{max}/\Delta\ \varepsilon_1$ values can be then converted to values of $\Delta\ \gamma_{max}/\Delta\ \varepsilon_v$. By plotting a sequence of small $\Delta\ \gamma_{max}$ vs $\Delta\ \varepsilon_v$, increments on coordinates of $\bar{\gamma}_{max}$ vs $\bar{\varepsilon}_v$, it is possible to obtain values of $\bar{\gamma}_{max}/\bar{\varepsilon}_{v-ss}$ from ratios of the coordinates of points on the resulting plot (fig 2.174). For any given value of $\Delta\ \gamma_{max}/\Delta\ \varepsilon_1$, it is also possible to obtain from figure 2.133 a related value of σ_1/σ_3. With appropriate conversions, the selection of a series of γ_{max} values could thus lead to series of $\bar{\gamma}_{max}/\bar{\varepsilon}_{v-ss}$ values paired with related τ_{max}/σ_1 values (fig 2.175), indicating that the proposed generalizations of Johnson and Bailey (1990) and those of Chancellor and Korayem (1965) are equivalent methods of representing the stress-strain behavior of soil samples in a cylindrical stress state. The latter method, however, involves two subgeneralizations, one between natural shear strain and Poisson's ratio, and the other between σ_1/σ_3 or τ_{max}/σ_1 and Poisson's ratio.†

---

* In this case, since γ_{max} was computed as $\Sigma\Delta\ \gamma_{max}$, therefore, in figure 2.132, $\gamma_{max} = \bar{\gamma}_{max}$.

† The term $\Delta\ \tau_{max}/\Delta\ \varepsilon_1$ is equal to 1 + Poisson's ratio.

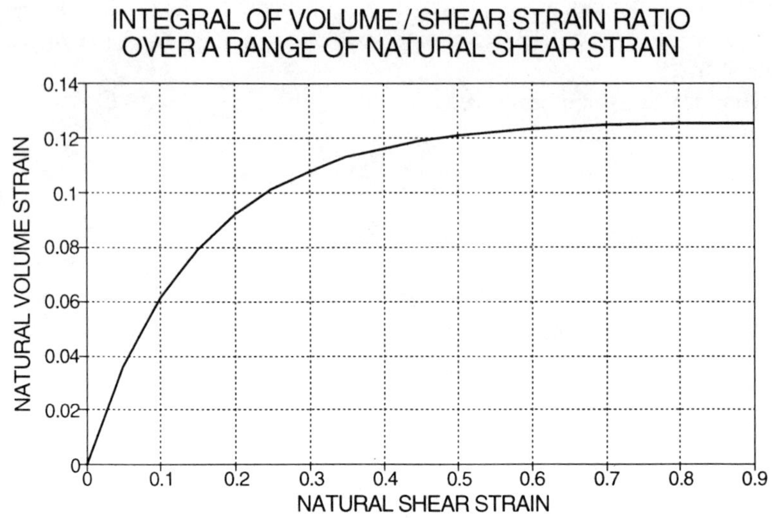

INTEGRAL OF VOLUME / SHEAR STRAIN RATIO
OVER A RANGE OF NATURAL SHEAR STRAIN

NATURAL VOLUME STRAIN

0.14
0.12
0.1
0.08
0.06
0.04
0.02
0

0 0.1 0.2 0.3 0.4 0.5 0.6 0.7 0.8 0.9

NATURAL SHEAR STRAIN

Figure 2.174–Integral of $\Delta \varepsilon_v / \Delta \gamma_{max}$ values (derived from fig 2.132) relative to a range of $\overline{\gamma}_{max}$ values for a composite of data for cylindrical samples of two soils at several moisture contents and initial dry bulk densities. The ratios of coordinates $\overline{\gamma}_{max} / \overline{\varepsilon}_{v-ss}$ for various points on the curve constitute the y-coordinates of points plotted in figure 2.175.

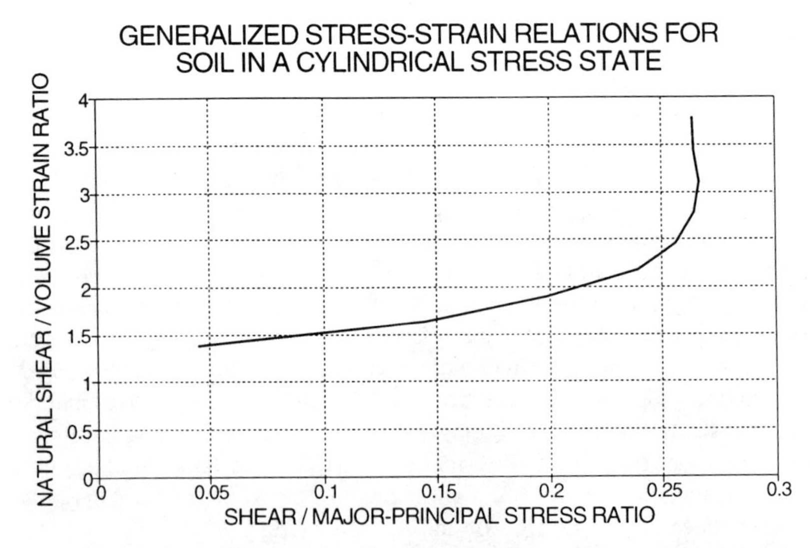

Figure 2.175–Plot of $\overline{\gamma}_{max} / \overline{\varepsilon}_{v-ss}$ values derived from ratios of coordinates of points of the curve in figure 2.174 vs corresponding values of τ_{max} / σ_1. Values of τ_{max} / σ_1 were derived from the generalized relationship in figure 2.133. Figure 2.175 represents an equivalent set of data to those presented in figure 2.172 but for different soils loaded with a different cylindrical stress process.

Viscoelastic, Viscoplastic, and Rheological Properties

In many cases, the stress-strain data obtained at quasistatic speeds in laboratory and field tests with agricultural soils are considered to be applicable to traction and tillage processes at conventional field speeds. However, many researchers have investigated the effects of time in the stress-strain relationships for agricultural soils. These time effects have usually been measured in three types of tests: (1) relaxation tests in which a given level of strain is imposed and the decay of stress with time is measured, (2) creep tests in which a given stress level is applied and the strain is observed over time, and (3) conventional tests in which variable strain rates are imposed. Results of these tests have frequently been categorized according to three different models. The first is that of a Newtonian fluid (fig 2.176) (Koolen and Kuipers, 1983), in which:

$$\tau = \eta \frac{d\gamma}{dt} \qquad (2.144)$$

where τ = shear stress, η = viscosity, γ = shear strain, and t = time.

The second is that of a viscoplastic Bingham body (fig 2.177) (Kézdi, 1974):

$$\tau = \zeta + \eta' \frac{d\gamma}{dt} \qquad (2.145)$$

in which ζ = yield stress in shear = $c + \sigma_n \tan \phi$, and η' = viscosity pertaining to shear stresses above ζ.

The third model is a Burger model (fig 2.178) (Ji et al, 1986):

$$z = \frac{\beta P}{\sqrt{A}} \left[\frac{1}{E_M} + \frac{1}{E_k} \left(1 - e^{-\frac{E_k}{\lambda_k} t} \right) + \frac{t}{\lambda_M} \right] \qquad (2.146)$$

in which

β	=	a constant for a given process
P	=	applied pressure to the top surface of a semi-infinite body
A	=	area on which the pressure is applied

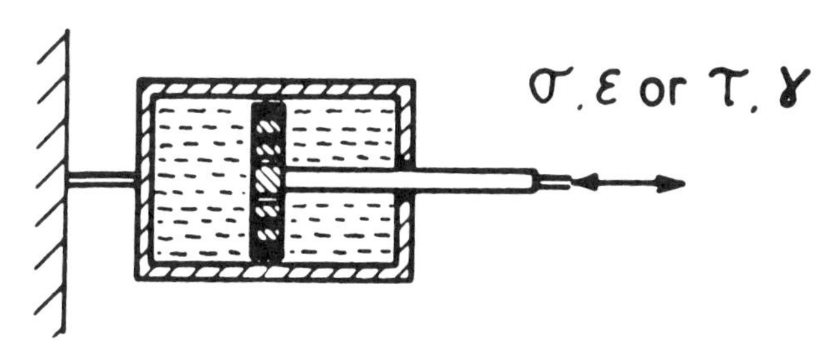

Figure 2.176–Representation of a viscous element, the force on which is directly proportional to the deformation rate (from Koolen and Kuipers, 1983).

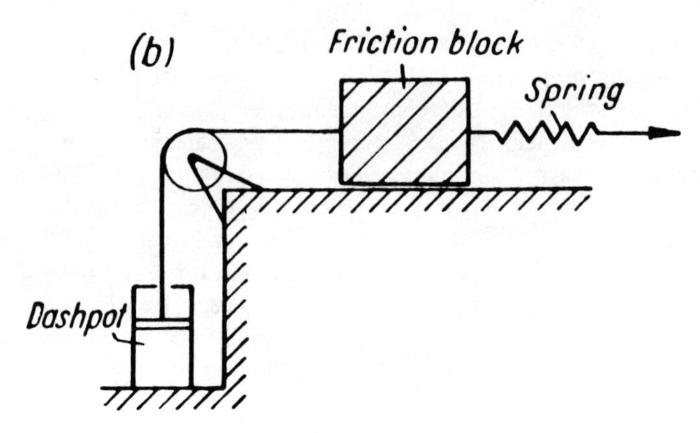

Figure 2.177–Representation of a Bingham body with a viscous element and a coulomb friction element in series (from Kézdi, 1974).

E_M and E_k	=	values related to the stiffnesses of the respective spring elements
λ_M and λ_k	=	values related to the viscosities of the respective dashpot elements
t	=	time
z	=	depth of impression of the loaded area into the semi-infinite body

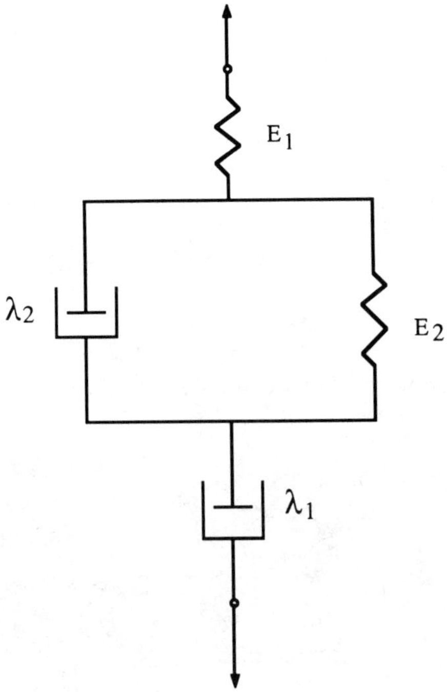

E_1

λ_2 E_2

λ_1

Figure 2.178–Representation of a Burger model with both spring and viscous elements (from Ji et al, 1986).

$$E_M = \frac{E_1}{1-\mu^2}, \qquad E_k = \frac{E_2}{1-\mu^2}$$

$$\lambda_M = \frac{\lambda_1}{1-\mu^2}, \qquad \text{and} \quad \lambda_k = \frac{\lambda_2}{1-\mu^2}$$

in which μ = Poisson's ratio for the soil.

In foundation engineering, stress-strain properties of soils over extended periods of time (as much as several years) are related to the consolidation process in which water is caused to flow from the soil pore space. These time effects are not considered here. However, the time aspects of stress-strain relationships for agricultural soils have most commonly been considered for clay soils at high moisture contents as are frequently found in wet rice culture.

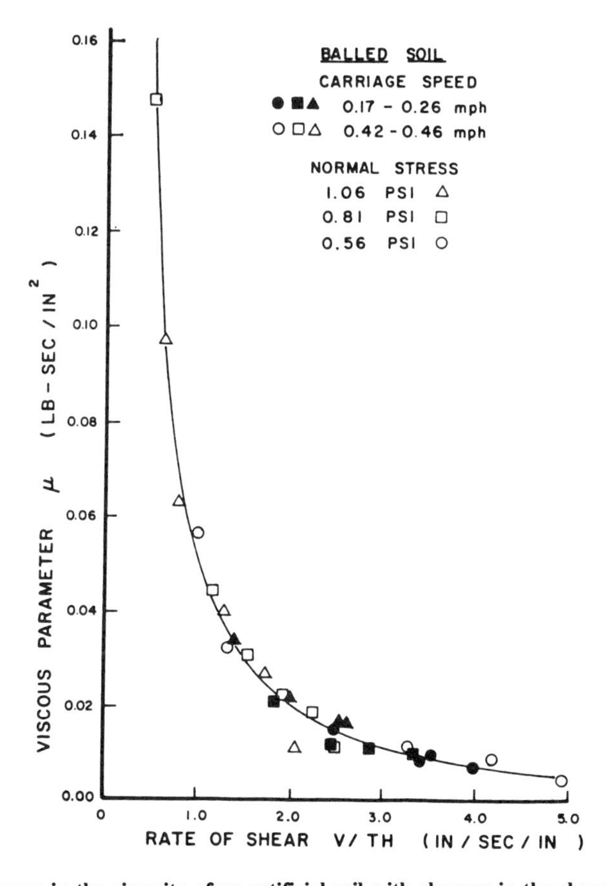

Figure 2.179–Changes in the viscosity of an artificial soil with changes in the shear rate (from Kunz, 1971).

Rowe and Hegedus (1959) tried to characterize a wet clay soil as a Newtonian-like fluid. They found that the viscosity of a super-saturated Volclay-Bentonite (typically about seven times that of water or about 0.007 $N\text{-}s/m^2$) decreased linearly with increases in the time-rate of shear strain, indicating that the material was not a true Newtonian fluid.

When a tracked carriage was run in a soil bin, at varying slip levels, over a ridge of artificial soil composed of sand, clay, and mineral oil, it was again found that the viscosity of the material could be described by a single curve for all speeds and normal stresses tested (fig 2.179). The curve indicated decreasing viscosity with increasing time-rate of shear strain (Kunz, 1971). Viscosity at a shear rate of 4 s^{-1}, was on the order of 50 $N\text{-}s/m^2$.

In another test in which a wet mud film was tested in a rotating viscometer (fig 2.180) (Persson and Chang, 1966), it appeared that shear stress tended to increase with the time-rate of shear strain and that there was a certain amount of cohesion that prevailed, unaffected by the time-rate of shear strain or by the normal stresses applied (fig 2.181). Typical viscosity values were in the range of 0.35 $N\text{-}s/m^2$.

The Bingham body model that represents the characteristics of a viscoplastic material (fig 2.177) is one of the most common forms used to represent the stress-strain-time properties of agricultural soils which are not necessarily saturated clays. This form is compatible with the $\tau_{max} = \sigma_n \tan \phi + c$ representation of the quasistatic shear strength of soils by the Mohr-Coulomb failure criterion. To this quasistatic shear strength is added a the term $\eta' \, d\gamma / dt$ to represent a further shear resistance associated with the time-rate of

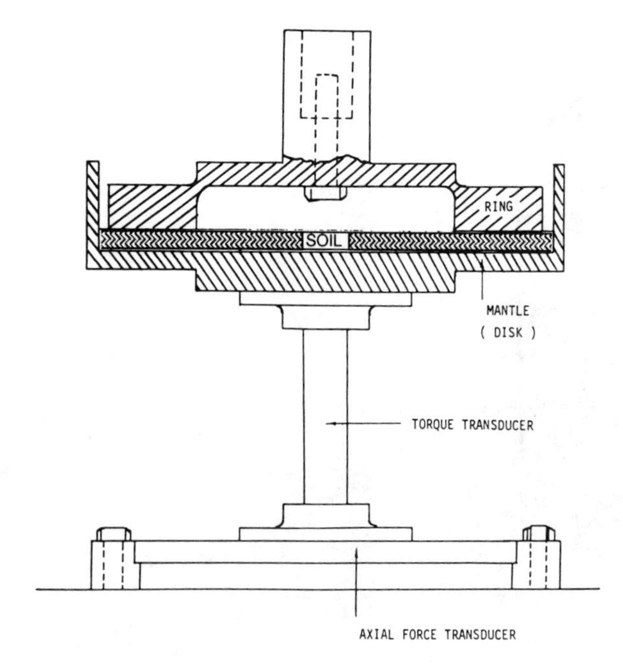

Figure 2.180–Annular apparatus for measuring viscous properties of soil (from Persson and Chang, 1966).

shear strain. Figure 2.68 (Hassan and Chancellor, 1970) illustrates the characteristics of a saturated clay soil that was of a viscoplastic nature when subjected to triaxial tests run at various levels of controlled strain rate. In this case, the superplastic* viscosity, η', appeared to be the same for all the levels of preconsolidation or initial porosity for the one soil used. However, when shear-vane measurements were made at various rotational velocities in a wet clay soil, superplastic viscosity before puddling was different than that after puddling (fig 2.182) (Awadhwal and Singh, 1985). In this thixotropic soil, the yield strength of the soil was greatly reduced by puddling, but when the "degree of puddle" was quantified by the term:

$$\text{degree of puddle} = 1 - \left(\text{puddled yield stress} / \text{unpuddled yield stress}\right) \qquad (2.147)$$

* Superplastic viscosity refers to viscosity characteristics based on shear stress levels measured above that required for quasistatic yield in shear.

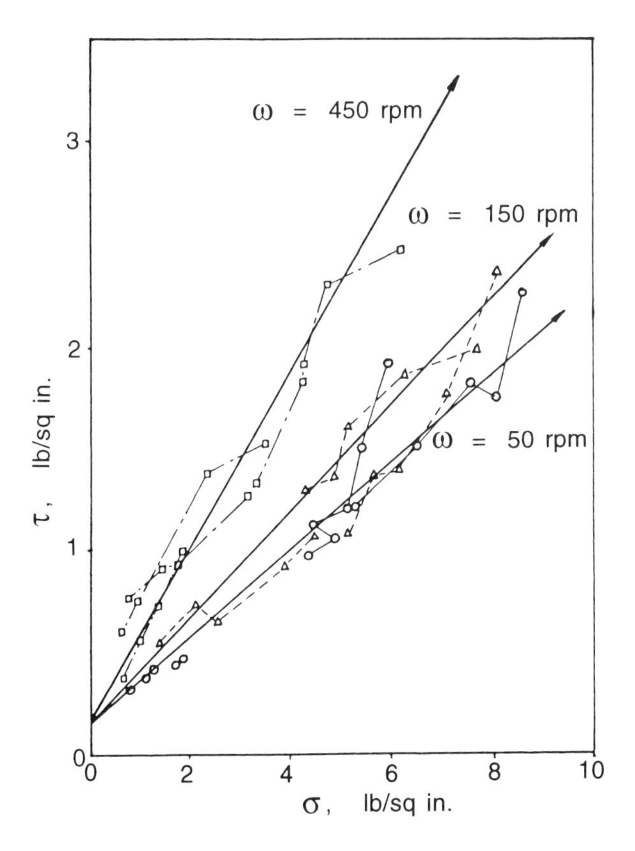

Figure 2.181–Changes in soil viscous shear stress, τ, with changes in normal stress, σ, and shear rate (from Persson and Chang, 1966).

it could be seen (fig 2.183) that much of the yield strength could be recovered in 100 h, particularly if the puddling action was not extensive.

When direct shear tests using a shear-box device were made at different speeds (Brandon et al, 1986), it was found that shearing stresses at failure increased with shearing velocities. These shear stress changes were due to increases in apparent cohesion and were not related to any significant change in the coefficient of internal friction with shear velocity.

Cylindrical monolithic samples (to simulate the characteristics of nonremolded field soils) of a clay loam soil were tested in confined and unconfined compression at rates of strain ranging from 0.018 min^{-1} to 0.254 min^{-1} over a range of moisture contents (Aref et al, 1975). The axial stress at failure, σ_1, was related to both the axial strain and the time-rate of axial strain by:

$$\sigma_1 = q_0 \, \varepsilon_1 + q_1 \, \dot{\varepsilon}_1 \qquad (2.148)$$

in which ε_1 = strain in the axial direction, $\dot{\varepsilon}_1$ = strain rate in the axial direction, and q_o and q_1 = coefficients.

Values of q_o and q_1, shown in figure 2.184, decreased with increasing moisture content. The general magnitude found for q_o and q_1 indicate that if the strain rate was such as to produce failure in 3 s, there would be a doubling of failure stress compared to that obtained with quastistatic loading. Thus, if extrapolated to field circumstances in

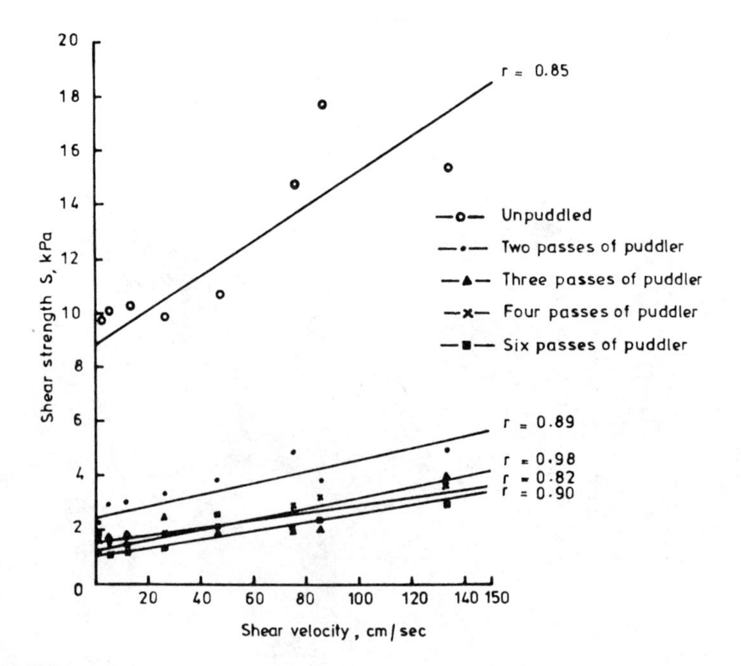

Figure 2.182–Effects of shear velocity on shearing strength of a puddled soil. Note the effect of puddling on the general reduction of shear strength levels for this thixotropic soil (from Awadhwal and Singh, 1985).

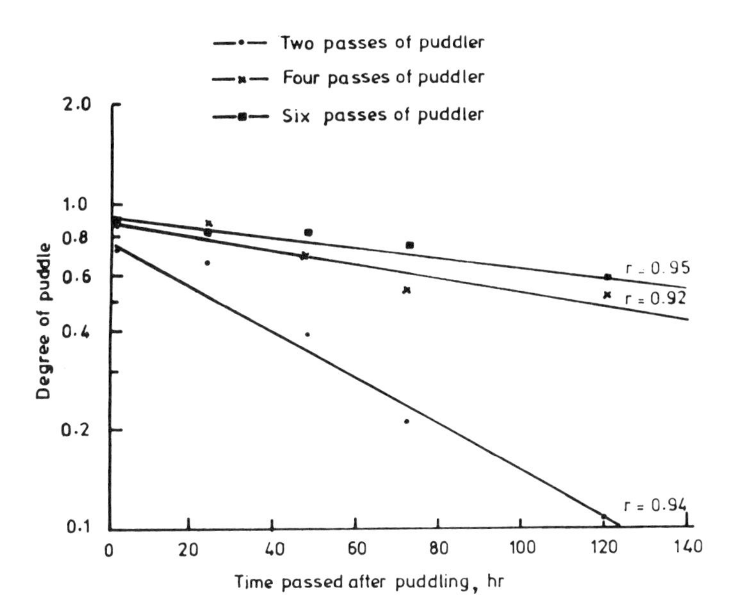

Figure 2.183–Illustration of the process by which a puddled thixotropic soil regains its strength by the process of floccule formation (from Awadhwal and Singh, 1985).

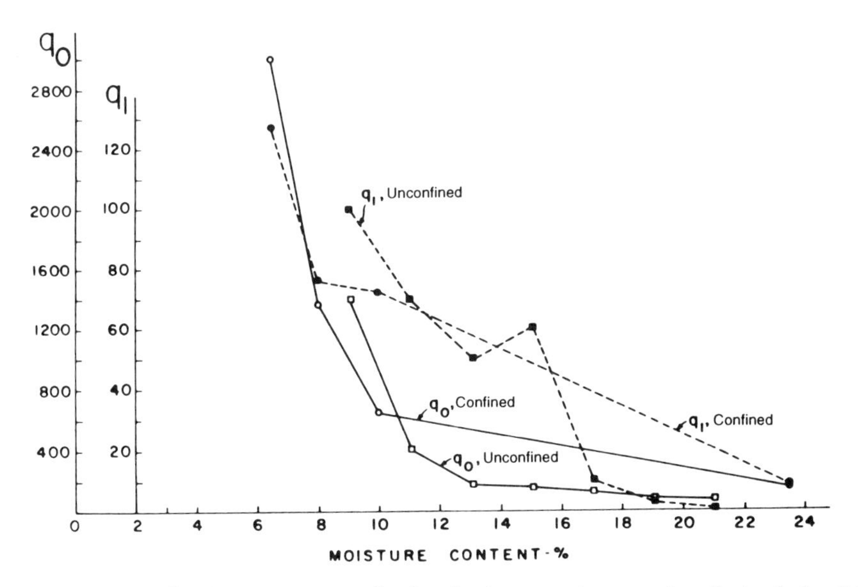

Figure 2.184–Effects of moisture content on the viscoelastic parameters q_0 and q_1 (inch min^{-1} units) for monolithic soil samples tested in confined and unconfined compression (from Aref et al, 1975).

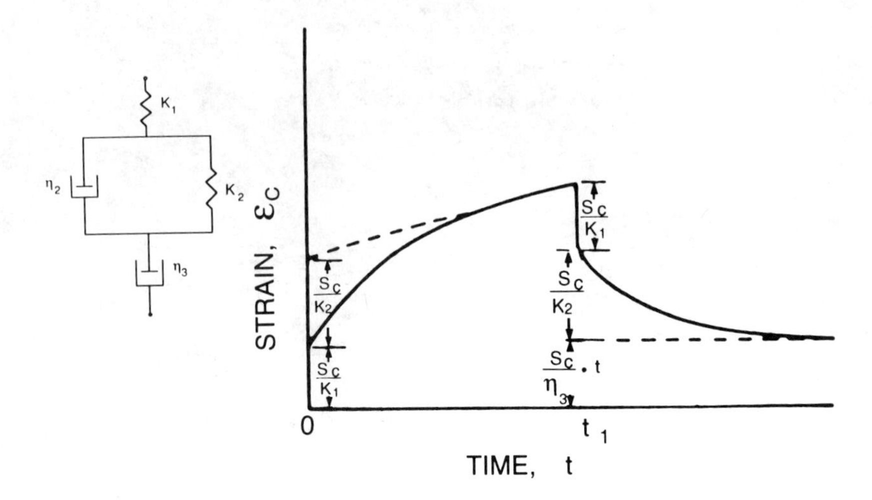

Figure 2.185–Illustration of both creep and relaxation characteristics of a soil having properties that could be represented by a Burger model. S_c = compressive stress, K values are constants for elastic compressive properties, and η values are constants for viscous compressive properties (from Ram and Gupta, 1972).

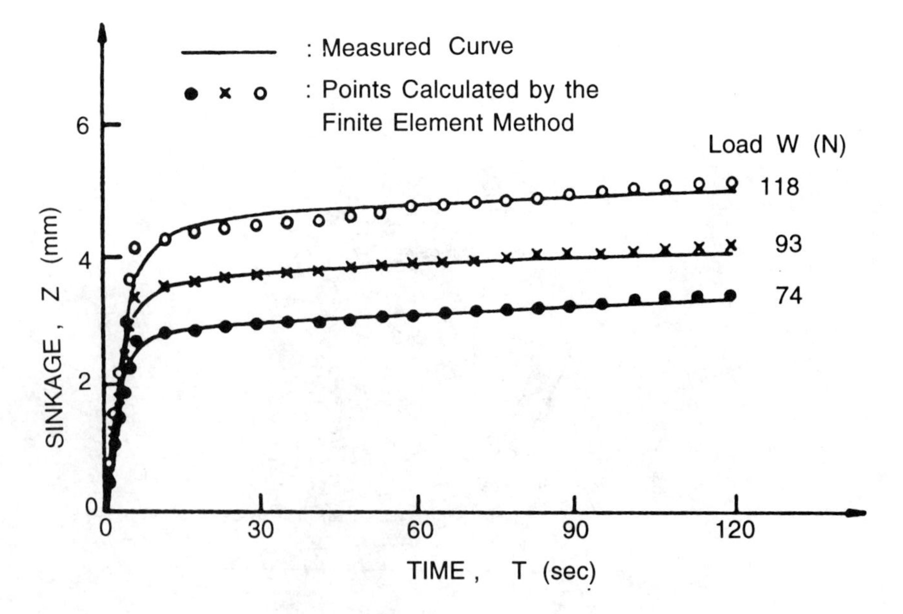

Figure 2.186–Calculated and measured values for wheel sinkage in a sandy loam soil with a moisture content of 21%. Calculated values were obtained using a finite-element model in conjunction with viscoelastic soil properties (from Oida and Tanaka, 1981).

which loading to failure occurs in less than 1 s, the time-rate-dependent soil strength characteristics would represent the major component of soil strength.

In figure 2.185 (Ram and Gupta, 1972) is illustrated a Burger's body model and an example of both a creep phenomenon when a constant stress level is applied over time and strain is measured (left side of plot) and a relaxation phenomenon when a fixed level of strain is applied over time and stress is measured. Tests of this sort may be applied both in a triaxial cell and by controlling the vertical load on, and position of, a plate of a given area resting on the soil surface (Ji et al, 1986; Lu et al, 1982; Pan et al, 1983). From the resulting data, the values of K_1, K_2, η_2, and η_3 may be determined. Such parameters have been used in a finite element analysis to successfully predict the sinkage (with time) of a wheel in a moist sandy loam soil (fig 2.186) (Oida and Tanaka, 1981). With the use of other types of analysis, the sinkage (with time) was predicted for tracked vehicles in soft, wet clay soils used for rice (fig 2.187) (Pan, 1984).

The effects of soil moisture content and clay content on the values of these four parameters for five rice-field soils are presented in figure 2.188 through 2.194 (Lu et al., 1982). In general, all parameters decreased in value with increasing moisture content and increased in value with increasing clay content.

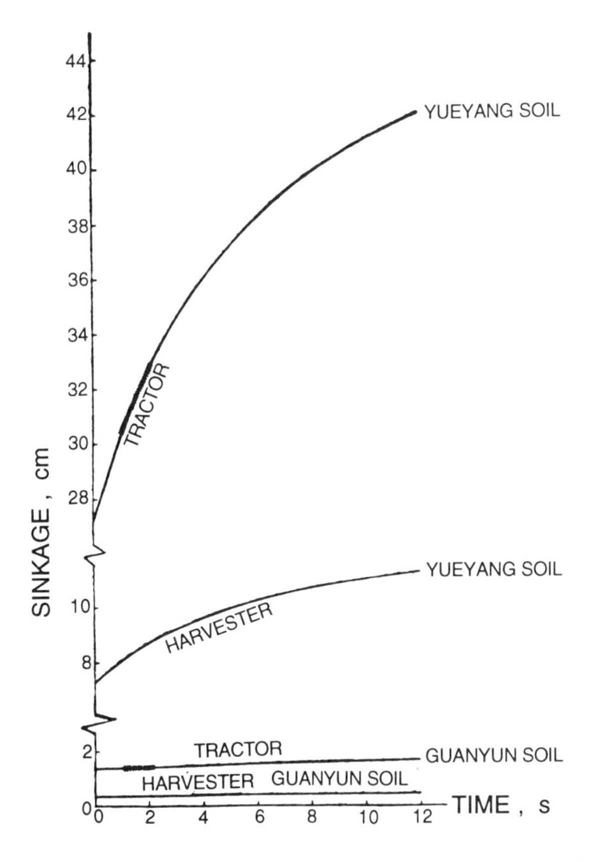

Figure 2.187–Tracked vehicle sinkage as a function of time in a wet rice-land soil (from Pan, 1984).

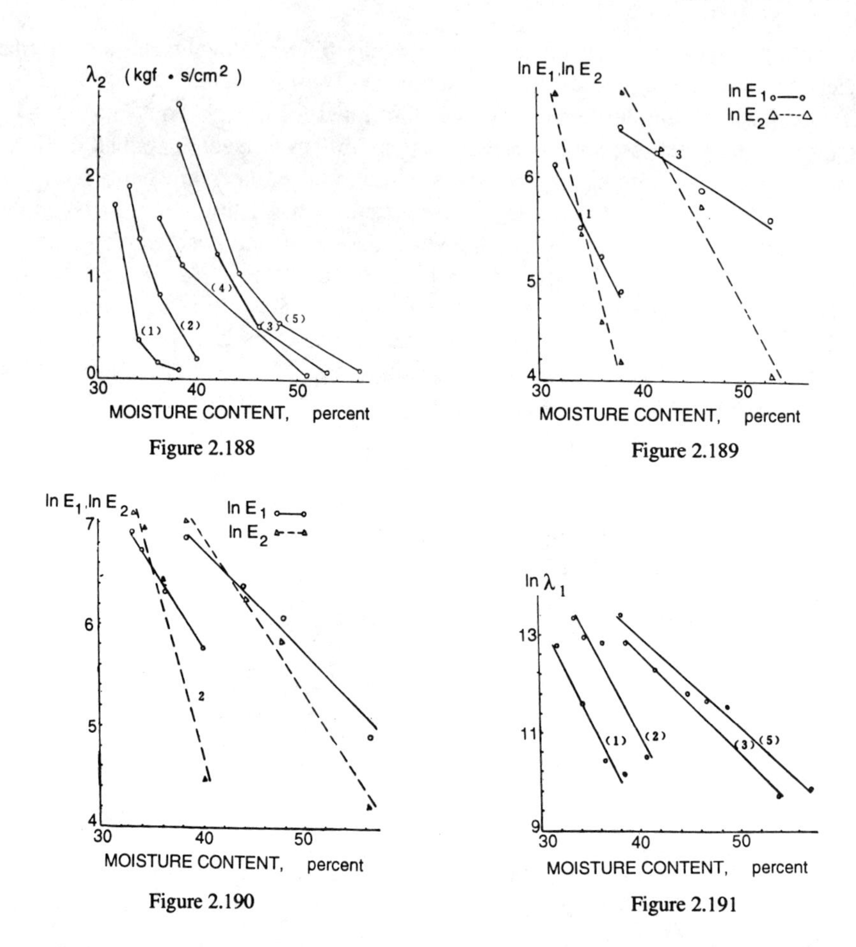

Figure 2.188

Figure 2.189

Figure 2.190

Figure 2.191

Figures 2.188 to 2.191–Burger model parameters E_1, E_2, λ_1, and λ_2 (see fig 2.178) as a function of moisture content for five rice-land soils. Units of E are kg f/cm^2 (from Lu et al, 1982).

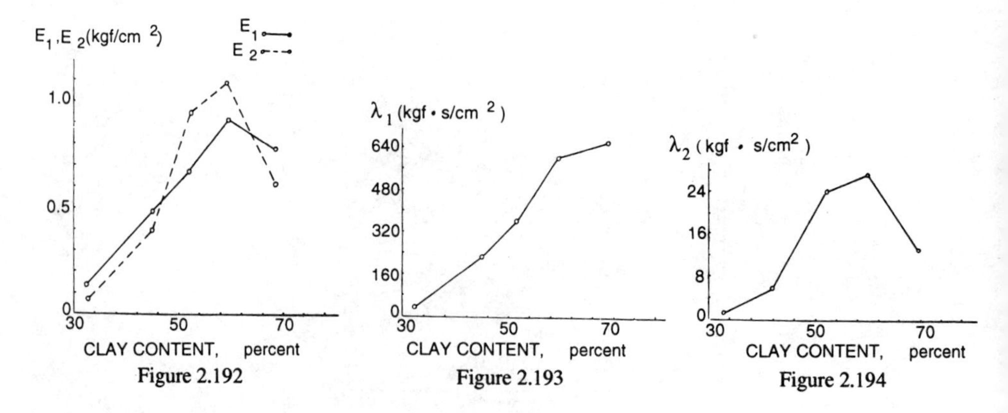

Figure 2.192

Figure 2.193

Figure 2.194

Figures 2.192 to 2.194–Burger model parameters, E_1 E_2, λ_1, and λ_2 as a function of clay content (abscissa) for five rice-land soils (from Lu et al, 1982).

It is for wet clay soils that the visco-elastic and viscoplastic relationships among stress, strain, and time most commonly find applicability. An illustrative diagram of a typical relationship (Pan et al, 1983) is provided in figure 2.195.

Sitkei et al (1992) compressed unsaturated soil (as well as other materials) axially in a confining cylinder at varying rates of deformation. A model using three spring-in-parallel-with-dashpot elements in series, all loaded through a single spring element, was found capable of representing the soil response characteristics in compression followed by either relaxation or creep. There was some scattering of response parameters — some of which was attributed to phenomena such as the differences of soil particle packing as a function of loading velocity — that caused end creep after sudden loading to be somewhat smaller than that with lower loading velocities.

Energy Adsorption, Storage, and Release Properties

When a volume of soil undergoes strain of any sort while subjected to stress, energy is transferred to or from the soil volume. Studies have been undertaken of the characteristics of the processes by which energy is transferred to or from a soil volume and of the forms in which energy absorbed by soil is converted to heat or stored for later release in mechanical form. Further, information about the amount of energy required, per unit volume of soil, to cause mechanical failure represents some sort of parameter linking stress and incremental strain, as well as provides some information about the

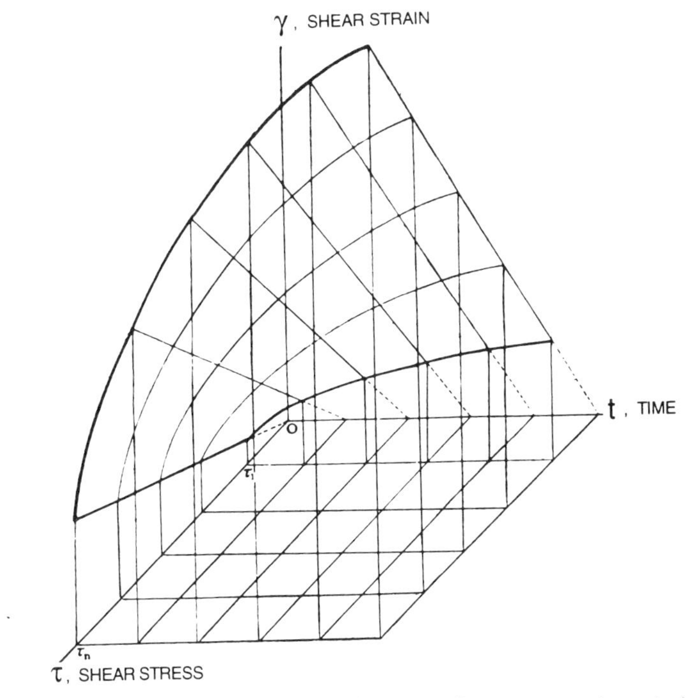

Figure 2.195–Hypothesized three-factor relationship among shear stress, τ, shear strain, γ, and time, t, for wet clay soils (imagine that the characteristic surface is a band shell opening up toward the viewer) (from Pan et al, 1983).

Table 2.24. Energy required per unit of soil surface increase

Soil texture	Soil moisture	Method and apparatus used to produce soil fracture	Energy consumed per unit new surface formed (N•m)/m²	Source*
Silty clay loam	14.5%	Blocks of 2.85 x 104 cm3 dropped 76 cm and allowed to shatter	7.38×10^2 (reduction to 2.54 cm m.w.d.)	1
Silty clay loam	14.5%	Small plow taking 2.54 cm cut	11.8×10^2 (3.73 cm m.w.d.)	1
Silty clay loam	28%	Soil block forced (slowly) over metal cutting grid	6.4 (2.54 cm m.w.d.)	2
Sandy loam	17.5%	Wire-cutting tool moved through a stationary soil ridge	1.96×10^2 (0.203 mm wire) 4.82×10^2 (1.04 mm wire) (slow speed only)	3
Clay	23.8%	Wire-cutting tool moved through a stationary soil ridge	1.99×10^2 (very slow speed)	4
Loam	11.4%	Wire-cutting tool moved through a stationary soil ridge	3.5×10^2 (very slow speed)	4
Sand	6.5%	Wire-cutting tool moved through a stationary soil ridge	1.92×10^2 (very slow speed)	4
Clay loam	21.0%	0.8-mm blade with 60° edge forced through soil	2.66×10^2 (very slow speed)	5
Sandy loam	69.1%	0.8-mm blade with 60° edge forced through soil	2.68×10^2	5
Clay	32.7%	0.8-mm blade with 60° edge forced through soil	8.57 x 10	5
Clay loam	21.0%	Unconfined compression of cylindrical sample	1.51×10^2†	5
Sandy loam	69.1%	Unconfined compression of cylindrical sample	7.7×10^2†	5
Clay	32.7%	Unconfined compression of cylindrical sample	1.98×10^2†	5

* 1. Gill and McCreery, 1960.
2. Bateman et al, 1965.
3. Hendrick and Buchele, 1963.
4. National Tillage Machinery Laboratory, 1961.
5. Kitani, 1965.

† Based on an assumed failure plane angle of 60° to base of samples.

linking stress and incremental strain, as well as provides some information about the process used to produce failure.

In cases for which the objective is to produce failure, energy inputs are frequently related to the amount of new surface produced upon failure (see sections on shatter resistance and on cutting resistance). Table 2.24 (Vomocil and Chancellor, 1969) presents some data on the energy requirements per unit new surface produced by failure.

Energy requirements per unit volume take the units of stress, or of the product of the units of stress times the units of strain:

$$\frac{\text{energy}\left(\text{N}\cdot\text{m}\right)}{\text{volume}\left(\text{m}^3\right)}=\frac{\text{force (N)}}{\text{area}\left(\text{m}^2\right)}\times\frac{\text{length (m)}}{\text{length (m)}} \tag{2.149}$$

Figure 2.83 and 2.84 show the effects of moisture content on the energy-per-unit volume required to produce failure by tension and unconfined compression, respectively. More moist soils had lower failure stresses and higher failure strains than drier soils, so there was not a particularly great change due to moisture content in energy-per-unit volume. Comparison of figure 2.196 (Vomocil and Chancellor, 1969) with figure 2.84 illustrates the over tenfold increase in energy-per-unit volume to cause failure when the soil material was supported by a confining stress of only 74 kPa as opposed to when

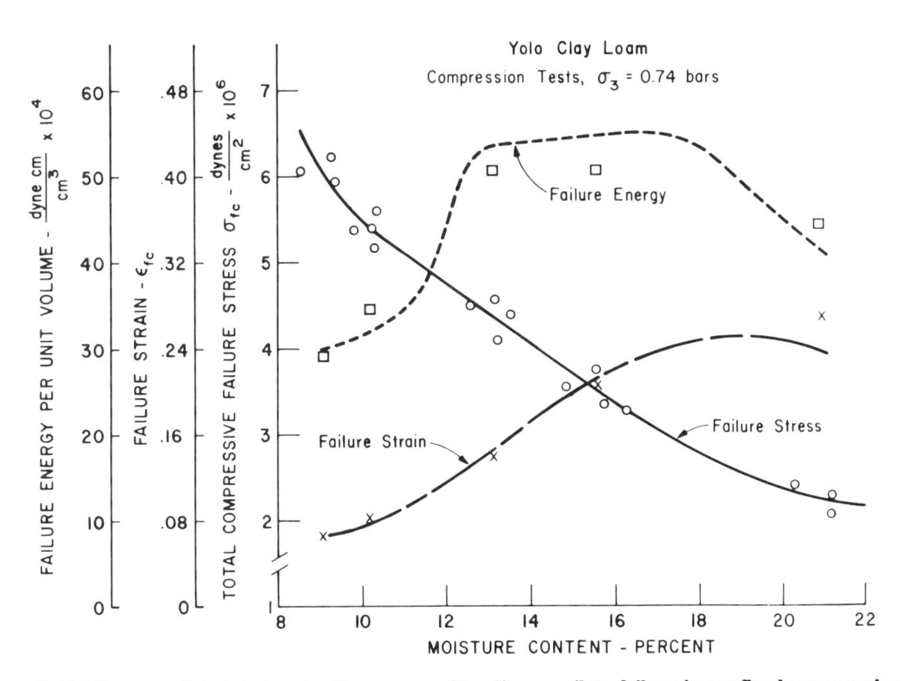

Figure 2.196–Energy-related factors in the process of loading a soil to failure in confined compression (from Vomocil and Chancellor, 1969).

unconfined compression was used. This was due to increases in both failure stress and failure strain. Figure 2.197 illustrates for a single soil the broad range of values of energy-per-unit volume to cause failure, that can prevail depending on the failure process used. Table 2.25 (Vomocil and Chancellor, 1969) presents some data on energy-per-unit volume required to produce failure of various soils using various loading procedures. It was found (Hendrick and VandenBerg, 1961) that increases in loading rate of monolithic soil samples resulted in significant decreases in the energy per unit volume to cause failure (fig 2.198). Soil density increases were also found to have a major effect, causing increases in energy-per-unit volume to produce failure (fig 2.199) (Panwar and Siemens, 1972).

When the volumes used in the computation of energy per unit volume are the incremental volume displacements resulting from incremental strains, the ratio of energy to volume is directly proportional to the stress causing the volume displacements. These considerations permit the analysis of various deformation processes for soil according to the disposition of energy in a given process (Chancellor and Korayem, 1965).

Energy Balance Formulations. Considerations will be restricted to the case of a three-dimensional, cylindrical soil sample in which energy is applied only in the direction of the major principal stress, σ_1. This direction is considered to be the axial direction of the cylindrical sample, as usually is the case in a triaxial cell. Both radial

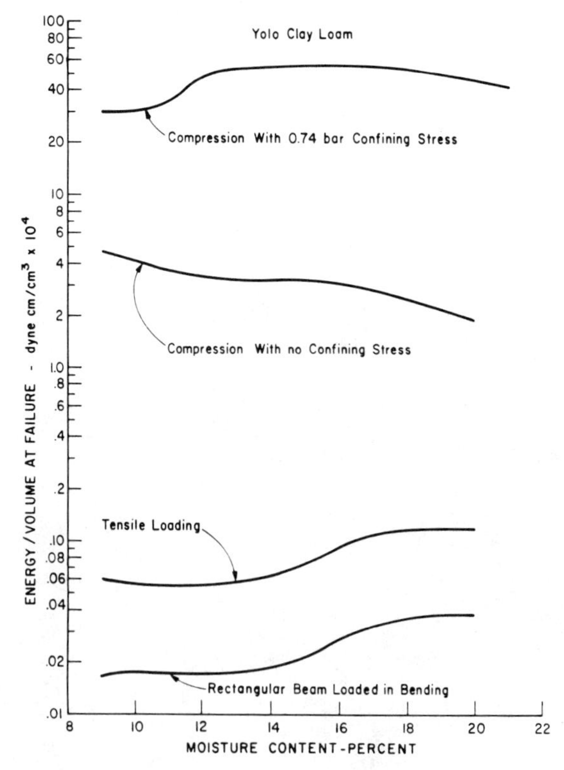

Figure 2.197–Energy per unit volume values supplied to cause soil sample failure using four different loading processes (from Vomocil and Chancellor, 1969).

dimensions are considered to sustain the minor principal stress σ_3 (fig 2.200) (Chancellor and Vomocil, 1985).

The basic three-stage procedure used to formulate the energy balance for a sample undergoing volume strain (fig 2.201) is as follows:

1. Volume strain is considered to take place isotropically.
2. Shear strain is considered to take place with the sample at constant volume until the original radial dimension (prior to the assumed isotropic volume strain) is reached. The energy involved in this portion of the process is the shear energy that must be added to the isotropic compression energy from the process in (step 1) to represent the case of cylindrical compression.
3. Additional shear strain (beyond that to restore the original radial dimension) is considered to take place with the sample at constant volume. Energy is thus

Table 2.25. Energy input per unit soil volume at failure

Soil Texture	Soil Moisture	Method and Apparatus used to produce Soil Fracture	Energy consumed per unit volume (N•m)/m^3	Source*
Clay	18, 23	Tensile loading of hourglass-shaped soil briquets	3×10^3 (18% MC) 9.5×10^2 (23% MC) (Slow speed only)	2
Clay loam	21.0	Tensile loading through adhesive connections	1.02×10^2	1
Sandy loam	69.1	Tensile loading through adhesive connections	7.66×10^2	1
Clay	32.7	Tensile loading through adhesive connections	1.99×10^2	1
Clay loam	21.0	Torsional loading of cylindrical sample (no axial load)	5.38×10^2	1
Sandy loam	69.1	Torsional loading of cylindrical sample (no axial load)	5.45×10^2	1
Clay	32.7	Torsional loading of cylindrical sample (no axial load)	2.14×10^2	1
Clay loam	21.0	Unconfined compression test of cylindrical sample	4.93×10^3	1
Sandy loam	69.1	Unconfined compression test of cylindrical sample	2.57×10^3	1
Clay	32.7	Unconfined compression test of cylindrical sample	6.50×10^3	1

* 1. Kitani, 1965.
2. Hendrick and VandenBerg, 1961.

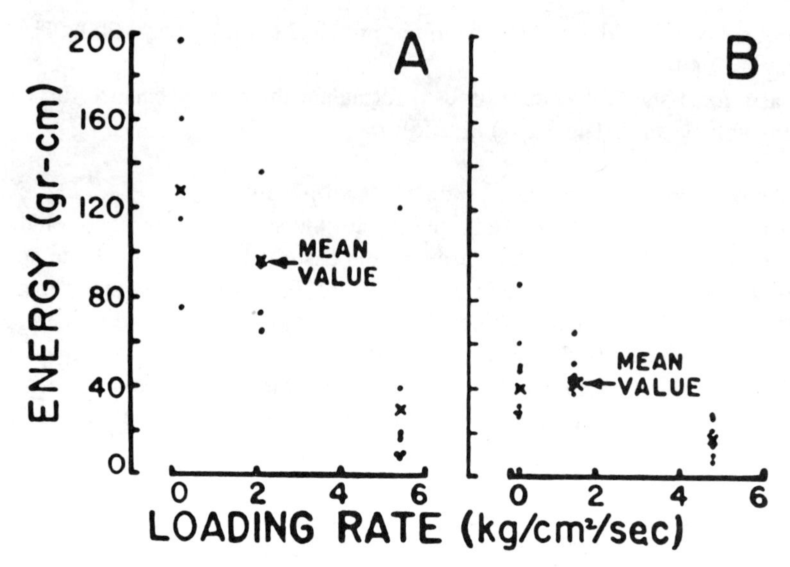

Figure 2.198–Energy inputs per unit volume to cause failure of monolithic soil samples loaded in tension as a function of the rate of stress application (from Hendrick and VandenBerg, 1961).

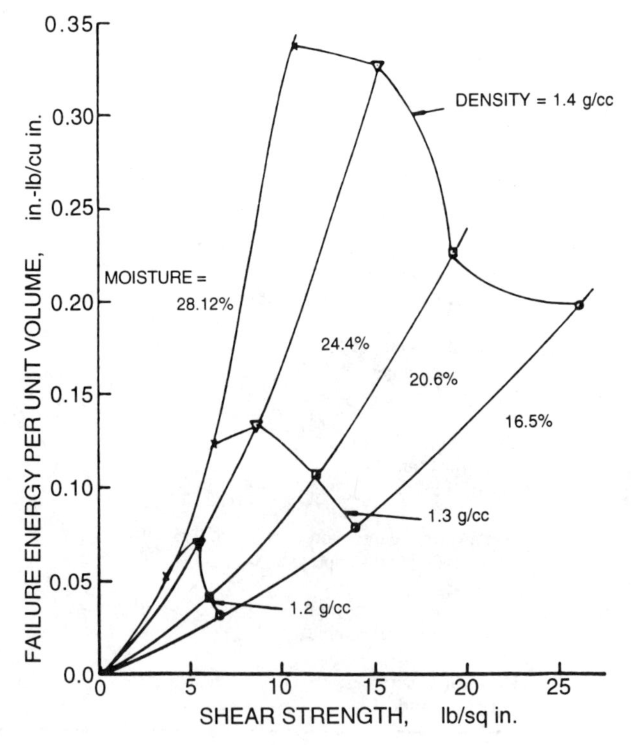

Figure 2.199–The effects of soil moisture and density on the energy per unit volume to cause failure with a shearing tool (from Panwar and Siemens, 1972).

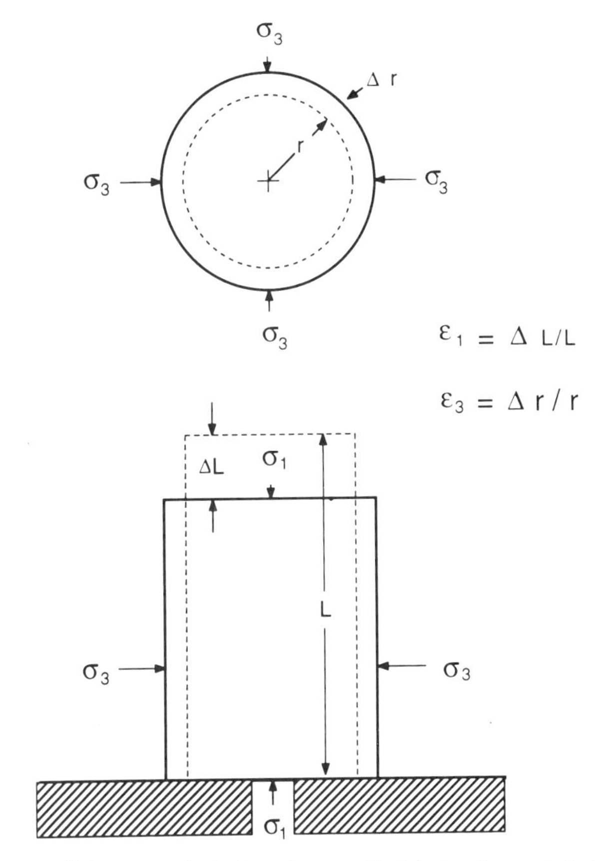

Figure 2.200–Diagram of stresses and strains when a cylindrical soil sample is loaded axially and supported radially (from Chancellor and Vomocil, 1985).

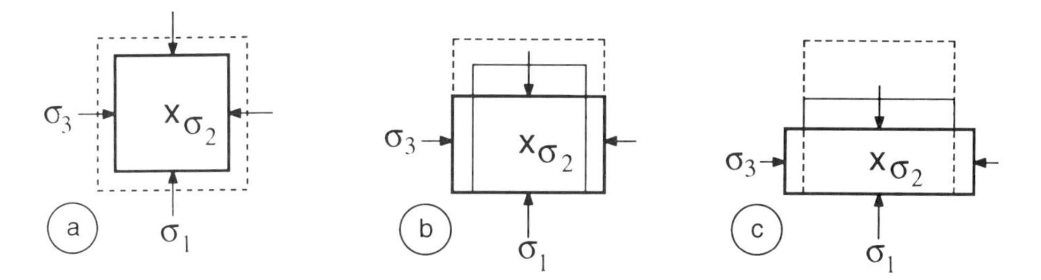

Figure 2.201–Diagram of the three stages of deformation into which a compression process is divided for energy analysis (from Soehne, 1965). The volume represented by the heavy-line box in sections a, b, and c is the same in each case. The process in section a is that of volume change alone. The change from a to b is that of shear strain to bring the lateral dimension of the sample back to its original value. The process in section c is additional shear strain.

3. Additional shear strain (beyond that to restore the original radial dimension) is considered to take place with the sample at constant volume. Energy is thus transmitted to the lateral surroundings of the sample volume.

In the following equations, increments of strain are considered to be very small, and values of stress are considered as mean values during the strain increment:

$$\text{Input energy} = \sigma_1 \, \Delta\varepsilon_1 \tag{2.150}$$

$$\text{Volume strain} = \Delta\varepsilon_1 + 2\Delta\varepsilon_3 \text{, in which}$$

compression considered as positive strain

$\Delta\varepsilon_1$ = increment of axial strain, and

$\Delta\varepsilon_3$ = increment of radial strain (2.151)

$$\text{Energy transmitted radially to sample surroundings} = -2\,\sigma_3\,\Delta\varepsilon_3 \tag{2.152}$$

$$\text{Volume that must be pumped to produce shear strain 2, above} = 2/3\,(\Delta\varepsilon_1 + 2\Delta\varepsilon_3) \tag{2.153}$$

$$\text{Volume that must be pumped to produce shear strain 3, above} = -2\Delta\varepsilon_3 \tag{2.154}$$

$$\text{Total volume of soil that must be pumped to produce shear strains 2 and 3, above} = \left[\frac{2}{3}\,(\Delta\varepsilon_1 + 2\Delta\varepsilon_3) - 2\Delta\varepsilon_3\right] \tag{2.155}$$

$$\text{Pressure drop through which volumes of soil must be pumped to produce shear strains 2 and 3 above} = \sigma_1 - \sigma_3 \tag{2.156}$$

Energy required for volume strain 1 then is the input energy minus the sum of the energy transmitted to the surroundings and that lost in the pumping procedure. The amount of this volume strain energy-per-unit volume strain is:

$$\frac{\sigma_1\,\Delta\varepsilon_1 + 2\sigma_3\,\Delta\varepsilon_3 - (\sigma_1 - \sigma_3)\left[\frac{2}{3}\,(\Delta\varepsilon_1 + 2\Delta\varepsilon_3) - 2\Delta\varepsilon_3\right]}{\Delta\varepsilon_1 + 2\Delta\varepsilon_3} \tag{2.157}$$

This reduces to:

$$\frac{\sigma_1}{3} + \frac{2\sigma_3}{3} = \sigma_{mean} = \sigma_m$$

and shows that this method of formulation is in accordance with the definition that the volume strain energy-per-unit volume strain is the mean stress.

The pressure drop $(\sigma_1 - \sigma_3)$ is twice the value conventionally called maximum shear stress (τ_{max}). With the change in maximum shearing strain, $\Delta\gamma_{max}$, defined as $\Delta\varepsilon_1 - \Delta\varepsilon_3$ the shear energy in the three-dimensional case considered here, is equal to:

$$\tau_{max} \times \frac{4}{3}(\Delta\varepsilon_1 - \Delta\varepsilon_3) = \frac{4}{3}\tau_{max}\,\Delta\gamma_{max} \tag{2.158}$$

In this case, in which all energy is applied in the direction of σ_1, the system that requires the least possible input energy-per-unit volume strain is that in which compression occurs axially with no radial strain (cylindrical compression). Under these circumstances, it is not possible (except with multiaxial compression) to produce volume strain without producing some shear strain. The sum of the energy for this shear strain (2), and the energy for volume strain (1) indicates the minimum amount of energy that must be supplied per unit volume strain.

$$\frac{\sigma_m(\Delta\varepsilon_1 + 2\Delta\varepsilon_3) + \frac{2}{3}(\sigma_1 - \sigma_3)(\Delta\varepsilon_1 + 2\Delta\varepsilon_3)}{(\Delta\varepsilon_1 + 2\Delta\varepsilon_3)} = \sigma_1 \tag{2.159}$$

Although there is no alternative relationship for the constant radius cylindrical case, this amount of energy is the minimum that must be supplied in all cases where items (1) and (2) above necessarily are linked (by the requirement, in this case, for all energy input to be in the σ_1 direction). However, item (3) above is not considered linked to item (1); therefore, even when radial expansion does take place the amount of energy associated with each unit of volume strain remains at σ_1.

The proportioning of input energy may be done as follows:

Volume strain energy/input energy =

$$\frac{\sigma_m(\Delta\varepsilon_1 + 2\Delta\varepsilon_3)}{(\sigma_1\,\Delta\varepsilon_1)} = \left(1 + 2\frac{\sigma_3}{\sigma_1}\right)\left(1 - \frac{2}{3}\frac{\Delta\gamma_{max}}{\Delta\varepsilon_1}\right) \tag{2.160}$$

Energy transmitted to surroundings/input energy =

$$\frac{-2\sigma_3\,\Delta\varepsilon_3}{\sigma_1\,\Delta\varepsilon_1} = \frac{2\sigma_3}{\sigma_1}\left(\frac{\Delta\gamma_{maax}}{\Delta\varepsilon_1} - 1\right) \tag{2.161}$$

Shear energy/input energy =

$$\frac{(\sigma_1 - \sigma_3)\left[\frac{2}{3}(\Delta\varepsilon_1 + 2\Delta\varepsilon_3) - 2\Delta\varepsilon_3\right]}{\sigma_1\,\Delta\varepsilon_1} = \frac{2}{3}\left(1 - \frac{\sigma_3}{\sigma_1}\right)\frac{\Delta\gamma_{max}}{\Delta\varepsilon_1} \tag{2.162}$$

In the above three equations, the proportioning of input energy is related to two parameters, (σ_3/σ_1) and $(\Delta\gamma_{max}/\Delta\varepsilon_1)$. The latter of these may be interpreted as follows (See fig 2.202):

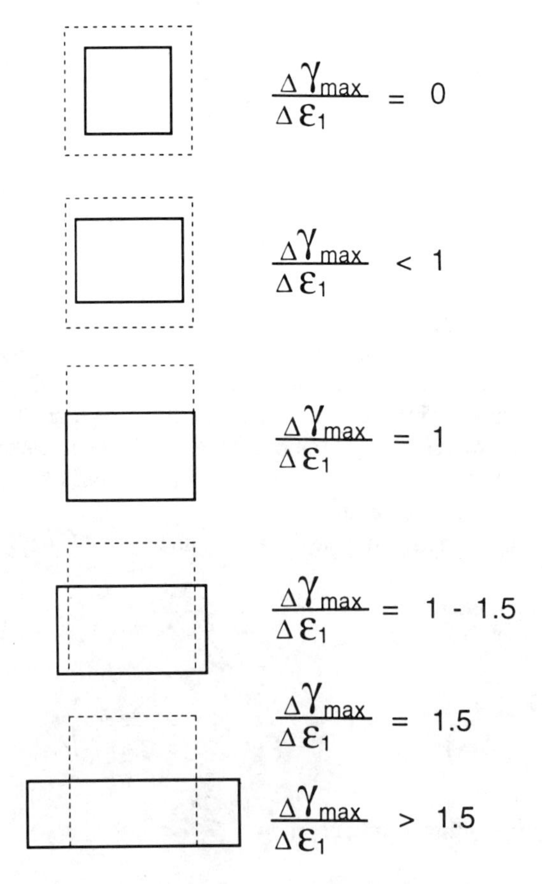

Figure 2.202–Five patterns of deformation that a soil sample may follow at any given stage when loaded axially in compression (from Soehne, 1965).

1. When $(\Delta\gamma_{max}/\Delta\varepsilon_1)$ is less than 1.0, energy is being put into the soil in all directions.
2. When $(\Delta\gamma_{max}/\Delta\varepsilon_1) = 1.0$, the soil is undergoing volume strain while being compressed in the direction of σ_1 with $\varepsilon_2 = \varepsilon_3 = 0.0$.
3. When $(\Delta\gamma_{max}/\Delta\varepsilon_1)$ is between 1.0 and 1.5, the soil is undergoing volume strain while being compressed in the direction of σ_1 and being expanded in the σ_1 and σ_3 dimensions.
4. When $(\Delta\gamma_{max}/\Delta\varepsilon_1) = 1.5$, the soil does not undergo volume strain but expands in the σ_2 and σ_3 dimensions while being compressed in the direction of σ_1.
5. When $(\Delta\gamma_{max}/\Delta\varepsilon_1)$ is greater than 1.5, the soil volume is increasing while being compressed in the direction of σ_1.

In some cases, this sort of energy accounting was used to analyze the results of triaxial tests with sands in such a way as to partition the sample strength into two components. One component was related to the coefficient of internal friction and another to the angularity of interparticle contacts, which would require sample dilation prior to failure (Rowe et al, 1964).

When short triaxial samples were prevented from undergoing catastrophic failure because of the failure planes intersecting the sample bases, stresses and strains for two very different soils over a broad range of moisture, density, and confining stress conditions could be represented by the characteristics illustrated in figures 2.131, 2.132, 2.133, and 2.154. The partitioning of energy during the axial compression process could

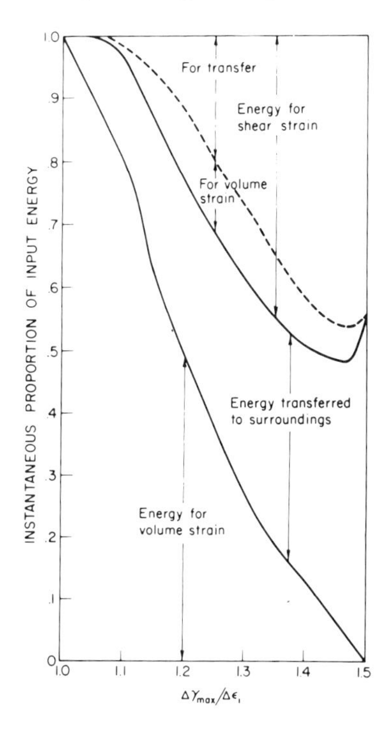

Figure 2.203–Energy disposition as a short, wide soil sample was loaded axially and plotted according to the deformation parameter $\Delta\gamma_{max}/\Delta\varepsilon_1$ = (Poisson's ratio + 1.0) (from Chancellor and Korayem, 1965).

be generalized for all samples and is presented in figures 2.203 and 2.204 (Chancellor and Korayem, 1965).

Cylindrical, monolithic soil samples (length twice diameter) of varying moisture content and texture were subjected to axial loading, with and without lateral confining stress. Stress strain characteristics of these samples are illustrated in figure 2.128. The energy-related responses of a moist soil sample are shown in figures 2.205 and 2.206. For some samples, energy continued to be absorbed in volume strain until the state of failure was reached (fig 2.207), while for other samples, energy absorption in volume strain stopped well before failure occurred (fig 2.130). In some samples (fig 2.208) not only was there a cessation of volume strain energy input, but prior to failure the sample underwent dilation and released large portions of volume strain energy to the surroundings of the sample in the form of lateral expansion.

It was found that with confined compression, lower proportions of the input energy were absorbed in shear strain and higher proportions of energy were absorbed in volume strain than was the case for unconfined compression (Aref et al, 1975). When confined compression samples were loaded to subfailure stress states and the applied stress levels were then relaxed, much of the energy absorbed in volume strain was released in the form of volumetric expansion (fig 2.209). This proportion of recovered energy ranged from 40 to 85%, increasing with increasing moisture content.

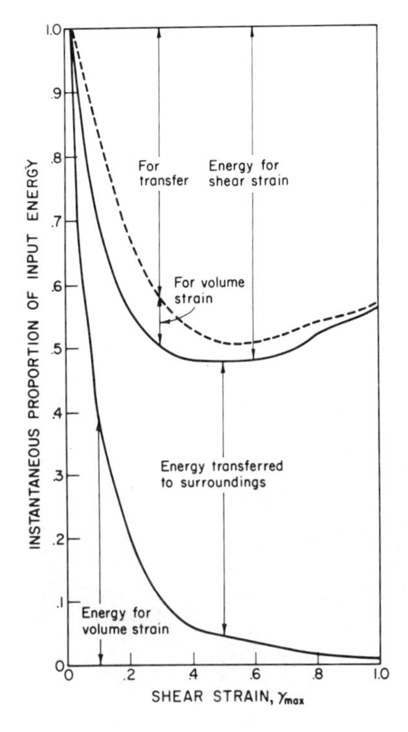

Figure 2.204–Energy disposition as a short, wide soil sample was loaded axially and plotted according to the level of shear strain sustained (from Chancellor and Korayem, 1965).

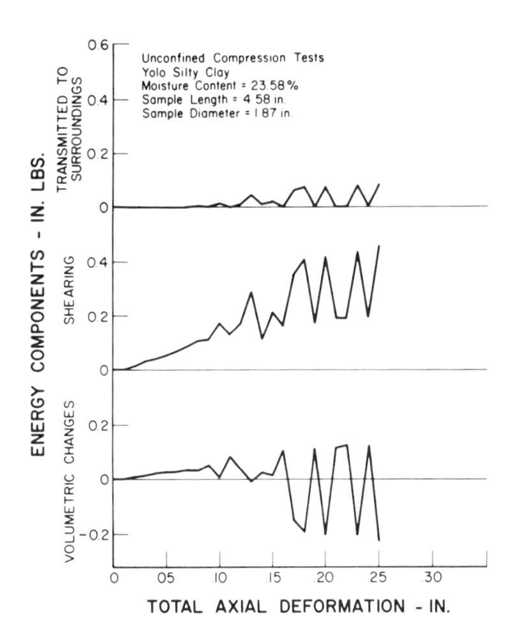

Figure 2.205–Unsmoothed energy analysis response curves as long monolithic soil samples were loaded axially to failure. Note the step-by-step shifts between shearing and volumetric energy absorption (from Chancellor et al, 1969).

The use of energy accounting relative to the division of energy absorption in shear strain, volumetric strain, and transmission to surroundings was a key element in a system to anticipate volumetric and shear strains in a soil matrix beneath surface loads (Chancellor, 1966) (see section on Surface Load-Stress Distribution).

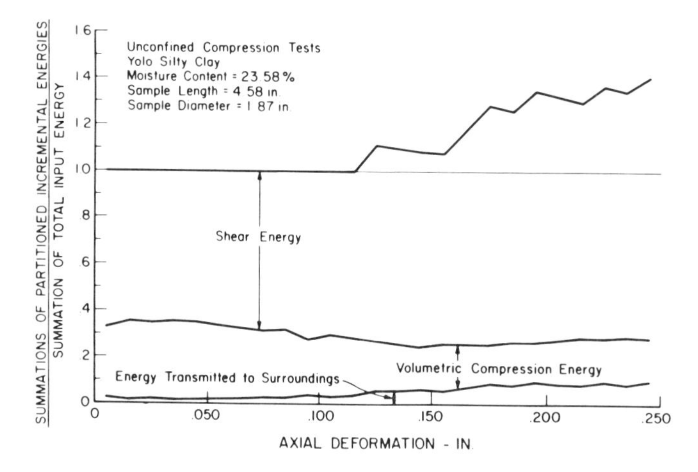

Figure 2.206–Cumulative energy absorption characteristics as monolithic soil samples were loaded axially to failure (from Chancellor et al, 1969).

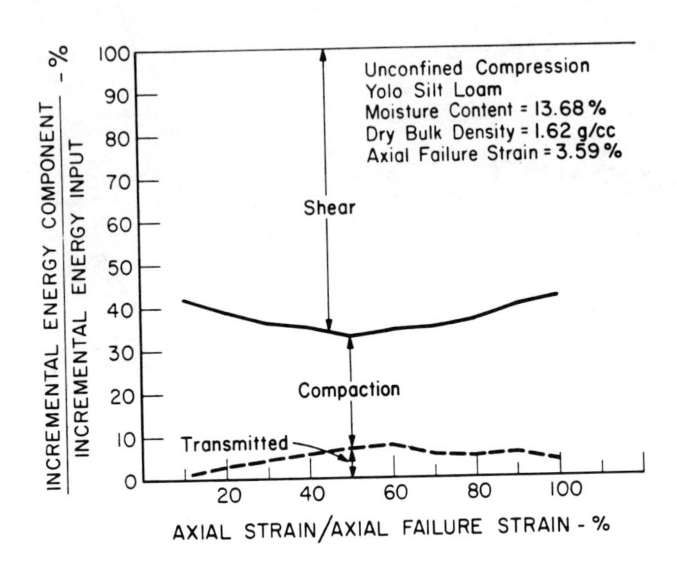

Figure 2.207–Energy component allocation as a long monolithic soil sample was loaded axially to failure. Note, that energy was being directed to volume strain throughout the entire process (from Chancellor et al, 1969).

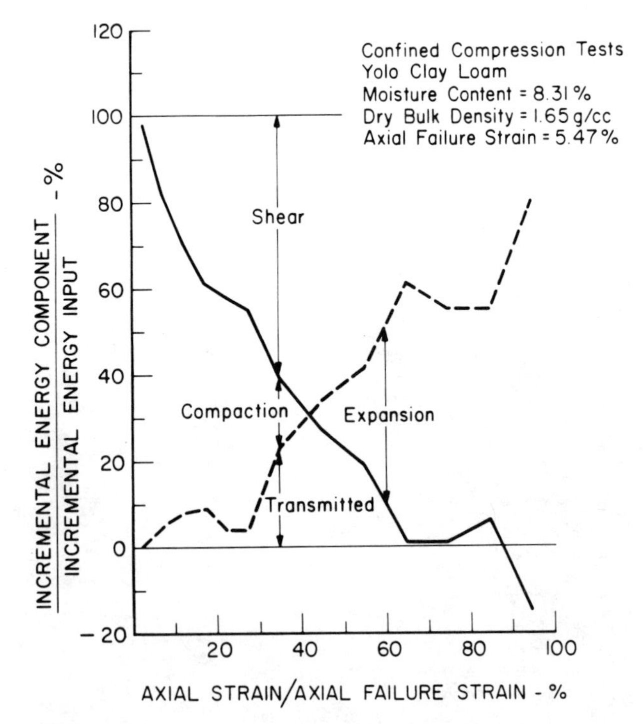

Figure 2.208–Energy disposition for a long monolithic soil sample loaded axially to failure. Note that for this dry sample, not only did the sample stop absorbing energy in volume strain well prior to failure, but that volumetric expansion energy was released as failure was approached (from Chancellor et al, 1969).

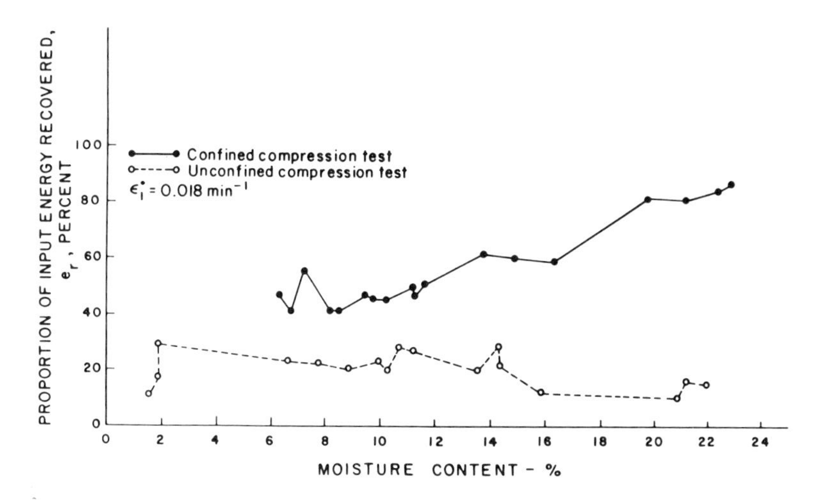

Figure 2.209–The proportion of compression energy returned upon stress release from a sub-failure level load. Confined samples returned a much higher proportion than unconfined samples (from Aref et al, 1975).

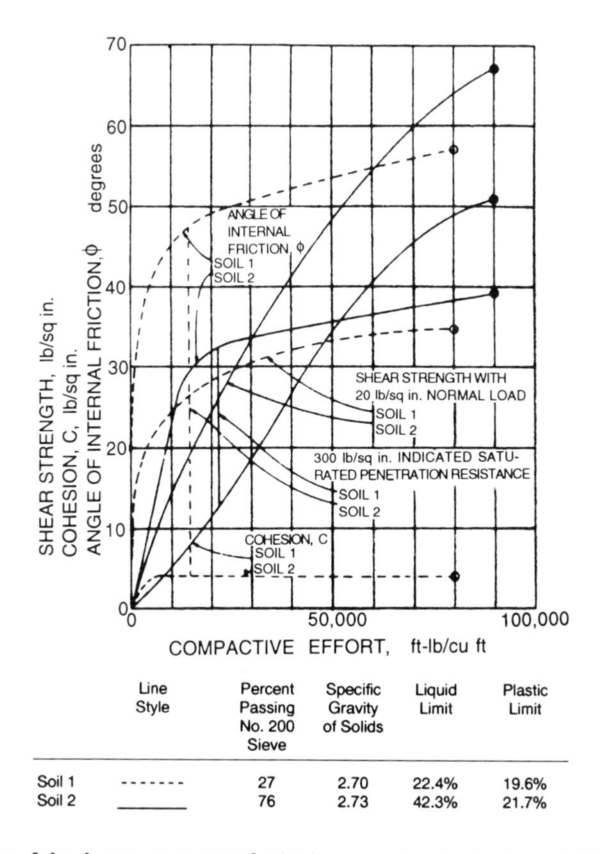

	Line Style	Percent Passing No. 200 Sieve	Specific Gravity of Solids	Liquid Limit	Plastic Limit
Soil 1	- - - - - - -	27	2.70	22.4%	19.6%
Soil 2	———	76	2.73	42.3%	21.7%

Figure 2.210–Values of the shear parameters of cohesion, c, and angle of internal friction, ϕ, as affected by the amount of compactive energy put into a given volume of an unsaturated soil (from Proctor, 1948).

In some cases, various soil properties can be related to the volumetric strain energy input to a soil of a given moisture content during a compaction process (fig 2.210) (Proctor, 1948).

Surface Load-Stress Distribution Characteristics

Two general approaches — definable within a system of mechanics — have been employed for predicting the stress that will be manifested within a soil half-space as a result of surface-imposed loads. These are the use of the finite-element method and the use of the classical Boussinesq and Cerruti equations for isotropic, elastic, and homogeneous materials. For each of these, data for different soil physical properties are required.

When the finite-element method, based on linear elastic behavior, is used, the two properties of the material that must be supplied are Poisson's ratio, μ, and the modulus of elasticity, E (Raper and Erbach, 1988a and 1988b; Duncan and Chang, 1970; Yong and Fattah, 1976) if the soil material is assumed to be elastic, isotropic and linear. In contrast with materials like steel, neither of these properties can be represented by a constant for soil, and both vary with changes in the degree and stage of loading, as well as with changes in moisture content, clay content, porosity, and other material parameters (see foregoing sections on Poisson's ratio and on the modulus of elasticity). Nevertheless, the use of iterative computer methods and associated techniques for working with materials with nonlinear properties can be employed (Duncan and Chang, 1970; Yong and Fattah, 1976). Generally some criteria are established for using various values of μ and E for each elemental calculation, usually in the form of the tangent modulus (see Duncan, 1980). Frequently, however, cases investigated are confined to conditions in which

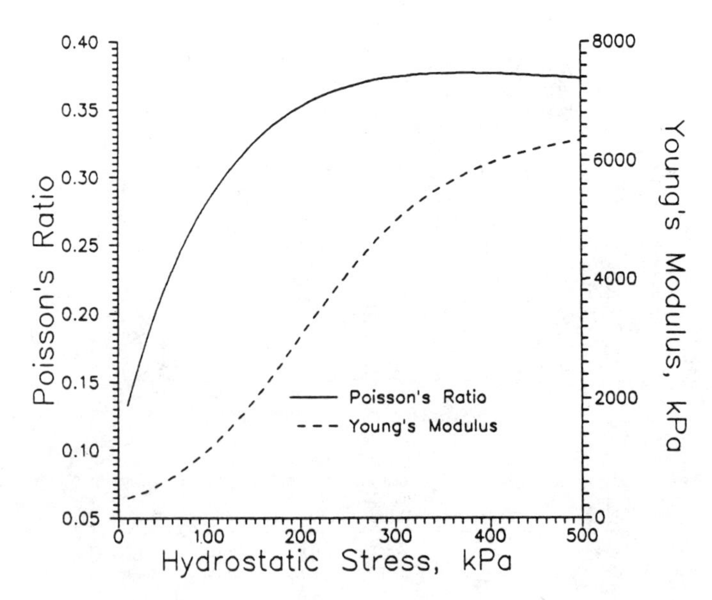

Figure 2.211–Range of values of Poisson's ratio and Young's modulus incorporated in a finite-element model of stress distribution and of volumetric strain under a loaded area on the soil surface (from Raper and Erbach, 1988a).

Poisson's ratio is not expected to change appreciably, and a single value is used for this parameter. In such cases it is implied that either the bulk modulus or shear modulus will change with changes in the level of the mean normal stress (Wood, 1990).

An investigation of the effects of making an incorrect assumption for Poisson's ratio (Raper and Erbach, 1988b) showed that predictions of vertical stress under a surface load similar to that which might be imposed by a tractor drive wheel, varying Poisson's ratio from 0.13 to 0.38 caused only about a 9% increase in stress at the more shallow depth and an 11% increase at a deeper depth over a broad range of possible E values.

Variation of E values from 506 to 5236 kPa resulted in even smaller changes in predicted vertical stresses over a range of μ values. Thus, there appeared to be some justification for the assumption of Poisson's ratio as a constant in the prediction of stress distributions within the soil. However, it was found that predicted vertical stresses for which constant values of μ and E were used were generally less than half the values actually measured in a soil bin. Incorporation of varying values of μ and E (fig 2.211) into the finite-element method calculations resulted in predicted stresses that were reasonably close to those measured in the soil bin (Raper and Erbach, 1988a).

When the finite-element method was extended to predict volume strains within the soil, it was found that these strains were sensitive to the values of μ used and particularly sensitive to the values of E used (Raper and Erbach, 1988b), indicating that the use of methods developed for linear elastic materials may not be satisfactory for use with soils.

The use of the Boussinesq equations for predicting soil stresses under tractor wheels (Soehne, 1958) has been accompanied by concerns for the fact that these equations were intended for use with homogeneous, isotropic, elastic materials, ie, materials having characteristics that might not apply to agricultural soils. To meet these concerns, the use of a soil-dependent "concentration factor" was considered for incorporation in the equations.

The basic equations are:

$$\sigma_r = \frac{3Q}{2\pi r^2} \cos\theta \tag{2.163}$$

and

$$\sigma_z = \frac{3Q}{2\pi r^2} \cos^3\theta \tag{2.164}$$

in which

σ_r = radial stress (major principal stress) (fig 2.212)
Q = vertical point load on the soil surface
r = radius from the point where the load, Q, intersects the surface to the point at which the stress, σ_r, is manifested
θ = angle between the vertical and the radius, r
σ_z = vertical stress on the volume element at radius, r (figs 2.212 and 2.213)

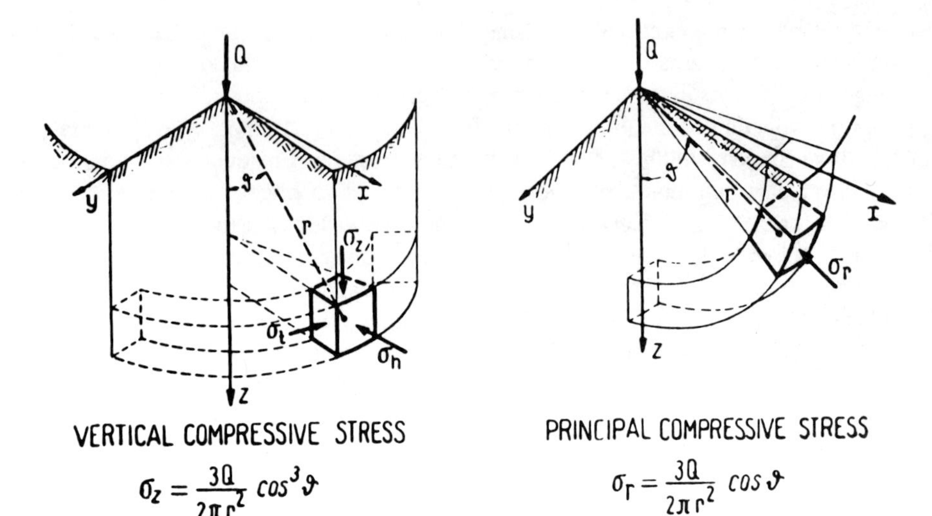

Figure 2.212–Principles of stress distribution from a point load on the surface according to the Boussinesq formulas (from Soehne, 1958).

Incorporation of the "concentration factor" as suggested by Froehlich (1934) caused the equations to take the following forms:

$$\sigma_r = \frac{\nu Q}{2\pi r^2} \cos^{\nu-2} \theta \tag{2.165}$$

and

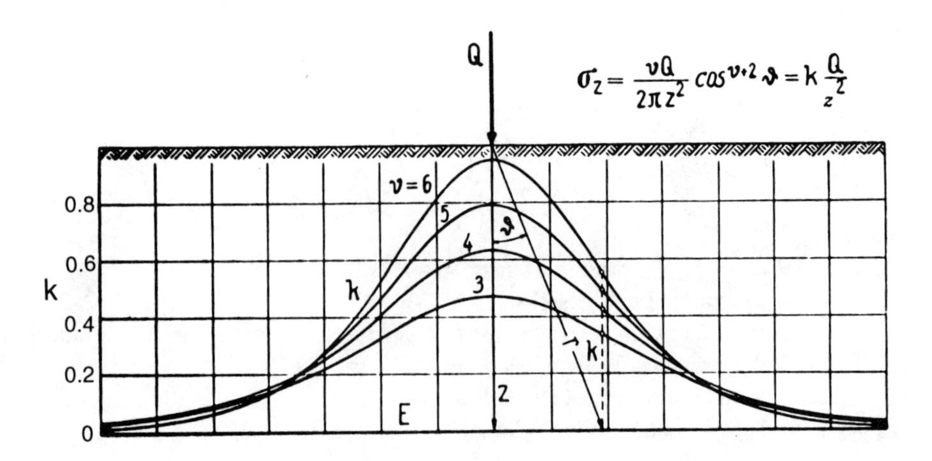

Figure 2.213–Vertical stress, σ_z, distribution at a given depth, z, with various values (ranging from 3 to 6) of ν, the concentration factor, applied to the Boussinesq formula (from Soehne, 1958).

$$\sigma_z = \frac{\nu Q}{2\pi r^2} \cos^\nu \theta \qquad (2.166)$$

in which ν = the concentration factor, the value of which would range from 3 to 6.

The purpose of the concentration factor was to cause the resultant stress distribution to be more concentrated near the center line of the load axis as a result of the very compliant coupling of the soil material on the center line to the material at some radial distance from the center line. Thus, the higher values of the concentration factor were to be associated with less rigid, or more compliant soil conditions. The general basis for selecting values of the concentration factor (Gameda et al, 1984) is as follows:

ν = 3 for isotropic materials that obey Hooke's law
ν = 4 for hard, dry soils subject to only elastic deformations
ν = 5 for agricultural soils with normal moisture contents and densities
ν = 6 for very wet, yielding soils

The result of various values of ν on the vertical stress at a given depth is illustrated in figure 2.214 (Soehne, 1958).

The concentration factor of 3 causes the equations to have the original form given by Boussinesq. In some cases the use the original Boussinesq form (ν = 3) in combination with a relationship between soil porosity and the major principal stress have given reasonable approximations of soil compaction under surface loads in experimental tests (Chancellor, 1966; Gupta et al, 1985b).

When correlations were made between field soil density changes upon surface loading, predicted using various values of the concentration factor and measured density values (Gameda et al, 1984), it was found that there was generally good correlation

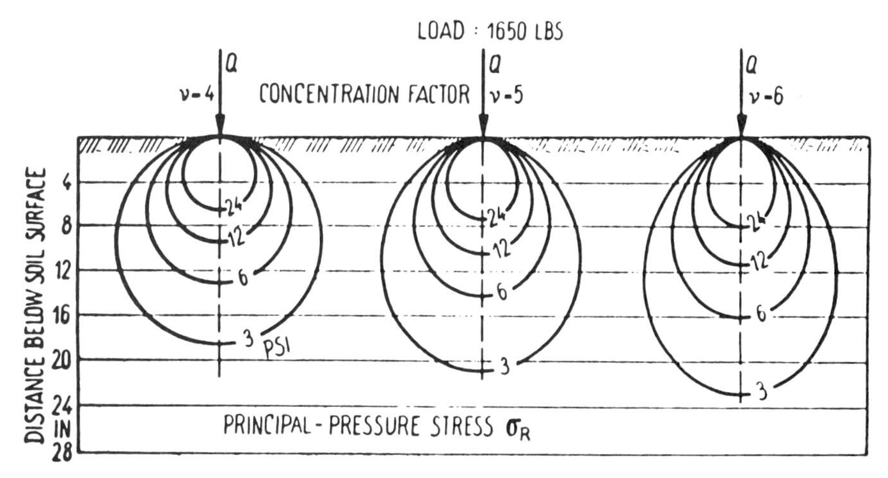

Figure 2.214–The effects of various concentration factor, ν, levels on distribution of the major principal stress, σ_r, within the soil half-space (from Sohne, 1958).

irrespective of the concentration factor used, except in the case of wet clay soil, in which ν values of 5 and 6 produced the best correlations. Generally, computed values of vertical stress did not correlate well with cone penetrometer readings irrespective of the values of ν used in making the computations.

The effect of using various values of the concentration factor on the depth of soil affected by compaction under a pneumatic tire (Jakobsen and Dexter, 1989) is illustrated in figure 2.215, indicating that higher values of ν result in compaction extending to greater depths.

When pressures were measured in the soil using deformable spherical transducers while varying loads were applied by a tractor tire (Blackwell and Soane, 1978), it was found that for accurate prediction of soil stress, concentration factors below ν = 4 were appropriate at depths greater than 30 cm, while predictions made using ν values of 6 were comparatively good at a depth of 10 cm (fig 2.216).*

In another investigation in which wafer-shaped stress transducers were used to measure the vertical stress, σ_z'', stress values of σ_z' were calculated using the Boussinesq equations and parameters for three different track-type tractors which

* Cone index values increased from 1100 kPa at 10 cm depth to 6000 kPa at 40 cm depth.

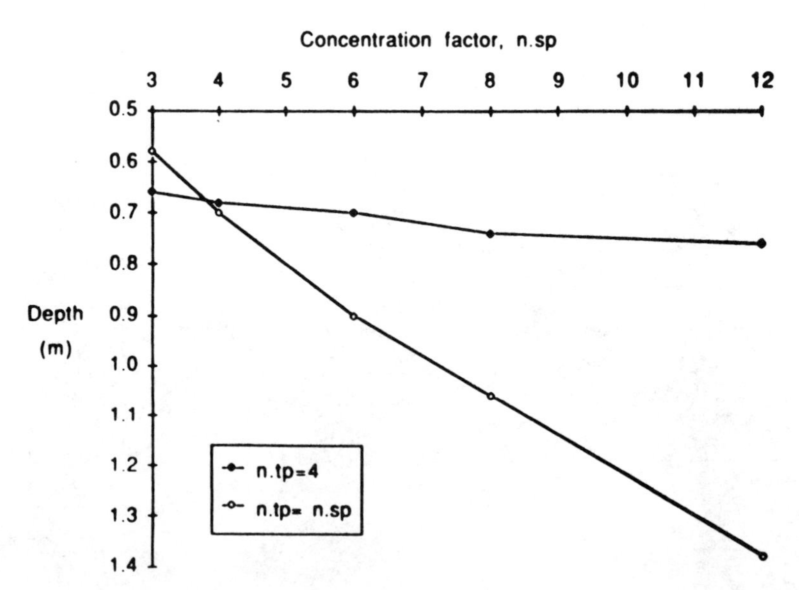

Figure 2.215–Depth of soil affected by compaction, depending on the value of the concentration factor, ν, chosen. The term "tp" applies to uniformly distributed surface pressure, while "sp" refers to a parabolic surface pressure distribution. Reprinted with permission of *Journal of Terramechanics* 26(2), Jakobsen, B. F. and A. R. Dexter, Prediction of soil compaction under pneumatic tires. Copyright © 1989, Pergamon Press, Ltd.

traversed the soil (Smirnov and Gorbunov, 1966). In this case the value of the concentration factor, ν, was computed from:

$$\nu = 3 \frac{\sigma_z''}{\sigma_z'} \tag{2.167}$$

The results are shown in table 2.26.

A similar method of evaluating the concentration factor, ν, was used by Bolling (1987) while operating tires under varying conditions on a sandy loam field soil.

The data in table 2.27 indicates that more dense soils are associated with lower concentration factor values. Increased speeds of load traverses also appeared to be linked to lower values of the concentration factor.

When a concentration factor of ν = 6 was used in combination with data about the pressures exerted on the surface by wide and narrow drive wheel tires, both carrying the

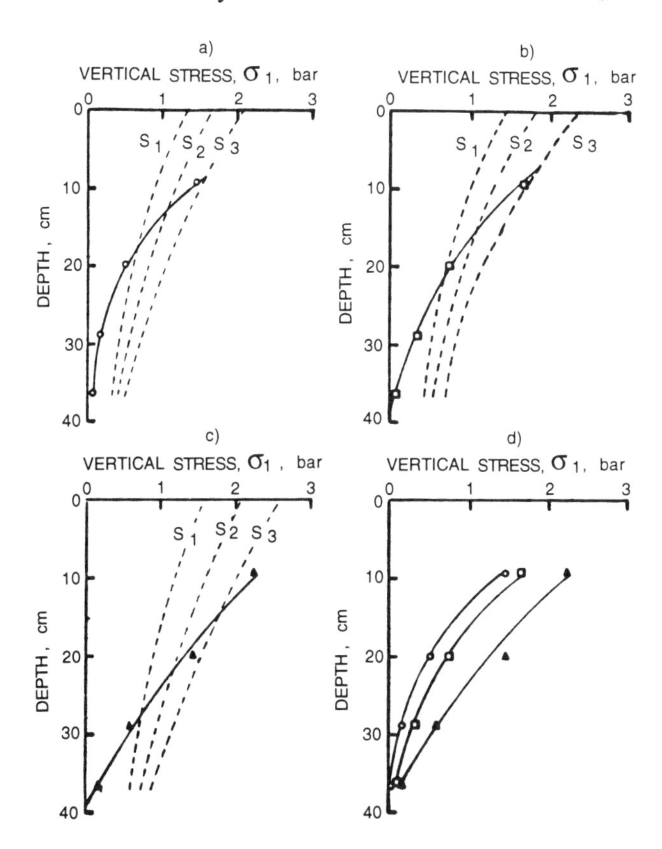

Figure 2.216–Comparison of observed (solid lines) and computed (dashed lines) stresses under a wheel with three different loads. Computed values were obtained using the Boussinesq equation with concentration factors, ν, of 4 (S1), 5 (S2) and 6 (S3). Reprinted with permission of *Journal of Terramechanics* 15(4), Blackwell, P. S. and B. D. Soane. Deformable spherical devices to measure stresses within field soils. Copyright © 1978, Pergamon Press, Ltd.

Table 2.26. Measured values for the concentration factor, ν*

Tractor Designation	Track Width (mm)	Average Ground Pressure (bars)	Values† of the Concentration Factor, ν at Depth of				
			10 cm	20 cm	30 cm	40 cm	Mean
DT-54A	390	0.46	3.2	3.9	5.4	2.0	3.8
DT-55A	533	0.26	4.2	4.7	5.8	2.1	4.2
S-100	500	0.50	3	4.6	7.5	2.5	4.4
Average			3.5	4.3	6.2	2.2	4.1

* Data from Smirnov and Gorbunov (1966).

† Moisture contents ranged from 26% at 10 cm depth to 18% at 40 cm depth with dry bulk density increasing from 1.55 g/cm^3 at 10 cm to 1.85 g/cm^3 at 40 cm depth.

same load (Johnson and Burt, 1986) to compute soil pressures, values of peak vertical stress obtained were approximately half of those measured in a soil bin with corresponding tire, soil, and load conditions.

It should be realized, however, that irrespective of what peak stress levels may be predicted or measured, the integral of vertical stresses over a horizontal plane at any depth in the soil should total the applied surface load (plus the soil weight). Thus, the effect of varying values of the concentration factor, ν, only applies to how that total vertical load will be distributed (in terms of stress levels and the areas over which any stress level will prevail) and not to the total load that will be supported by increased vertical stress at any given level in the soil.

Empirical Dynamic Properties

In many cases the dynamic properties of soil that are definable within a system of stress-strain mechanics are complex to measure and difficult to use because of their tendencies to change during various dynamic processes of tillage and traction. For

Table 2.27. Measured values for the concentration factor, ν*

Forward Speed (km/h)	Soil Density g/cm^3	Tire Size	Vertical Load (kN)	Contact Area (cm^2)	Concentration Factor, ν
	1.5	16.9–R30	15	1474	5.5
	1.5	16.9–30	12	1620	5.5
	1.5	16.9–30	12	1603	5.5
	1.7	12.4–32	8.5	517	3.15
	1.7	18.4–38	16.3	1006	3.15
	1.7	23.1–26	34.5	1811	3.15
2	1.28	9.5–36	8.2	455	4.7
6	1.28	9.5–36	8.2	463	3.9
10	1.28	9.5–36	8.2	541	3.5

* Data from Bolling (1987).

engineering design and analysis purposes empirical versions of dynamic processes of elemental form have been defined. Soil response characteristics as they interact with these processes have been categorized, and soil parameters have been generalized according to the specific processes employed.

Shear Displacement – Shear Stress Characteristics

When a normally loaded grouser plate is placed on the soil surface and loaded horizontally until it begins to move (fig 2.217) (Pavlics, 1958), stress-deformation characteristics of one of two general types tend to appear: shear stress asymptotic to the maximum shear stress as deformation continues, and shear stress reaching a maximum value at a certain level of deformation and then dropping to a lower and nearly constant level as deformation continues (fig 2.218) (Yong et al, 1984).

The second type of response characteristics are more difficult to represent. Two different methods have been proposed for defining soil parameters to describe this response. Both require the definition of three soil-related values in addition to the determination of $\tau_{max} = c + \sigma_n \tan \phi$.

In one case (Pavlics, 1958) the amount of grouser plate displacement when the maximum value of shear stress, τ_{max}, is obtained, is designated as d_{opt}. With the use of these two parameters to normalize the data from the grouser plate test, the data are plotted on the diagram shown in figure 2.219. The three soil parameters needed are d_{opt}, K_1, and K_2. The soil response is represented by:

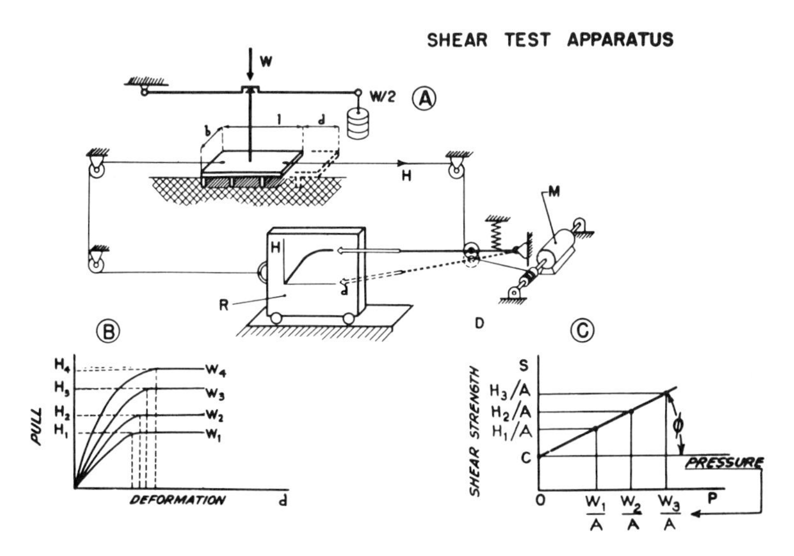

Figure 2.217–Schematic diagram of an apparatus to measure the shear-stress vs horizontal displacement characteristics on a soil surface (from Pavlics, 1958).

$$\frac{\tau}{\tau_{max}} = \left\{ \frac{\left[\exp\left(-K2 + \sqrt{K_2^2 - 1}\right) K_1 d - \exp\left(-K_2 - \sqrt{K_2^2 - 1}\right) K_1 d\right]}{\left[\exp\left(-K_2 + \sqrt{K_2^2 - 1}\right) K_1 d_{opt} - \exp\left(-K_2 - \sqrt{K_2^2 - 1}\right) K_1 d_{opt}\right]} \right\} \quad (2.168)$$

The parameter K_2 can be obtained by determining which of the precomputed curves in figure 2.219 is best fitted by the plot of grouser-plate shear stress-horizontal deformation data. The K_1 value can then be obtained by referring to the K_2 scale at the top of figure 2.219, tracing vertically downward to intersect the curve representing K_2 vs K_1 d_{opt}. By referring to the right hand scale of K_1 d_{opt}, the value of K_1 can be determined by division of K_1 d_{opt} by d_{opt}.

Another system for representing the humped stress-deformation characteristics (Yong et al, 1984) is by use of the equation:

$$\frac{\tau}{\sigma} = fm\left[1 + \frac{a}{\cosh\left(\frac{\delta}{k_\tau}\right)}\right] \tan h\left(\frac{\delta}{k_\tau}\right) \quad (2.169)$$

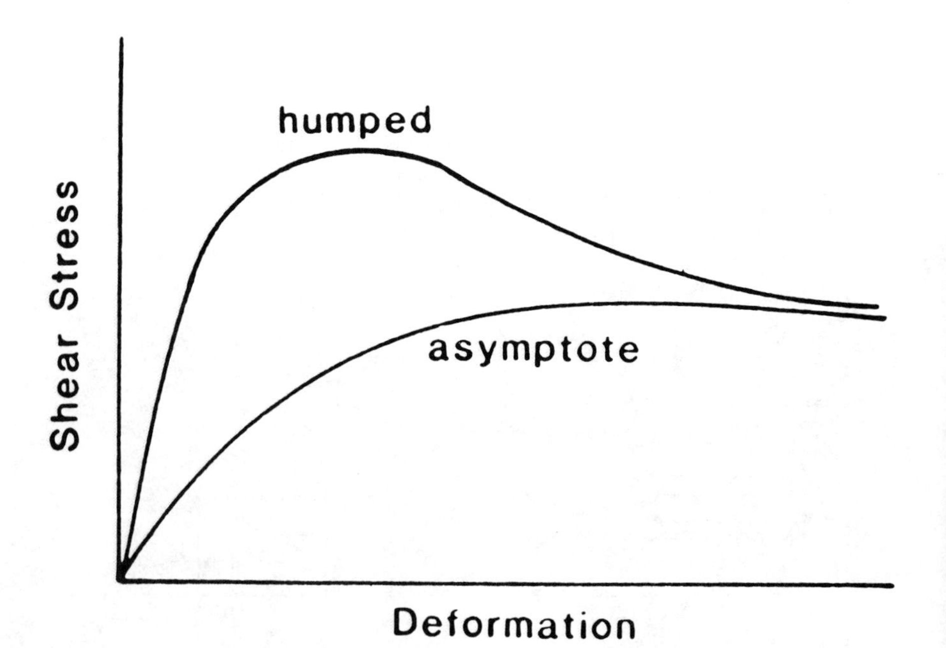

Figure 2.218–Two typical sorts of response characteristics displayed by shear-stress vs horizontal displacement test results (from Yong et al, 1984).

in which τ = shear stress, σ = normal stress, δ = shear deformation, and k_τ = displacement when τ is maximum. This function is illustrated in figure 2.220. The three soil parameters needed to describe soil response are:

1. The ratio, fm, of the residual shear strength to the contact stress, σ, at large displacements.*
2. The displacement, $k_{\tau,}$ required to reach the peak shear stress.
3. The constant, a, which depends on the ratio of residual to peak strength. The magnitude of a is given by:

$$a = 2.55 \left(\frac{fs - fm}{fm}\right)^{0.825} \tag{2.170}$$

Of the two methods used to define the shear stress-horizontal deformation characteristics of soil in the asymptotic form, one (Taylor and VandenBerg, 1966) uses two parameters, T and n, while the other (Janosi, 1962; Wills, 1963) uses a single parameter, k.

For the two-parameter case (Taylor and VandenBerg, 1966), the equation used to describe the process is:

* The definitions of fm and fs are illustrated in figure 2.220.

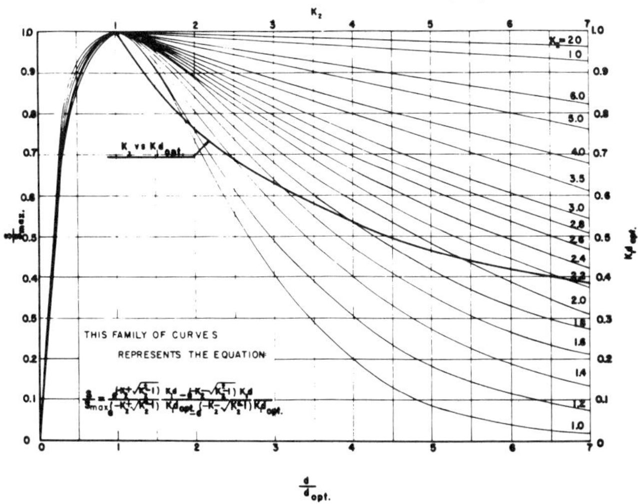

Figure 2.219–Diagram for determining values of K_1 and K_2 from the results of a shear-stress vs horizontal displacement test (from Pavlics, 1958).

$$\tau = c + P\left(\frac{J}{P}\right)^{n} T \qquad (2.171)$$

in which τ = shear stress, c = cohesion, P = normal stress, and J = shear displacement. By plotting values of $(\tau\text{-}c)/P$ vs J/P on log-log coordinates and then constructing a straight line through the test data points (see fig 2.156), n can be determined by measuring the physical slope of the line, and T can be determined by noting the value of $(\tau\text{-}c)/P$ at the point where the line intersects the value of J/P = 1.0 irrespective of the units used for J/P. Typical data and computed values are shown in figure 2.221.

The most commonly used procedure is the single-parameter method, which describes soil stress-deformation behavior in shear by:

$$\frac{\tau}{\tau_{max}} = 1 - e^{-\frac{j}{k}} \qquad (2.172)$$

in which τ = shear stress, $\tau_{max} = c + \sigma_n \tan \phi$, e = natural logarithmic base, j = shear deformation, and k = soil shear length parameter. In this case, the data are represented by the same general form used to describe charge on a capacitor vs time for an R-C circuit. The value k is the shear displacement necessary for the shear stress to reach about 63% (1 – 1/e) of its maximum value (similar to the time constant of an R-C circuit). Alternatively, k may be defined by the slope of the shear-displacement curve at the origin (fig 2.222) (Janosi, 1962). Field test data seldom fit this general stress-displacement form exactly. By constructing a precomputed diagram of straight lines representing a range of k values on a semi-log plot (fig 2.223) (Wills, 1963), and then plotting the experimental data on the diagram, the value of k may be determined by deciding which of the preplotted lines best represents the experimental data. Nonlinear regression techniques may also be used for this purpose.

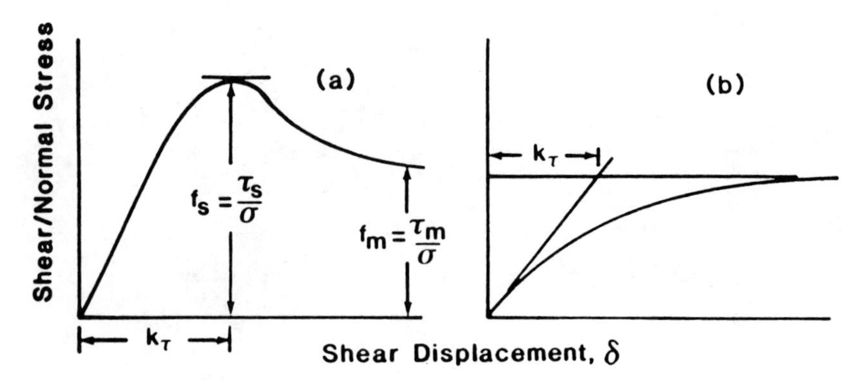

Figure 2.220–Diagram for interpreting the parameters of a humped response characteristic from a shear-stress vs horizontal displacement test (from Yong et al, 1984).

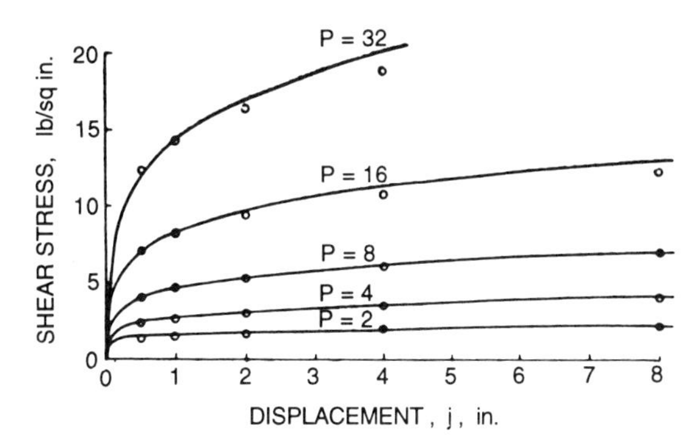

Figure 2.221–Typical response characteristics (circles) from shear-stress vs horizontal displacement tests at varying levels of normal stress, P. Predicted values are shown as solid lines (from Taylor and VandenBerg, 1966).

It can be seen that all of the four above-mentioned methods force the field test data to be represented by one form or another. Although this sort of procedure prevents the field data from being represented exactly, it allows a group of one or more standardized parameters to serve as a generalized middle ground between the great variability of individual field data points and the total generality of a predictive equation.

Tests were conducted (Upadhyaya, 1989) using five different sizes of grouser plate and three different vertical loads on a single dry loam soil in tilled and firm (tilled then rolled) condition. It was found that neither grouser plate size nor vertical load had a significant effect on the value of k. However, the k value for the tilled condition (k =

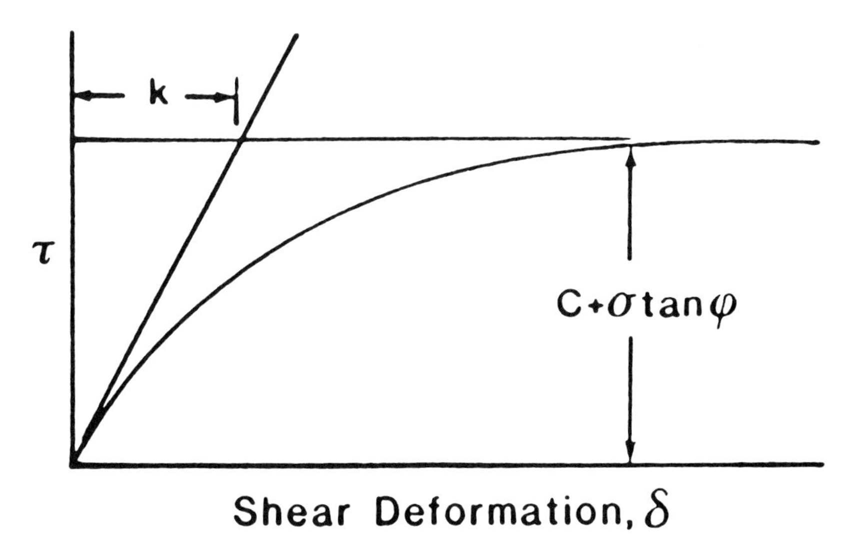

Figure 2.222–Typical response form of an asymptotic result characteristic for a shear-stress vs horizontal displacement test. The shear-displacement stiffness parameter, k, is shown as the inverse of the initial slope of the response curve (from Yong et al, 1984).

11.63 mm, sd = 5.55 mm) was somewhat higher than for the firm soil (k = 9.26 mm, sd = 5.84 mm). These values for k are only slightly lower than those shown in figure 2.223. In additional tests on moist soil it was found that at a given level of normal stress k decreased with increases in plate length. Also values of k decreased with increases in vertical load-per-unit plate width. For either fixed normal stress or fixed vertical load, k values increased with increases in plate width (Upadhyaya et al, 1993). Values of k on moist soil were approximately 30% greater than those found on dry soil. For a Michigan farm soil at 21.3% moisture (remolded) a value of k = 8.2 mm (sd = 1.3 mm) was found (Janosi and Hanamoto, 1961). Values found for two soils (remolded) at varying moisture and density (Taylor and VandenBerg, 1966) are as shown in table 2.28.

It is, thus, difficult to discern any general relationship between k values and other soil physical conditions, other than the general expectation that a stiff, rigid soil will have lower k values than a more compliant soil which tends to develop strength as it deforms.

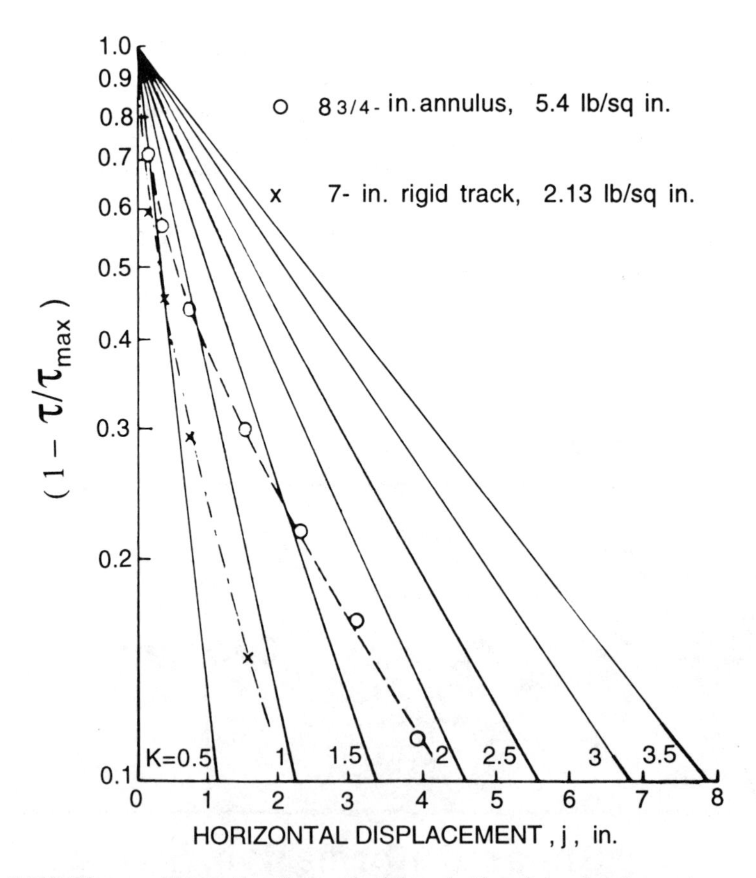

Figure 2.223–Diagram illustrating the method for plotting data of a shear-stress vs horizontal displacement test for comparison with precomputed lines for specific k values. The k value for the test is determined by estimating which of the precomputed lines most nearly represents the test data (from Wills, 1963).

Load-Sinkage Parameters

The sinkage of loads such as tractor wheels into the soil surface has been a matter of interest to persons in both agriculture and cross-country mobility and transport. Investigations into the nature of such phenomena and of the soil parameters involved have arrived at findings in two different categories — situations in which time is considered an important factor and those where it is not.

In areas in which flooded rice culture is commonly practiced, there has been a considerable amount of work on time-dependent sinkage, which is frequently related to viscous properties of the soil (see previous section on viscoelastic, viscoplastic, and rheological soil properties). One time-dependent approach that does not involve viscosity of the soil directly (Ito, 1974) expresses sinkage by:

$$h = h_0 \left(1 - e^{-K_1 t}\right) \tag{2.173}$$

in which h = sinkage at any time, t, h_0 = sinkage at t = ∞, and $K_1 h_0$ = the tangent to the slope of the sinkage-time curve at t = 0. From this:

$$\frac{dh}{dt} = K_1 \left(h_0 - h\right) \tag{2.174}$$

A second parameter, K_2, was also defined by:

$$\frac{dh}{dt} = K_2\, h \left(h_0 - h\right) \tag{2.175}$$

It was found that not only K_1 and K_2 but also h_0 were dependent on slippage.

For the case when time is not considered to be a factor, one of the earlier approaches used was that of the bearing-capacity relationships used in foundation engineering (Terzaghi, 1942) in which a critical pressure p_{crit} for soil failure was determined from:

$$p_{crit} = cN_c + \gamma \left(zN_q + 0.5\, bN_\gamma\right) \tag{2.176}$$

Table 2.28. Values for the soil shear length parameter, k*

Soil Type	Moisture Content (%)	Dry Bulk Density (gm/cm^3)	k (mm)
Hiwassee Sandy Loam	9.2	1.31	22.9
Hiwassee Sandy Loam	14.3	1.69	17.8
Hiwassee Sandy Loam	8.2	1.43	25.4
Lloyd Clay	22.3	1.00	15.2
Lloyd Clay	24.4	1.26	5.1

* Data from Taylor and VandenBerg (1966).

in which c = cohesion, γ = soil weight per unit volume, z = depth of sinkage, b = width of loaded surface, N_c, N_q, and N_γ are dimensionless coefficients with values shown in figure 2.224.

Another relationship used was the Berstein equation (Gill and VandenBerg, 1968):

$$P = k\,z^n \tag{2.177}$$

in which P = pressure, z = depth of sinkage, and k and n are constants.

One of the most-used forms of this equation (Bekker, 1960) was that of:

$$P = \left(k_c / b + k_\phi\right) z^n \tag{2.178}$$

or

$$\text{Load} = P\,A = P\,\ell b = \left(\ell k_c + b\,\ell k_\phi\right) z^n \tag{2.179}$$

in which b is the width of the loaded area and ℓ is its length.

The two parameters k_c and k_ϕ separated the sinkage resistance parameter, k, into two components — one linking load at any depth to loaded area and the other to the length of the loaded surface.* Thus, three parameters were required to describe the sinkage phenomenon, k_c and k_ϕ and n. These parameters are determined using surface penetration tests with loaded areas of at least two different widths† (fig 2.225) (Pavlics, 1958). The values of the test data were plotted on a log-log form and the values of a_1 and a_2 were those of soil pressure for load widths b_1 and b_2, respectively, when the depth of sinkage, z, was 1.0 irrespective of the units used. The value of n was equal to tan α, in which α was the physical slope (on the log-log plot having equal cycle lengths for both

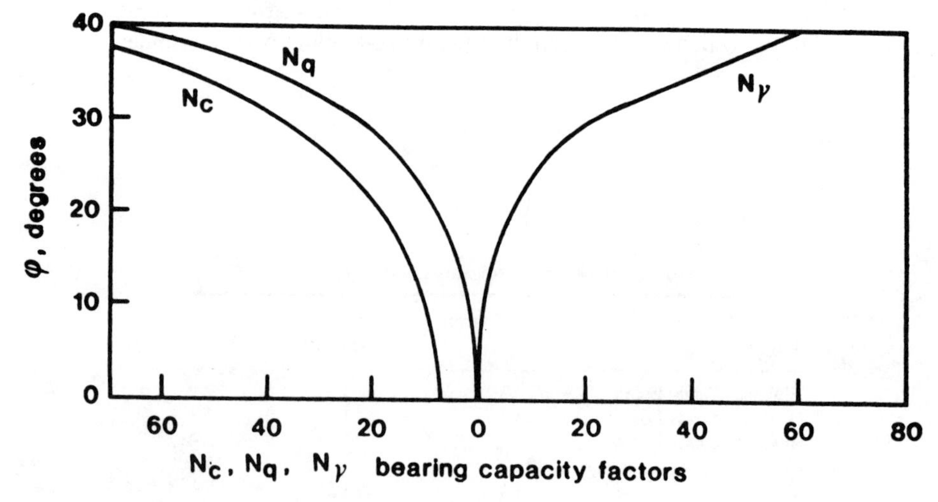

Figure 2.224–Dimensionless coefficients, N_c, N_q, and $N\gamma$ for the bearing capacity equation commonly used in foundation engineering (from Yong et al, 1984).

the z and P scales) of the two parallel lines assigned to represent field data for the two load widths. A statistical method to formulate these lines when only two load widths were used would be to construct a multiple linear regression of the type:

$$\log P = A + B \log z + C\,(\text{dummy}) \tag{2.180}$$

in which A, B, and C are coefficients, and dummy = 0.0 for one load width and 1.0 for the other.

Values for k_c and k_ϕ could be determined (Pavlics, 1958) from:

$$k_c = \frac{(a_1 - a_2)\, b_1\, b_2}{(b_2 - b_1)} \tag{2.181}$$

* Traditional bearing-capacity formulations frequently ascribed a portion of bearing capacity to the loaded area and a portion to the perimeter of the loaded area.

† It has been shown (McKyes, 1985) that using three or four different plate widths and averaging the results could result in markedly reduced coefficients of variation for k_c and k_ϕ, as opposed to when only two plate widths were used.

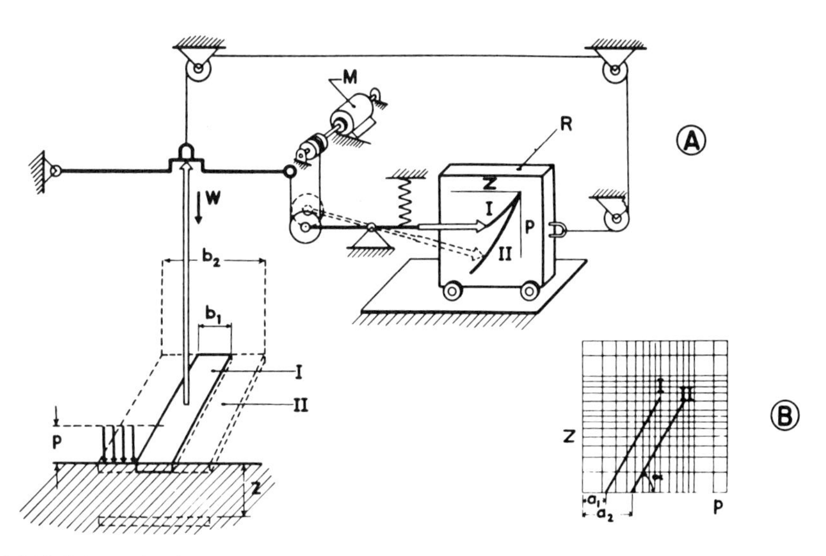

Figure 2.225–Schematic diagram of the apparatus and result plot used for determining the sinkage parameters, k_c, k_θ, and n (from Pavlics, 1958).

$$k_\phi = \frac{(a_2 b_2 - a_1 b_1)}{(b_2 - b_1)} \tag{2.182}$$

Because it is very difficult to transpose values for k_c and k_ϕ from one system of units to another, the values are presented in table 2.29 in the units with which they were determined.

In figures 2.226 and 2.227 (Hanamoto and Hegedus, 1958) it can be seen that both k_c and k_ϕ increase rapidly when the moisture content of a typical loam soil decreased. Similar characteristics were found for clay-silt mixtures (figs 2.228 and 2.229) (Trask and Klehn, 1958), and this same tendency also prevailed for values of the parameter n (fig 2.230) (Trask and Klehn, 1958).

A modified form of the above sinkage equation was proposed (Reece, 1964) in which the depth, z, was normalized by the width, b, of the impression surface:

$$P = \left(c\,k_c' + \gamma \frac{b}{2} k_\phi'\right)\left(\frac{z}{b}\right)^n \tag{2.183}$$

in which c = cohesion and γ = soil weight-per-unit volume.

In this case, the parameters k_c' and k_ϕ' are dimensionless. For one example soil, values of $k_c' = 11$, $k_\phi' = 123$, and n = 1.0 (lb-in. units) were found (Hegedus and Liston, 1966).

A further modification of this equation form (Upadhyaya, 1989; Upadhyaya et al, 1990) to:

Table 2.29. Values for soil sinkage parameters, n, k_c, and k_ϕ

Soil	Units	n	k_c	k_ϕ	Source*
Dry sand	lb-in.	1.07	0	3.3	1,3
Sand	lb-in.	0.8	0	7.0	2
Wet sand	lb-in.	1.06	1.0	6.6	3
Michigan farm soil	lb-in.	1.04	–15	8	4
Michigan farm soil	lb-in.	0.93	16.5	2.8	4
Michigan farm soil	lb-in.	1.2	25	–7	4
Michigan farm soil	lb-in.	1.04	–1.5	2.9	4
Fine Sandy Loam (14.5% moisture)					
γ = 1.2 g/cm³	N-cm	0.51	45.64	2.11	5
γ = 1.48 g/cm³	N-cm	0.51	65.20	13.31	5
γ = 1.68 g/cm³	N-cm	0.51	277.10	21.05	5

* 1. Czako, 1958.
2. Bekker and Janosi, 1958.
3. Czako and Bekker, 1958.
4. Sloss, 1966.
5. Bolling, 1987.

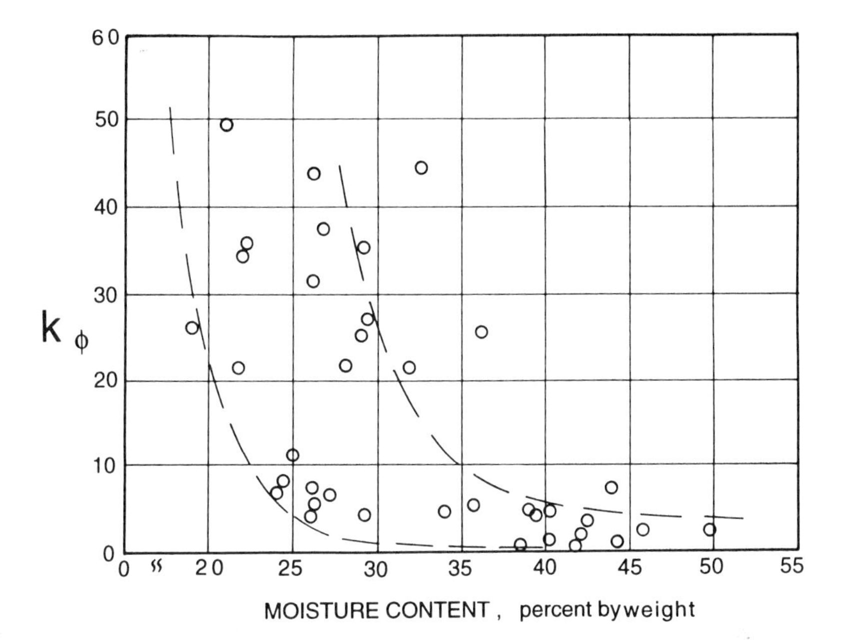

Figure 2.226–Effects of soil moisture content on values of the sinkage parameter, k_ϕ (from Hanamoto and Hegedus, 1958).

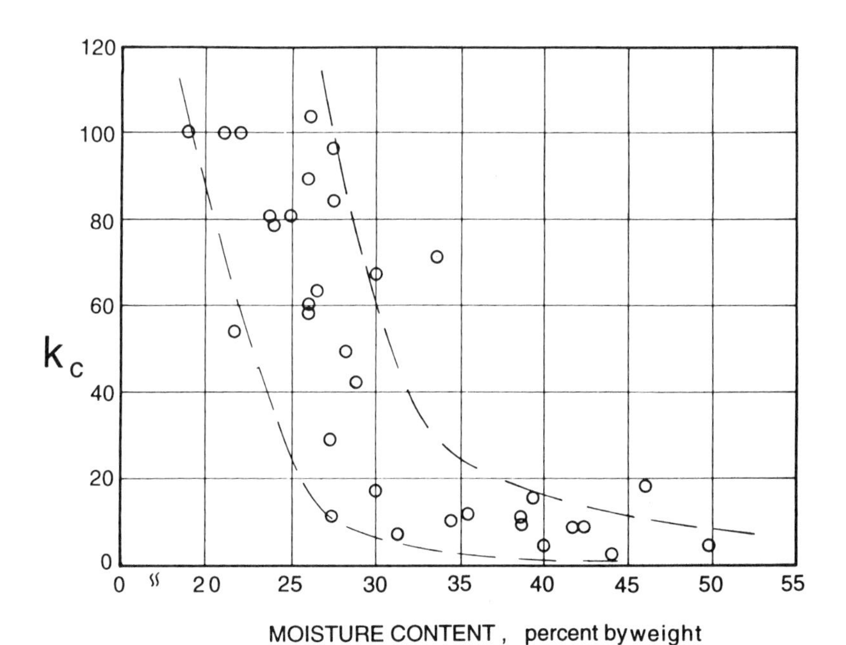

Figure 2.227–Effects of soil moisture content on values of the sinkage parameter, k_c (from Hanamoto and Hegedus, 1958).

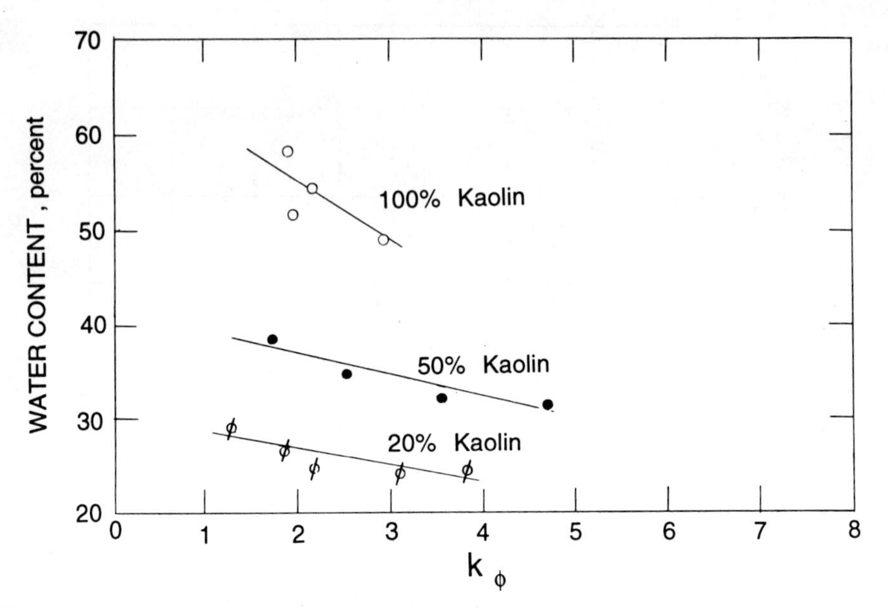

Figure 2.228–Effects of moisture content and clay content on values of the sinkage parameter, k_ϕ (from Trask and Klehn, 1958).

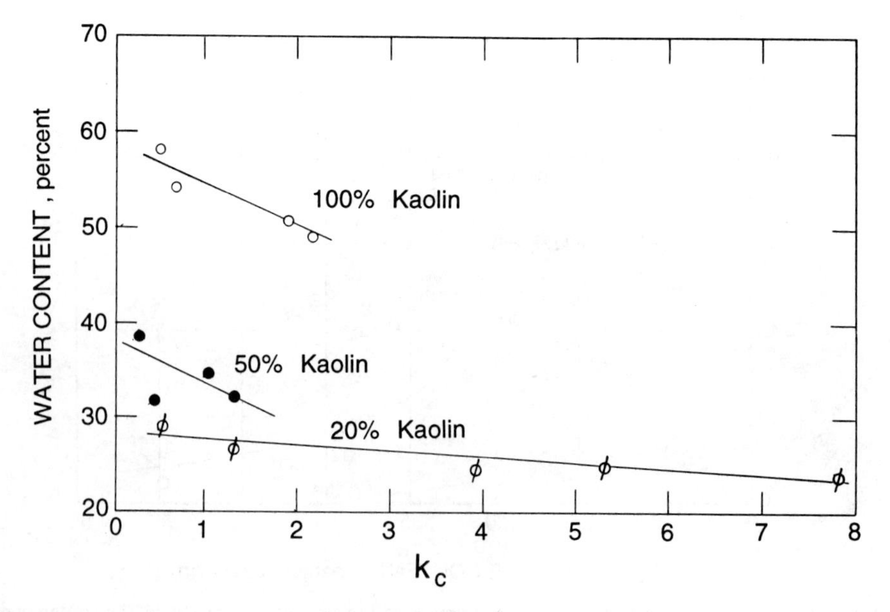

Figure 2.229–Effects of moisture content and clay content on values of the sinkage parameter, k_c (from Trask and Klehn, 1958).

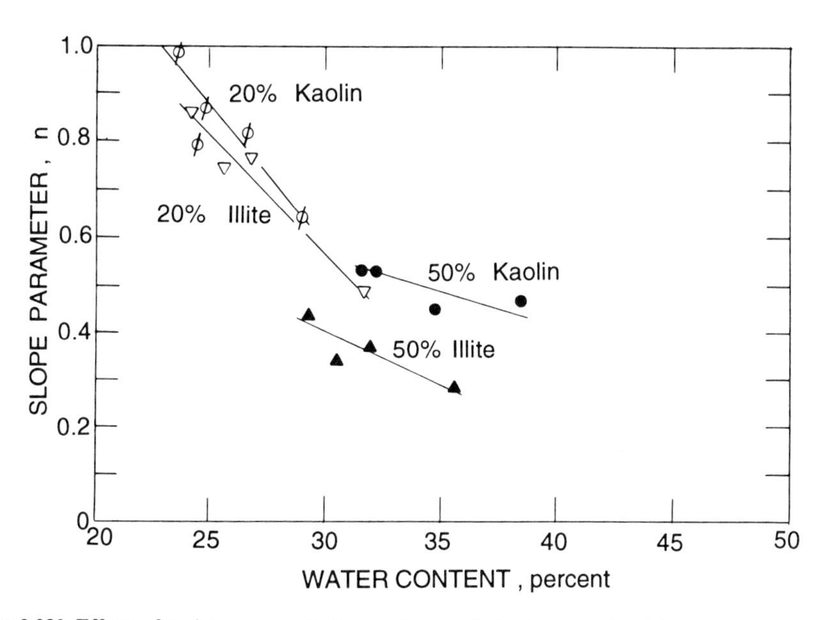

Figure 2.230–Effects of moisture content, clay content, and clay type on the sinkage parameter, n (from Trask and Klehn, 1958).

$$P = (k_1 + k_2\, b) \left(\frac{z}{b}\right)^n \tag{2.184}$$

allowed the two parameters, k_1 and k_2 to have units of force/area and pressure/width, respectively, and thus be convertible between systems of units. This form also obviated having to measure cohesion and density in development of a load-sinkage process description. Values of k_1, k_2, and n for a dry loam soil in tilled and firm (tilled and rolled) condition were as shown in table 2.30.

Another formulation for a load-sinkage relationship, the parameters of which are dimensionless (Hegedus, 1965) is of the form:

$$P = \frac{\gamma\, \ell}{1\,/\,(A^m B)} (z/s)^{1/m} \tag{2.185}$$

Table 2.30. Values for sinkage parameters, n, k_1, and k_2*

Soil Condition	k_1	k_2	n
Tilled	268 kPa	0.516 kPa/mm	0.616
Firm	–348 kPa	20.774 kPa/mm	0.601

* Data from Upadhyaya (1989).

in which γ = weight of soil per unit volume, ℓ = characteristic dimension of the impression surface, s = perimeter of the impression surface, and A, B, and m are dimensionless parameters associated with the soil properties and surface geometry

This relationship appeared best suited to cohesionless soils, for some of which a linear relationship on a log-log plot could be found between $\gamma\ell/P$ and z/s (Murphy, 1967), provided z/s values were less than 0.10.

A simple pressure-sinkage relationship (Kogure et al, 1983) with which vehicle motion resistance values can be easily developed is of the form:

$$P = K_s \left(\frac{z}{D}\right)^n \tag{2.186}$$

in which D = diameter of circular impression surface, and K_s and n are soil parameters. The characteristics of such a relationship are illustrated in figure 2.231.

Another version of the same representation of the pressure-sinkage relationship (fig 2.232) (Bolling, 1987) can be quantified in the form:

$$P_{(z)} = A\left(1 - e^{-Bz}\right) + Cz \tag{2.187}$$

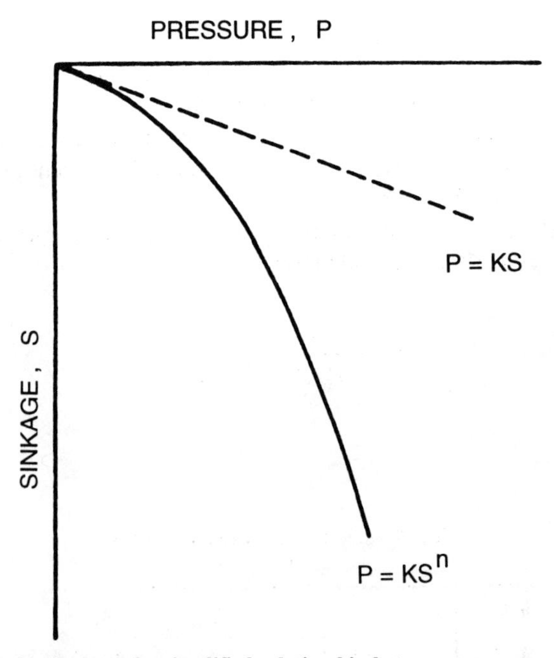

Figure 2.231–Schematic diagram of a simplified relationship between pressure and sinkage. Reprinted with permission of *Journal of Terramechanics* 20(314), Kogure, K., Y. Ohira, and H. Yamaguchi, Prediction of sinkage and motion resistance of a tracked vehicle using plate penetration test. Copyright © 1983, Pergamon Press, Ltd.

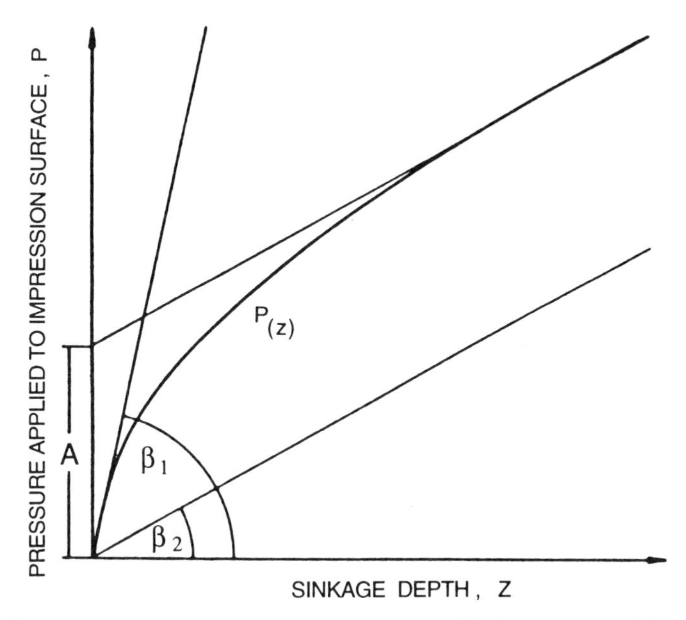

Figure 2.232–Diagram representing the parameters of a simplified pressure vs sinkage relationship (from Bolling, 1987).

in which A, B, and C are parameters of the soil and of the impression area, C = tan β_2, and B = (tan β_1 – tan β_2)/A. See figure 2.232 for determination of A and specifications of β_1 and β_2.

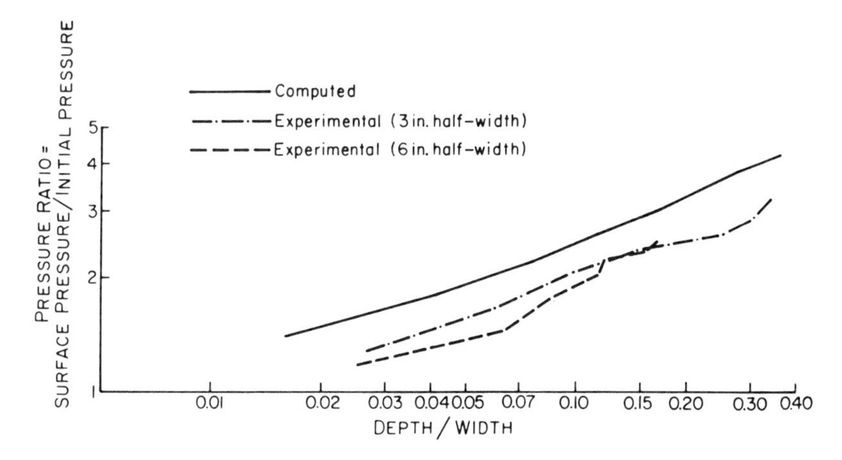

Figure 2.233–Comparison of computed and experimentally determined pressure vs sinkage relationships. Sinkage was computed from a soil compaction model in which rut volume was equated with pore volume lost upon compaction (from Chancellor, 1966).

One somewhat complex approach to predicting load-sinkage characteristics in agricultural soils (Chancellor, 1966) involves using: (1) the Boussinesq stress distribution characteristics, (2) a semi-logarithmic relationship between major principal stress and soil porosity, (3) a system of energy conservation within the soil half-space, and (4) the assumption that all volume absorbed in volumetric strain throughout the soil half-space appears as impression volume. This procedure, when applied on a two-dimensional basis (long, narrow impression surface), is scaled in dimension by the width of the impression surface. For a small diameter circular impression, it would be approximately scaled by the square of the diameter of the impression. Laboratory tests showed that this method tended to underpredict sinkage (fig 2.233) in part because the volume of soil pushed above the original soil surface, some distance from the impression surface, was not considered.

In this last case knowledge of the soil parameters of density, cohesion, angle of internal friction, compressibility constant, and compressibility coefficient would be required in addition to information on the size, shape, and load of the surface to be impressed. The system would not apply to saturated or near saturated soil conditions. Sinkage values obtained with this system, however, are based on a defined soil deformation process.

Chapter 3

Advances in Soil-Plant Dynamics

Henry D. Bowen Thomas H. Garner David H. Vaughn

Soil-plant dynamics may be defined as the interaction and effects of forces and motions between soils and plants. The effects of primary interest would be those that influence plant growth and development, since it is from plants that most returns of beneficial products are obtained.

This broad subject can be narrowed somewhat by recognizing that the primary means of soil-plant interaction is through plant roots, except in the case of seedling emergence, which involves hypocotyl/cotyledon — or coleoptile — soil interaction. In the text that follows, discussion will focus heavily on soil-root interactions, because what happens to roots has a significant influence on plants. Russell (1977) discussed research on hormonal mechanisms that control plant growth. This research led to the realization that the absorption of water and nutrients is only part of the function of root systems: they are also the source of compounds that are involved in regulating development throughout the entire plant. He further stated that it is now evident that the explanation of many responses of the above-ground parts of plants to environmental factors can be adequately explained only if the performance of their roots is considered. This work by Russell also deals with many other aspects of plant root development. Rendig and Taylor (1989) give a basic treatment of soil-plant interrelationships.

History of Plant Dynamics/Soil-Root Interaction

The classical work of Pfeffer (1893) on measurement of plant growth forces is cited in most references as being the first successful attempt to document the amount of pressure a plant root could exert. According to Pfeffer (translated by Gill and Bolt, 1955) corn roots exerted an average of 1.47 MPa in an axial direction and 0.67 MPa in a lateral direction. Russell (1977) indicated that for more than 60 years after Pfeffer reported his experiments, the subject of root growth pressures received little attention. However, Zobel (1989) indicated that there had been an almost continuous interest in root growth and development throughout the past 100 years. Bowen (1981) cited three events that occurred in the late 1940s and early 1950s that intensified work in the area of mechanical impedance of roots by soils. These events were:

1. Veihmeyer and Hendrickson (1948) showed that increase in bulk density reduced root growth even in soils where aeration should not have been a problem.

2. Lutz (1952) stressed that major gaps existed in our knowledge about mechanical impedance and plant growth.
3. Gill and Bolt (1955), reviewed Pfeffer's (1893) work on root growth pressures.

Both Russell (1977) and Bowen (1981) cite the work of Stolzy and Barley (1968) and Taylor and Ratliff (1969) that confirmed maximum longitudinal growth pressures of 0.9 to 1.5 MPa for several crops. These data have also been used as reference points for predicting emergence force of seedlings.

Soil – Seed/Hypocotyl – Coleoptile Interaction

Morton and Buchele (1960) stated that the design of precision planters was hampered by lack of adequate information concerning ideal environment for emergence of seedlings. They performed experiments that indicated that emergence energy increased with seedling diameter, soil compaction pressure, initial soil water content, depth of planting, and surface drying. They suggested that applying pressure at the seed level and preventing evaporation from the soil surface reduced emergence energy. Williams (1963) measured emergence force of forage legume seedlings by subjecting seeds to the vertical force of weighted glass rods. These rods were sized precisely to a mass of 10, 20, 30, or 40 g. They were inserted in glass tubes and allowed to rest on the germinating seeds. When a rod was pushed up a distance of 2 mm or more by the arching hypocotyl, the seedling was considered to have exerted a force proportional to the mass of the rod supported. They obtained force values ranging from 0.158 to 0.625 N for different cultivars of legumes. Jensen et al (1972) obtained values that were slightly lower but of the same order of magnitude for similar plant species.

Arndt (1965a) stated that the mechanics of seedling emergence was too variable and complex to be expressed by a single index such as tensile strength of free samples as in modulus of rupture determinations. He preferred the direct measurement of mechanical impedance and used a simulated seedling to measure forces required to push through soil seals (crusts). He measured forces on the order of 1.47 N required for a mechanical probe to emerge through a crust. He also attempted to measure the emergence force of pea seedlings by having them push against different amounts of weighted lead washers. He found that lateral support was important for a hypocotyl to be able to develop its maximum emergence force. In other studies he found that a pea seedling would grow in a spiral under a soil crust for a period of 30 days and to a length of 20 cm. He indicated that emergence force potential of seedlings might be better utilized if lateral soil resistance could be increased by agronomic practices. Arndt (1965b) in another study, found that the load-bearing potential of a cotton hypocotyl 1.27 cm long was four times greater than a hypocotyl 3.81 cm long. From this result, he inferred that deeper planting would increase buckling tendency.

Drew and Buchele (1962) measured the forces exerted by corn shoots with a "seedling thrust meter". They found a mean value of 2.66 N. Garner and Bowen (1966) measured emergence force of cotton seedlings for one temperature and soil condition and obtained an average value of 2.03 N for the maximum emergence force. Drew et al (1971) measured emergence forces for cotton for temperatures of 23.9, 26.7, 29.4, and 32.2° C. Maximum thrust was developed at 23.9° C but a longer time was required to develop this thrust (135 h compared to 100 h at 32.2° C). The range of average maximum emergence force developed was from 2.26 to 2.73 N, but the difference due to temperature was not significant statistically. Prihar and Aggarwal (1975) measured emergence force of corn seedlings and recommended a firm base below the seed to enhance the emergence capability of the growing plumule by providing better anchorage.

Goyal et al (1980) measured emergence forces of soybean seedlings at different ambient temperatures. They obtained forces of 2.92, 3.53, 3.88, and 3.75 N for temperatures of 15, 20, 26, 32° C, respectively.

Wanjura et al (1965) performed a study in which the pressures exerted by various planter press wheels were measured in a laboratory. The same surface press wheels were then used to plant cotton (*Gossypium hirsutum* L.) in the field on an irrigated Amarillo loam soil. Pressures in the horizontal and vertical directions were measured at the soil surface and at the 5.08-cm depth in the laboratory studies for soils at 7 and 12% soil water content. Vertical pressures measured at the surface for soil at 7% soil water content varied from 0.003 to 0.023 MPa and at the 5.08 cm depth these pressures varied from 0.003 to 0.019 MPa. For the 12% water content, pressures trended slightly lower than for the 7% water content. From the field tests it was concluded that, under normal conditions, surface press wheels decreased cotton emergence in an Amarillo loam soil. Surface press wheels provided a slight increase in maximum seed zone temperature and a reduction in soil water loss. However, the high soil strength caused by surface pressing reduced cotton emergence. For the indicated press wheels they measured the following pressures required to insert a hand penetrometer 0.64 cm into the soil crust associated with that press wheel: 6 × 16 in. tire, 0.352 MPa; 7 × 18 in. tire, 0.144 MPa; 7 × 22 in. tire, 0.032 MPa; innertube, 0.018 MPa; no press wheel, 0.000 MPa. These authors recommended that reducing drying rate without creating high soil strength would be one means for increasing cotton emergence on light and medium soils.

Root Growth Pressure and Forces

Taylor and Gardner (1960) used wax substrates to measure the penetrating abilities of plant roots. These waxes were available in different penetration numbers with different hardness. Barley (1962) examined the ability of maize roots to overcome mechanical resistance. In one experiment the apex of the root was compressed in a way that permitted the cells to elongate and differentiate while subject to stress. Growth in length proceeded without interruption but at a reduced and continuously declining rate. He measured longitudinal and radial pressures for one species each of corn (*Zea mays* L.) and broadbean (*Vicia faba* L.). The respective longitudinal and radial pressures were 1.22 and 0.71 MPa for corn and 1.11 and 0.61 MPa for broadbean.

Barley also stated that oxygen shortage accentuated the effects of mechanical restraint.

Stolzy and Barley (1968) measured directly the force exerted by the radicle of pea (*Pisam sativum* L.) growing into soil. Maximum force exerted was approximately 0.59 N in about 15 to 20 h. After the root entered a soil core, approximately 0.2 N of additional force was required to overcome skin friction between the root and soil. They found that force required for a mechanical probe was greater than the force required for root penetration. They gave as reasons for the difference the tapered shape, smoother surface of the root, and the root's ability to grow along planes of weakness in the soil.

Barley and Greacen (1967) appealed for more basic understanding of the root growth-soil resistance problem. They stated that the fact that soil mechanics was the domain of the engineer had been a further handicap in applying the subject to agronomic problems. They stated that engineers solve problems of a practical nature without explaining the processes involved. Greacen et al (1968) reported that high mechanical resistance offered by a layer in the profile could hinder the vertical growth of main roots; similarly, highly resistant peds could hinder the horizontal growth of laterals. They discussed the difficulties of applying penetrometer data. These data were hard to interpret in terms of root growth forces. In soils where the growth of maize roots was stopped, penetrometer readings registered 2.2 MPa. This value was considerably greater than Pfeffer (1893) and Barley (1962) determined could be exerted by corn roots. Dexter and Hewitt (1978) discussed the deflection of plant roots. When a plant root meets an interface it either deflects or penetrates. If the root has just grown across a void it may buckle when it meets the interface. They discuss procedures for estimating the proportion of roots that could be expected to penetrate soil layers of different strengths. They presented an equation analogous to a column failure equation:

$$\sigma_b = \frac{4(20.2)EI}{\pi(dr)^2\ 1^2} \quad (3.1)$$

where σ_b = buckling stress (MPa), E = modulus of elasticity (MPa), I = moment of inertia (m^4), dr = root diameter (m), and l = length of air gap (m).

They measured E and dr for the roots of several different species. In a follow-up study, Whitely et al (1982) measured buckling stresses for root tips of seven plant species. They compared measured buckling stresses with calculated stresses based on elastic properties of roots. The values were of the same order of magnitude but there was no consistent relationship between calculated and measured stresses. Misra, et al. (1986) measured axial and radial growth pressures for roots of several species. They obtained the following values for axial growth pressures: Pea (*Pisum sativum* L.), 0.497 MPa, Cotton (*Gossypium hirsutum* L.), 0.289 MPa, Sunflower (*Helianthus annuus* L.), 0.238 MPa.

The rate of growth of roots under different soil physical conditions has received much attention for many years. Baldovinos (1953), using a traditional technique, scribed maize root tips with India ink marks spaced 1 mm apart. The movement of these marks was followed with time. The growth pattern he obtained is illustrated in figure 3.1. It is apparent that the elongation occurs within the first 4 mm of the root tip. Baldovinos reported that the first millimeter behind the root cap was a region containing slowly dividing meristematic cells. The second millimeter was a transition zone between cell division and cell enlargement, with rapid cell division and cell enlargement in the distal (end nearest the root tip) portion of the section and cell enlargement alone in the proximal half. Garner and Bowen (1966) measured displacement with time for both the top of a cotton (*Gossypium hirsutum* L.) hypocotyl and the root tip. With a controlled

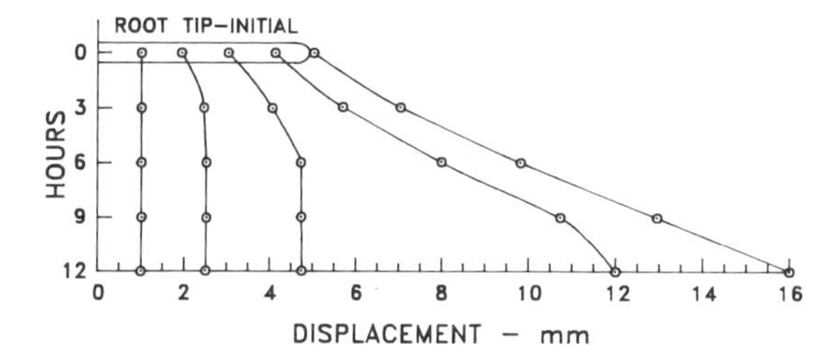

Figure 3.1–Displacement of India ink marks on a maize root tip due to the growth of the root tip. Measured by Baldovinos (1953).

observation apparatus they obtained curves such as the one shown in figure 3-2 which showed that after an initial period of approximately 35 h at 32° C., root tip displacement from the planted position was linear while the hypocotyl crook displacement followed an equation of the form:

$$y = k + at + bt^2 \tag{3.2}$$

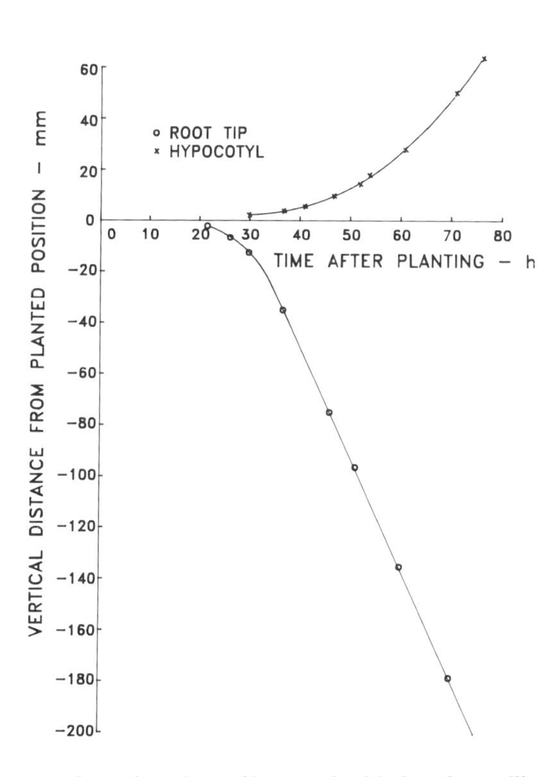

Figure 3.2–Displacement of root tip and top of hypocotyl, with time, for seedling grown in Ruston Loamy Sand; temperature = 31.6 ± 6° C, CO_2 = 0.1%, O_2 = 20%, soil compacting stress = 0 MPa (from Garner and Bowen, 1966).

where y = displacement from planted position (mm), t = time (h), and k, a, and b = constants. The constants varied with soil compaction stress above the planted seeds.

Effect of Soil Physical Factors on Rate of Root Extension

There have been many studies that documented the drastic effects of soil physical factors on the rate of root extension. Bowen (1966) discussed edaphic factors which affect seedling emergence. These factors also affect root extension. He emphasized that the total seed environment can be classified into three distinct parts for discussion and analysis: (1) the chemical environment, (2) the biological environment, and (3) the physical environment. The physical environment can only be described in terms of the histories of soil temperature, soil water, soil aeration, and soil impedance. Bowen then used "limiting factors" concepts to develop recommendations for planting conditions.

Drew (1979) stated that the primary edaphic factors that influence the growth and function of roots in soil can be identified as temperature, soil water potential, oxygen concentration, mechanical resistance, and concentrations of nutrient ions and other solutes. Russell (1977) said that the rate of root extension was subject to considerable variation depending on genetic and environmental factors. Taylor (1971) summarized the mechanisms for root penetration through high-strength pans. He stated that when a root grows from a loosened cultivated layer and encounters a soil pan it may be diverted horizontally, grow into the layer a short distance and cease further elongation, or it may continue to elongate in the horizontal direction but at a slower rate. He explained that the root tip would enter compact layers if a pore or crack larger than the root cap diameter was available. The root is capable of circumnutational movements that may aid in finding soil voids. Whitely et al (1981) tentatively concluded that weak tensile strength of soil could facilitate penetration by roots. Okada and Shimura (1990) stated that when root tips encounter obstacles in soil, they change the direction of their growth. They grew roots (*Arabidopsis thaliuna*) on a surface of agar that was slanted at 45°. The root tips grew in a waving pattern since the tips were unable to penetrate the agar. In the process the root tips also rotated. Prebble (1970) investigated the effect of smeared soil surfaces on wheat (*Triticum aestivum* L.) root penetration. At a soil water potential of –0.01 MPa, penetration into a horizontal surface was decreased by smearing (as would occur in pulling tools through wet soil). For a soil water potential of –0.1 MPa root penetration decreased 50%.

Barley et al (1965) stated that penetration and growth of roots is controlled chiefly by strength of the soil. Soil strength should be regarded as a property that has general influence on root elongation, rather than as a limiting condition encountered in unusual soils. Moreover, certain biological processes associated with roots may also play a key role in their growing through soils. Russell (1977) cited recent evidence that contact between growing roots and the solid matrix of the soil is established largely by the mucigel secreted by roots rather than by the walls of the root cells themselves. He emphasized that this substance could not be ignored in the study of root/soil relationships. Rovira et al (1979) described several materials that are associated with the soil-root interface that may be important to understanding of phenomena related to roots.

Collis-George and Yoganathan (1985) found that interfacial stresses in soil of 3.0, 2.3, 1.7, and 0.8 MPa were respectively limiting germination, root elongation, coleoptile elongation, and emergence of wheat (*Triticum aestivum* L.). Lachno et al (1982) found evidence of possible hormone control of root elongation with corn (*Zea mays* L.). They grew maize seedlings in "sand/garden loam" mixtures at normal bulk density of 1.29 Mg/m^3 and at a higher bulk density of 1.69 Mg/m^3. Roots from the compacted soil were about 40% as long as the controls and were much thicker. They measured abcissic acid (ABA) and indolacetic acid (IAA) levels in the root tips for each treatment. They found no difference in ABA levels in 10-mm root tip segments. IAA levels were increased from 32.4 to 176.3 ng/g fresh wt, and it was concluded that this response was likely to have been the main cause of the morphological and growth changes brought about by soil compaction. They suggested that soil resistances far smaller than those of cell turgor could dramatically impede growth. They further expanded on the hormone-mediated mechanism by listing three bases for the theory:

1. Cell volume is not necessarily reduced as demanded by the hypothesis that elongation response is governed by turgor pressure.
2. Mechanically impeded roots do not recover rapidly. It requires several days, a factor which is easily explained in terms of decay of a growth inhibiting hormone.
3. Pressures caused root growth inhibition if applied to the root cap, but not if applied to the meristem of decapped roots. This response implies that the hormone source is the root cap.

Lachno et al (1982) indicated that it was possible that IAA stimulated ethylene production which caused the growth responses. However, Moss et al (1988) concluded that, despite widespread views to the contrary, ethylene is not the cause of the morphological response of roots to physical impedance. Feldman (1984) stated that Lachno et al (1982) may have been premature in saying that IAA levels were likely to be the main cause of changes in roots subjected to soil compaction.

Brock and Kaufman (1991) discuss many plant hormones and their role in regulating plant development at molecular, cellular, organ, and whole plant levels. Itai and Biunbaum (1991) state that plant growth regulators (PGR) are involved in the regulation of all the functions and processes of roots including their response to the environment. They indicate that roots are capable of synthesizing all PGRs known today, without exception.

Abdalla et al (1969) subjected barley (*Hordeum vulgare* L.) roots to mechanical stress while they grew in glass beads enclosed within a modified triaxial apparatus. At a confining pressure of 0.0107 MPa there was very little effect on root growth. However, at pressures above this level, root elongation was rapidly reduced, and at 0.0621 MPa root growth was almost stopped. The roots were thickened under constraint. Richards and Greacen (1986) called attention to an error they had made in assuming that the pressure on a growing root was equal to the confining pressure of the growth medium. They found that when lateral pressure was of order 0.02 MPa the soil pressure on the root was 0.2 to 0.3 MPa. It is suspected that this same effect may have been experienced in the Abdalla et al (1969) experiments. Boone and Veen (1982) studied the influence of mechanical resistance and phosphate supply on the morphology and function of corn (*Zea mays* L.) roots. The influence of mechanical resistance on root distribution and morphology reduces the uptake of nutrients and concomitantly shoot growth if a nutritional factor is limiting. Their results further indicated that the number of lateral roots per main axis decreased with increasing mechanical impedance.

Relationship Between Soil Dynamics and Plant Dynamics

Foster et al (1983) gave a thoughtful presentation regarding the root-soil interface as follows: Few subjects related to the growth and development of higher plants attract the interest of as many separate disciplines as does the subject of the root-soil interface.

They refer to the interests of (1) soil microbiologists, (2) plant physiologists, (3) soil physicists, and (4) structuralists. They could well have included the engineers who, working with professionals from the other disciplines, have the responsibility of manipulating or not manipulating soils and seeds to achieve desired physical conditions or relationships. Foster et al (1983) describe soils as perhaps the most heterogeneous of all natural materials. They consist of minerals, organic matter, water, salts, in solution, gases, roots of plants, small animals, and microorganisms. Besides breaking down soil organic matter, bacteria and fungi secrete carbohydrate slimes and gels. Such polysaccharides constitute about 15 to 20% of the soil organic matter and hold the soil together to form microscopic aggregates that, in turn, are loosely bound by the hyphae of fungi and actinomycetes to form soil crumbs. The organic acids and chelating compounds secreted by roots modify nearby particles, accelerating the weathering of soil minerals and releasing nutrients in the rhizosphere. These physical and chemical changes induced by root growth and metabolism make the soil near the root very different from the bulk soil.

It is interesting also to consider the macroscopic effects of roots on the soil-root matrix. Russell (1977) stated that one rye plant could have 13 million root axes and laterals with a total length of 500 km and a surface area of greater than 200 m^2. Wheat roots are abundant at the 1.0-m depth, with some reaching depths of 2 m. Powell and Day (1991) stated that roots are a large, dynamic component of an ecosystem and a significant portion of net primary production (NPP). Person (1978) cited research that on apple trees, fine root death and renewal was a natural cyclic process and bears resemblance to leaf shedding in evergreen plants. There is considerable organic matter from dead fine roots. Fine root mass decaying per year is about twice the annual litter fall.

It is now emphasized that soil dynamics includes the effects of applied forces on soil water, soil temperature and soil aeration as well as effects on soil strength, although these factors are all interrelated and may be affected by tillage and traffic. Letey (1985) stressed that in managing soils for optimum plant productivity, attempts should be made to manage so that water potential, aeration, mechanical resistance, and soil temperature are in a balanced range for plant performance .

Soil Conditions as a Medium for Plant Roots

Russell (1977) mentioned three types of stresses that are experienced by roots in soil: chemical — shortage of nutrients, unbalanced supply of nutrients, presence of toxic substances; physical — inadequate water, mechanical impedance to root penetration, anaerobic conditions or unfavorable temperature; biological — caused by flora or fauna of the soil, plant pests, diseases.

Soil Physical Conditions for Meeting Plant Needs

Cannell (1977) discussed the effects of soil compaction, aeration, and management on root growth. He stated that in well-aerated soils the composition of the soil air approaches that of the atmosphere. An external pressure of 0.05 MPa would appreciably restrict root extension, the sizes of the pores would restrict roots from entering and the root system would be stunted.

Russell (1977) reported that roots cannot penetrate rigid pores with a diameter less than that of the extending zone of the root. They cannot decrease their diameters to enter pores; in fact, they get larger when elongation is restricted. Mechanical impedance causes the zone in which the cells are actively extending to be much shorter. Cells may attain their maximum length within 1 mm of the root tip. Root hairs often develop in these locations and anchor the root. Extraction of water by these systems may further complicate the problem by increasing the mechanical resistance to penetration. Russell (1977) indicated again that roots can exert longitudinal pressures of 0.9 to 1.3 MPa, although considerably smaller pressures can severely impede root extension. For this reason we should focus our attention on the minimum pressures that can appreciably restrict root elongation.

Many researchers have attempted to develop useful measures of mechanical impedance to root growth. Soil bulk density and cone index as measured by various types of penetrometers are the most used parameters. It is generally recognized that soil water content must be considered in the interpretation of the effects of each of these variables on root extension. Jones (1983) stated that the US Soil Conservation Service (SCS), National Soil Survey Laboratory, routinely measures bulk density at –0.033 MPa soil water potential in all soil layers in soil Pedon descriptions. However, the relationship between root growth and this measure is not clear. The point at which root growth stops depends upon bulk density and soil water content. This relationship is also affected by soil texture; for example, as the percentage of clay or of silt plus clay increases, the critical bulk density decreases. Eavis and Payne (1969) succinctly pointed out that root growth and function are affected by a large number of factors of which soil water appears to be of particular importance since it acts both directly on growth and function, and indirectly by affecting other relevant factors such as aeration, soil impedance, and soil temperature. They found no difference in penetrometer force between penetration speeds of 4 mm/min and 1 mm/h, and stated that forces experienced by roots were considerably less than those encountered by a root-shaped probe of the same size. Russell and Goss (1974) explained three aspects of roots that cannot be duplicated by penetrometers: (1) the capacity of the root apex to deform in response to external pressure, (2) ability of the root to curve around obstacles, and (3) the possible lubricating effect of the mucigel sheath typically on the root cap. Tollner and Verma (1984) reported on the development of a lubricated penetrometer. This penetrometer had less resistance and did not cause vertical displacement of layers of tissue placed in soil. Standard penetrometers did vertically displace the tissue. Okello (1992) stated that even though the penetrometer is the only technique that can assess variation in soil resistance with depth, neither the cone index nor its gradient with depth is uniquely related to soil cohesion or density, but varies with water content and structural state. He called attention to the fact that soil bodies form ahead of the cone, which changes its geometry. Dexter (1987) presented an empirical model for characterizing the interrelated effects of soil water potential and soil penetrometer strength on root growth. Development of this model resulted in the equation:

$$\frac{R}{R_{max}} = \frac{-\psi_o}{\psi_w} + e^{-0.6931\,(Q_P / Q_{1/2})} \qquad (3.3)$$

where

R = rate of cell elongation, mm/day
R_{max} = maximum rate of cell elongation, mm/day
ψ_o = external water potential = $\pi_o + \psi_m$, MPa
π_o = osmotic potential, MPa
ψ_m = soil matric potential, MPa
ψ_w = wilting point potential for the plant species, MPa
Q_P = soil penetrometer strength, MPa
$Q_{1/2}$ = soil penetrometer strength to reduce elongation rate by 1/2, MPa

Dexter (1987) prepared graphs to display the results of this equation as shown in figures 3.3 and 3.4. He conducted additional studies and determined the values of $Q_{1/2}$ and R_{max} for different plant species (table 3.1).

Dexter (1986) also studied root penetration relationships for plant species that have seminal roots. He stated that when elongating seminal roots of developing plants reach the base of a tilled seedbed, they often encounter a layer of dense, strong, untilled soil. At this interface they may be deflected horizontally, instead of penetrating, and may form a horizontal mat of roots at the base of the seedbed. If this occurs, the plants are

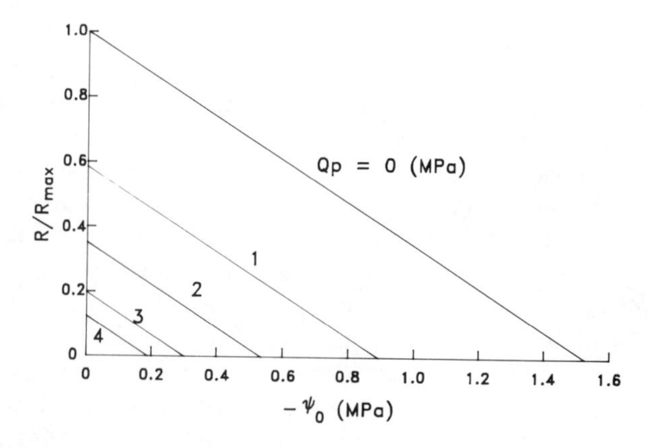

Figure 3.3–Values of relative rate of root elongation, R/R_{max}, under combinations of conditions of soil water potential, ψ_o, and penetrometer strength, Q_p, predicted from equation 3.3 (from Dexter, 1987).

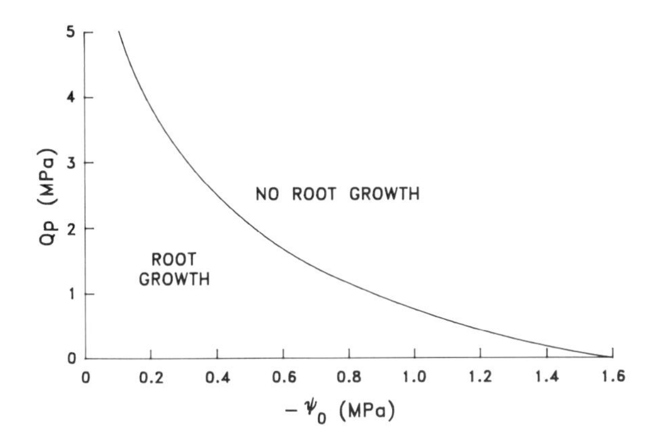

Figure 3.4–Curve indicates combinations of values of soil water potential, ψ_o, and soil penetrometer strength, Q_p, at which root growth will cease as predicted from equation 3.3 (from Dexter, 1987).

unable to absorb the reserves of water in the subsoil and become very sensitive to short periods of drought. He found that for wheat (*Triticum aestivum* L.) the proportion of roots penetrating this interface decreased exponentially as subsoil strength increased. It was concluded that the base of the seedbed should be composed of fine aggregates and that the penetrometer strength of the underlying untilled subsoil should not exceed 0.4 MPa for plants with a single seminal axis or 3.0 MPa for plants such as wheat with four seminal axes. Ehlers et al (1983) performed intense studies of penetration resistance, soil bulk density, soil water content and root growth of oats (*Avena sativa* L.) in tilled and untilled loess soil. Soil bulk density and penetration resistance were higher in the top layer of untilled soil but in the tilled soil a traffic pan formed at 0.25 to 0.30 m. This traffic pan had a higher soil bulk density and penetrometer resistance than any layer in the untilled soil. Root growth was linearly related to penetrometer resistance. The limiting penetrometer resistance for root growth was 3.6 MPa in the tilled Ap horizon, but 4.6 to 5.1 MPa in the untilled Ap horizon and in the subsoil of both tillage treatments. There was a buildup of a continuous pore system in the untilled soil created by earthworms and roots from preceding crops.

Table 3.1. Values of penetrometer strength to reduce cell elongation by one-half ($Q_{1/2}$) and maximum elongation rate (R_{max}) for four plant species*

Plant species	$Q_{1/2}$ MPa	R_{max} mm/day
Cotton	0.72	85
Corn	1.3	26
Pea	2.03	24
Peanut	1.91	65

* From Dexter (1987).

Cassel (1982) discussed how soil bulk density and mechanical impedance may be modified by tillage and the effects of these properties on plant growth. Mechanical impedance is related in a poorly understood manner to clay mineralogy, soil bulk density, soil texture, soil structure, soil water content and percent organic matter. He stated that the influence of soil bulk density was not well understood. At less than optimum bulk density, poor water relations may exist. At higher bulk density, poor aeration and high mechanical impedance may limit root extension. Cassel (1982) cautioned that many times the natural variation in soil bulk density is greater than the differences caused by treatments. Researchers must have a sound statistical scheme before beginning to collect samples. This scheme involves measuring the existing bulk density with sufficient replication to establish the standard deviation. Although he was very concerned about using cone penetrometer data to represent mechanical impedance of soil to root growth, he used the penetrometer to assess mechanical impedance differences affected by different tillage operations in a given soil. When taking cone index data, researchers must obtain data simultaneously on soil water content, soil bulk density, soil texture and organic matter. He further recommended measuring cone index at *in situ* field capacity. Blancher et al (1978) stated that root growth was more closely correlated to probe resistance than to soil bulk density. Root growth decreased as probe resistance increased from 1.0 to 2.0 MPa. Root growth stopped when probe resistance exceeded 2.0 MPa.

Erbach (1987) published a comprehensive review on the measurement of soil bulk density and moisture. He stated that bulk density affects soil strength for structural purposes and for trafficability of vehicles. Bulk density also influences plant growth, water infiltration, drainage, power required for tillage and performance of tillage. It is often used as a measure of effectiveness of tillage operations or of damage caused by the passage of wheels. He concluded that there is no quick, easy, accurate method for the determination of bulk density or soil moisture content. Raper and Erbach (1987) concluded that bulk density measurements obtained with a core sampler without an auger were significantly smaller and more variable than those obtained with an augered core sampler. The augered core sampler gave more accurate results.

It is perhaps well to cite results that give a different viewpoint on root growth since these results may help to explain the overall picture. Bamford et al (1991) studied the effects of soil physical conditions on root growth and water use of barley (*Hordeum vulgare* L.). Their experiments were performed on barley grown in containers. They found that increasing the number of large pores in sieved sandy loam soil at a depth of 0.2 to 0.4 m allowed roots to proliferate easily and at the same time, extract water held at relatively low suctions (high matric potentials). When no extra water was added these plants came under water stress sooner than those grown in containers with relatively compacted subsoil. Reduced rates of root growth meant that water was made available over a longer period. They conclude that, under certain conditions, subsoil loosening will increase crop water stress and reduce yield. Wolf et al (1981) experienced a situation with cotton (*Gossypium hirsutum* L.) in which subsoiling under the row allowed plants to grow large early in the season, then when a drought period came later in August, the large plants possibly experienced greater water stress than those in nonsubsoiled plots presumably because transpiration had utilized more of the water in the profile and the larger plant had a greater demand for water. It should be pointed out that in this experiment neither supplemental irrigation nor applied plant growth regulators were used, and that yields of neither subsoiled nor nonsubsoiled cotton were exceptionally high in that year. Campbell et al (1974) stated that with the coarse-textured soils of the Southeastern Coastal Plains, the beneficial effects of subsoiling decrease with drought duration because even the subsoil has a low water holding capacity. They also cite other relationships among soil management practices and soil physical properties affecting root and shoot growth and yield.

Goss et al (1977) stated that the rate of root elongation was considerably reduced by mechanical stresses less than 0.05 MPa. They also emphasized that soil compaction can impose considerable restraint on emerging shoots. Veen (1982) performed experiments in which it was shown that soil resistances much smaller than root growth pressure would induce significant reductions in root growth. He worked with barley (*Hordeum vulgare* L.) as the test plant showing that a mechanical resistance of 0.05 MPa brought about an 80% reduction in the root growth rate but for soil resistances between 0.1 MPa and the value of root growth pressure, the growth rate was practically independent of the level of applied impedance. These results lead to the hypothesis that it is unlikely that mechanical resistance retards root growth directly. Veen (1982) showed that with corn (*Zea mays* L.) mechanical impedance stimulated root branching, with the impeded roots having more laterals that were longer and showed tertiary branching. He proposed that roots may become shorter and thicker under mechanical resistance because of ethylene production.

Sands et al (1979) studied compaction of soils in radiata pine forests. They found that penetrometer resistance was largely dependent on soil water content and was directly related to soil bulk density. Soil strength at constant soil bulk density increased with depth because of overburden pressure and a decrease in organic matter. Soil under native scrub was less compact than that from adjacent radiata pine plantations of the same soil type. Soil from pasture was usually more compact in the surface 0.2 m than soil from pine plantations, but was less compact at greater depths. Radiata pine roots preferentially penetrated areas of lower soil strength. Root penetration was severely restricted above a critical penetrometer resistance of about 3.0 MPa. Soil compaction reduced porosity but had little effect on water storage capacity. Increased organic matter at constant soil bulk density also reduced porosity but greatly increased water storage capacity and unsaturated hydraulic conductivity. Sands et al (1979) offered the following suggestions for decreasing soil compaction: (1) avoid use of heavy logging equipment, (2) use deep plowing, (3) promote soil organisms favoring increased litter decomposition (not all that simple to do), (4) add Ca CO_3 to increase litter decomposition rate, (5) incorporate organic matter, (6) plant pasture between pine rotations, and (7) do not burn litter during site preparation.

Saini and Chow (1982) stated that soil condition affects root activity directly and also affects growth and yield of above-ground parts of plants by modifying root functions. They discussed some more complicated responses to shoot-root development. The pattern of shoot-root development as affected by compaction and water stress differs in shallow-rooted crops such as corn (*Zea mays* L.) from deep-rooted crops like alfalfa (*Medicago sativa* L.). For example, at a depth of 0.15 m for both compact and loose soils water matric potential was higher for alfalfa than for corn, whether in a wet or dry year. The drier surface condition for corn was attributable to higher transpiration of corn plants. On the other hand, the soil at a depth of 0.6 m had a higher moisture content for corn than for alfalfa. They suggested that in soils with a compact subsoil, a shallow-rooted crop such as corn may be adversely affected by an unfavorable soil moisture regime at certain stages of plant development. VanLoon and Bouma (1978) studied the effect of soil compaction on potato growth in a loamy sand. Strong compaction of the topsoil resulted in shallow rooting, low water availability, and relatively slow foliage and root growth in the first 60 days after emergence. Compaction of the subsoil (formation of a plow pan) initially resulted in rather rapid foliage growth, which was partly due to a high capillary flux from the water table. However, the foliage growth rate slowed as vertical root elongation became inhibited by the plow plan. Top soil compaction resulted in depressed yields and severe induction of second growth. At maturity, differences in total yields were small except for the treatment with the plow pan for which yield was low. Soil compaction had a negative effect on marketable yield.

Threadgill (1982) found that the soil-loosening effects of subsoiling were still present the next crop year after the treatment. However, for the second year after treatment the loosened areas were no longer distinguishable. Soils can consolidate under natural forces.

Sojka et al (1990) studied the effect of subsoiling and penetrometer resistance on sunflower (*Helianthus annuus* L.) production in the Southeastern Coastal Plains. They found that seed yield and quality were generally improved under nonirrigated conditions if adequate profile disruption was achieved. A soil strength corresponding to a penetrometer resistance of 2 MPa would cause some root restriction. Soils that had a penetrometer resistance of 3 MPa would become a total barrier to root elongation in the absence of macropore channels.

Lutz et al (1986) discussed the relationships between citrus root health and soil physical properties. They describe different types of soil conditions that restrict water movement. Some of the problems are pans, compact layers, and clay lenses. All pans can limit citrus productivity by limiting water, oxygen, and nutrients. Restriction of water movement is the major problem. Maximum root growth pressure for citrus is 1.5 MPa. Greater strength than this severely restricts root growth. Some pans are penetrable when wet but when the soil dries, roots can only enter through cracks. Perched water tables also cause limited oxygen, roots die (maybe from soil pathogens). In a good soil most feeder roots are in the top 0.6 m, with the majority being between 0.15 and 0.3 m deep. High water tables can force roots to the top 0.15 m. Feeder roots of navel and mandarin oranges will grow horizontally rather than vertically when they reach the water table. There is evidence that roots growing through compacted soils will leak nutrients, attracting phytophtora and other pathogenic fungi. Oxygen can diffuse through air-filled spaces 10,000 times as fast as through water-filled pores.

Schumaker and Smucker (1981) found that mechanically impeded roots were deformed, branched more frequently, were less porous and consumed more oxygen per unit of fresh weight. Oxygen required to prevent anoxia (oxygen stress) at the root surface can be predicted to be greater for mechanically impeded roots due to changes in morphology and metabolism of the stressed root system. Kirkham (1987) said that aeration of soil is necessary for the maximum absorption of nutrients by plant roots. Insufficient oxygen influences uptake of N, P, K, Ca, Mg, Cl, B, Zn, Cu, Mg, and Fe, all of which are essential organic elements. For adequate aeration, roots usually need at least 10% by volume air in the soil for survival.

The Plant as a Living Physical System

Characteristics of Root Systems

Hale and Moore (1979) stated that the root in its soil environment is part of a complex, interacting system. They discuss that the sphere of influence of the root (which was first called the *rhizosphere* by Hiltner in 1904) on its environment extends for a distance of at least 1 to 2 mm from the root surface. Fitter (1991) describes the two primary functions that root systems perform as: (1) acquisition of soil based resources (principally water and dissolved ions) and (2) anchorage of the plant. Other root system functions such as storage, synthesis of growth regulators, propagation, and dispersal can be seen as secondary. Fitter (1991) stated that there had been considerable discussion on trying to classify roots without much success. Roots grow differently in different environments, plus there are species differences even though these differences are not strictly characteristic, such as those of top growth.

Birdsall and MacLeod (1990) stated that roots can be broadly subdivided into two main types. The first is characteristic of dicotyledonous plants and during early growth is dominated by a single taproot that develops from the radicle and produces many laterals. If the radicle is short-lived it may be replaced by one or many equally vigorous laterals. The root system in the latter resembles that found in monocotyledonous plants. Root systems for monocotyledonous plants consist of many codominant adventitious roots, each of which bears many laterals.

Rendig and Taylor (1989) describe the essentials for root growth by stating that new cells must be formed and then expand. This expansion occurs when hydrostatic pressure (turgor) is sufficient to overcome cell-wall constraint plus the resistance to deformation of the surrounding soil. Osmotica in the root cells are responsible for causing water to flow from soil to the expanding cells. These osmotica can be maintained only if respiration can proceed. Root growth thus can continue only if continuing supplies of water are present for hydrostatic pressure, of oxygen for respiration, of hormones for cell-wall loosening, of calcium for cell-wall synthesis and of carbon and other metabolites for energy and cellular building materials.

According to Rendig and Taylor (1989) shoots are the initial source for most of the organic metabolites used in growth and maintenance. Roots are the initial source for nearly all the inorganic nutrients and for water. Roots also serve other functions: anchorage of plant, anchorage for shoot emergence, and anchorage for roots to force a path through a soil matrix. Roots and shoots are obviously interdependent. By themselves, however, these interdependent roles do not explain the close coordination of root and shoot development. When seeds first germinate much of the metabolite is directed toward root expansion. For a specific environment, genotype and growth stage, there will be a set fraction of the metabolites directed toward this expansion. If half of the root mass is removed from the seedling, root growth will accelerate, relative to shoot growth, until this set fraction is achieved. Similar responses happen for shoots. Chemical signals probably are interchanged between roots and shoots to control growth and development.

Rendig and Taylor (1989) gave information on the root growth characteristics of some plant species. Wheat (*Triticum aestivum* L.) roots in England grew at a rate of 6 mm/day between December and 8 April; between 8 April and 31 May they grew at a rate of 18 mm/day. They grew to a total depth of 2 m. In Germany wheat roots grew at a rate of 17 mm/day between 23 April and 18 June. Total depth of penetration was 1.95 m. In Colorado wheat roots were reported to have grown to a depth of 3 m. In England a total root length of 2.4×10^4 m was found to be present beneath a square meter of land area. In Africa 3.5×10^3 m were found beneath each square meter.

Rendig and Taylor (1989) also reported that the soybean radicle develops into a taproot that grows downward at 25 to 50 mm/day. This root can grow to depths of 1.5 m but often terminates at lesser depths because of adverse soil conditions. Lateral roots emerge at 90-degree intervals around the taproot circumference. A large proportion of the soybean root system consists of four to seven extensively branched roots that originate near the base of the taproot. Basal roots 2 to 3 mm in diameter grow outward at a characteristic angle that depends upon cultivar and soil temperature. They grow laterally for 0.2 to 0.36 m then abruptly turn downward and rapidly grow to a 1.8 m or greater depth. Their diameters even at 1.5 to 1.8 m depth are about 1 mm. They often grow faster than the taproot.

Masle et al (1990) studied the effects of soil penetration resistance and ambient partial pressure of CO_2. They concluded that increased soil resistance induces a factor that retards shoot growth, partly by decreasing its sensitivity to carbohydrate levels, making more carbon available to root growth. Roots growing under mechanical resistance were visibly thicker and the faster growth rate was based upon root dry weight rather than root length. Carmi et al (1983) also found that restricting root growth of bean (*Phaseolus vulgaris* L.) by growing the roots in small pots resulted in a suppression of shoot growth, in spite of the fact that there was a greater production of dry matter per unit leaf area. Ben-Porath and Baker (1990) found that taproot restricted cotton (*Gossypium hirsutum* L.) had a better partitioning of assimilates into yield components. Earliness of fruit occurred at the expense of frame. Their results indicated a potential for agronomic improvement of drip irrigated cotton via taproot restriction. They suggested one means of implementation of this effect would be through dense canopies such as would occur in narrow-row cotton.

There have been many references to the concept of root signals for controlling both root and shoot growth. An example of such a study is provided by the work of Gowing et al (1990). They found that leaf area expansion rate declined in a drying soil in spite of the fact that bulk water potential of the leaves was not detectably perturbed by the treatment. The growth data provide strong evidence that shoot growth is controlled by a positive chemical signal. Blum et al (1991) saw evidence in their research with wheat (*Triticum aestivum* L.) grown in a drying topsoil that a nonhydraulic root signal may have a role in reducing leaf area, preparing the plant for a forthcoming water deficit. Kramer (1988) cautioned against drawing sweeping conclusions from laboratory experiments. He stated that under field conditions the shoots usually wilt while the roots are still actively growing. Thus, it seems unlikely that in the field the roots are often primary sensors of water stress or that biochemical signals from the roots are as important as the direct hydraulic effects of shoot water stress.

Brown (1984) reported on the use of a fiber-optic scope with television microcamera to record soybean (*Glycine max* L.) root growth to a depth of 1.2 m. He described different stages of growth depending on age of the plants. He emphasized the need for more efficient root growth, better distribution, and longevity of the root system for continuous water and nutrient absorption. This need includes altering the soil environment to a greater depth. The surface and subsoil properties such as acidity, fertility, and tilth must be made more favorable for water and nutrient uptake.

The subject of root competition between species has been studied. Martin and Snaydon (1982) grew barley (*Hordeum sativum* L.) and field beans (*Vicia faba* L.) both by themselves and with either their roots or shoots or both roots and shoots intermixed. Nitrogen supply, total density, and relative time of planting were also varied. The relative total yields (RTYs) of mixtures were significantly greater than 1.0 only when the root systems intermixed. The RTYs were reduced when N fertilizer was applied. They concluded that the yield advantage of intercropping was due mainly to beans and barley using different N sources. The competitive ability of barley was greater than that of beans because of its fine root system. Sowing barley before beans further increased its competitive ability. Root competition had a much greater effect on the relative performance of the two species than did shoot competition. Bozoa and Oliver (1990) studied the competitive mechanism of cocklebur (*Xanthium strumarinus* L.) and soybean (Glycine max L.) during early growth stage. Soybean was approximately 1.5 to 2.0 times taller and was more competitive above ground than cocklebur during the first few weeks of growth. Cocklebur had an approximate 20 to 50% more fine root system and was more competitive below ground than soybean during early growth. This below-ground competitiveness of cocklebur may be attributed to its fine root system. These authors document in their literature review that full season cocklebur interference with soybean had caused yield reductions in soybeans of 10 to 80%.

Russell (1977) stated that roots can exert longitudinal pressures of 0.9 to 1.3 MPa but considerably smaller pressures can much reduce root extension. Rendig and Taylor (1989) give the range of root growth pressures to be 0.9 to 1.5 MPa. They also say that much smaller resistance will reduce elongation rate substantially. Impeded roots are thicker and distorted, a factor also demonstrated vividly by Veen (1982). Bennie (1991) cites Pfeffer's (1893) range of root growth pressures as 0.7 to 2.5 MPa and gives as the accepted values today for axial root growth pressure 0.24 to 1.45 MPa and for radial pressures values of 0.51 to 0.90 MPa. To express the root growth-mechanical impedance relation in terms that have practical field applicability, the previously cited model of Dexter (1987) has much merit because it takes into account soil water potential, penetration resistance, and plant species differences. There remains a need for much research on various crops regarding the most efficient level of root growth, taking into account vegetative plant growth vs fruit formation and retention, seasonal weather effects, irrigation vs nonirrigation and many other factors.

An emerging area of research that shows much promise is computer simulation. Jones et al (1991) have reviewed several categories of root models and have attempted to develop a model that considers the major soil properties and crop characteristics affecting root growth. Their model integrates several of the major factors affecting root system growth and death in soils. It is designed to be a component of simulation models of crop growth and development. Models such as this have the possibility of utilizing existing data bases on weather, soil properties, etc, which would enable better crop management strategies to be developed.

Effects of Various Cultural Operations on Soil Parameters that Influence Plant Growth

General Soil-Crop Relationships

Numerous research efforts on a laboratory scale have demonstrated that high soil resistance to root penetration results in less root elongation and thicker roots and also reduces shoot growth. Masle and Passioura (1987) working with wheat (*Triticum aestivum* L.) showed that leaf area, shoot dry weight, and root dry weight were negatively correlated with soil strength as measured by penetrometer resistance. The growth of roots was less affected than that of shoots. High soil strength also produced smaller stomatal conductance. All leaf, shoot, and stomatal effects were the same regardless of whether changes in soil strength were brought about by changes in soil water content or changes in soil bulk density. They suggested that growth of the shoot is primarily reduced in response to some hormonal message induced in the roots when they experience high soil strength. Tu and Tan (1991) found that plant biomass (roots and shoots) and plant height were reduced as degree of compaction increased. Uniformity of plant growth was also reduced and maturity was delayed at higher levels of soil compaction. Cornish et al (1984) in pot experiments with ryegrass (*Loliam perenne* L.) found that increasing bulk density of a sandy loam soil from 1.0 to 1.54 Mg/m^3 slightly increased root diameter and reduced root hair length. Increasing bulk density caused an increase in soil penetration resistance and a reduction in root length. A soil penetration resistance of 5.89 MPa reduced root elongation rate to zero. Carr and Dodds (1983) stated that experiments in England with several arable and vegetable crops have shown that loosening the subsoil, even when there is no obvious compact layer, can increase yields by as much as 30%. They worked with a structureless sandy loam packed in glass tubes to 1.25 and 1.50 Mg/m^3 with lettuce (*Lactuca sativa* L.). They found that root elongation rate in the higher-density soil was 7 to 9 mm/day compared to 20 mm/day in the lower-density soil. Lowery et al (1970) artificially established soil pans in metal drums placed in dug holes 0.65 m in diameter $\times$ 0.76 m deep. The drums were placed in the holes then refilled with soil after providing a drainage system. Pans were formed at 0.10 m, 0.2 m, and 0.3 m by compacting with a metal disk. At each depth-to-pan treatment, final plant height and yield of cotton (*Gossypium hirsutum* L.) were reduced as soil bulk density or penetrometer resistance increased. Growth rate and yield were less in the 0.10 m depth-to-pan treatment than in the 0.20 or 0.30 m depths. They concluded that the reduction in yields was caused by failure of roots to penetrate soil pans. The reduced yields were not due to a shortage in oxygen. Almost no water was extracted by roots from below the soil pans. Taylor et al (1964) stated that when soil pan strength is greater than 3.0 MPa, no cotton roots can penetrate Amarillo fine sandy loam. Growth rate and yield of both cotton and grain sorghum (*Sorghum vulgare* L.) were drastically reduced as soil pan strength increased to 2.5 MPa, but further strength increases above 2.5 MPa did not reduce yields. Yields of cotton and grain sorghum from high-strength pans were about half those from low strength soil.

Hardy et al (1990) recommended that improper management of topsoil can lead to chemical restrictions to root development in argillic horizons of ultisols. Ca, Mg, and K applied as lime, gypsum, and fertilizer move downward with time and accumulate in the subsoil. The enrichment of subsoil with bases, especially Ca, decreases exchangeable Al, which improves soil environment for root growth and increases volume of soil available to plants for extraction of water and nutrients.

Torbert et al (1990) studied four lines of fescues (*Festuca arundinacea* Shreb) plus a Ky 31 check to determine the relative ability of their roots to grow through compact soils. These plants were grown on soils known to have compacted zones at 0.4 to 0.6 m. Two of the fescue lines with large diameter roots penetrated the hardpan and extracted soil water to a depth of 1.2 m. These authors stated that in Alabama, for a rooting depth of 0.3 m, there will be 42 days of water stress from May through July, in 5 out of 10 years.

Traffic Effects on Soil Conditions Affecting Root Growth

Gaultney et al (1982) cited indications that the cumulative effects of subsoil compaction in Indiana can severely reduce corn growth, vigor, and yield. The natural forces of freezing and thawing do not appear to be reliable agents for breaking up soil pans. Some of the increases in soil bulk density appeared to be due to tillage implements, but the majority of compaction is due to pneumatic tires. Tractors, implements, wagons, trucks, nurse tanks, fertilizer spreaders, combines and other production machinery pass over almost the entire area of a field. Some of these field operations take place when soil moisture is high, which results in greater soil compaction. For tires normally inflated to 0.069 to 0.103 MPa, stresses applied to soils varied from 0.138 to 3.45 MPa. Both static and dynamic load can compact soil. Most compaction is caused by dynamic loading. The first wheel pass does about 80% of the total compaction. Gaultney et al (1982) documented over 50% yield reductions due to subsoil compaction in a relatively wet summer. Moderate subsoil compaction reduced yields by 25%.

Carpenter et al (1985) listed the factors affecting depth and degree of soil compaction: (1) vehicle weight, (2) weight distribution, (3) soil type, (4) soil water content, (5) amount of traffic, (6) wheel slip, (7) drawbar pull, and (8) vehicle speed. They recommended that future design criteria for equipment that travels on soil should include limited ground contact pressures. These authors called attention to the fact that subsoils may be wet when topsoil is at optimum moisture content for field work. Blackwell and Soane (1981) developed a simplified soil mechanical model to predict soil compaction beneath wheels when running on soils of certain characteristics. This model gave good results for soils with bulk densities greater than 1.1 Mg/m^3 and for soil penetration resistances greater than 0.5 MPa. They pointed out that contact pressure alone can be a misleading guide to soil compaction. Increases in soil bulk density below the 0.1 m depth are considerably influenced by total wheel load. The most effective way of reducing compaction requires the use of both minimum load and maximum tire contact area.

Raghavan and McKyes (1978) studied the effects of repeated tire passes on soil dry bulk density. They drove a tractor on assigned plots for 1, 5, 10, and 15 passes. They measured soil water content and dry bulk density (at 0.05 m increments to a depth of 0.40 m). They developed many equations for different soil types and conditions. For example, the following equation was presented for the prediction of dry bulk density of a sandy loam soil to include all levels of soil water content and for slip values of less than 20%:

$$\gamma_{dry} = 2.47 + 0.0052(y) - 0.00013(x) + 0.0024(s) + 0.0069\ln(np) - 0.415\ln(mc) \quad (3.4)$$

where y = depth (cm), x = horizontal distance from center of tire (cm), s = slip, np = number of machine passes, and mc = soil water content.

This equation was determined by stepwise regression techniques and had an R^2 value of 0.84. Fausey and Dylla (1984) compared yields of corn (*Zea mays* L.) and soybeans (*Glycine max* L.) on different rows that were produced with six-row equipment. Rows 1 and 6 had no adjacent wheel traffic while rows 2 to 5 had five passes of a tractor with 3.5 Mg rear axle load. Neither soybeans nor well-fertilized corn suffered yield reductions due to five passes of the tractor wheels along one side of the rows even though the interrow traffic compacted the soil significantly in the top 0.3 m of the soil profile. In a case where supplemental fertilizer was not added to the corn, yields in the trafficked rows were 11% less than from rows with no adjacent wheel traffic.

Lowery and Schuler (1991) reported on soil compaction caused by massive farm equipment. They used three different machines that had 4.5, 8.0, and 12.5 Mg/axle loads. Soil compaction from these machines extended to the subsoil and persisted for more than 4 years. Voorhees et al (1986) performed a study that exhibited some of the variable results that may be expected from compaction experiments. Their experiments included six treatments, a control with no compaction, 9 Mg/axle load, and 18 Mg/axle load each with and without annually applied interrow surface compaction of 4 Mg/axle load. They measured a 9 and 30% corn (*Zea mays* L.) yield reduction respectively. In the second year, the 18 Mg/axle load reduced yields 12%. High axle loads on dry soil caused little subsoil compaction. Grain yields were only reduced by 6% by these high axle loads. High axle loads on a relatively wet soil caused compaction to a depth of 0.6 m, but dry climatic conditions the following year negated potential adverse effects, yields were not affected. Surface layer compaction from annual interrow traffic did not cause significant yield decreases.

Gameda et al (1985) applied 10- and 20-Mg/axle loads to two soil types at two soil water contents. Topsoil and subsoil dry density levels were significantly increased by high axle loads. Moldboard plowing to 0.2-m depth did not fully relieve subsoil compaction. In a loam field the effects of 10-Mg axle load were minimal, but in a clay soil, corn yields were significantly reduced to 43 to 68% of the control plots.

In a continuing international study reported by Hakansson et al (1987) the effects of high axle-load traffic on subsoil compaction and crop yield are being measured. This study involves 26 fields in Europe and North America. It was concluded that axle loads greater than 6 Mg may cause subsoil compaction to depths greater than .4 m. In their other treatments of 10 Mg per single axle or 16 Mg per tandem axle, compaction was caused to a depth of 0.5 m. In Sweden vane shear strength of the soil was 17% higher after compaction. Six years later the soil strength was 15% higher even with freezings. Based upon presently available data, one of their conclusions was that axle loads greater than 10 Mg should not be used. Reporting on the North American part of this international study, Voorhees et al (1986) showed that when subsoils were relatively wet, bulk density increased about 0.08 Mg/m^3 from the 0.3- to 0.5-m depth. Saturated hydraulic conductivity decreased from 20 to 2 mm/h with an 18-Mg/axle load on a Webster clay loam. Evidence of subsoil compaction was measured 4 years after treatment, in spite of annual winter freezing, to a depth of 0.9 m. They recommended that a total description of the soil's pore structure would seem more appropriate than soil bulk density since the former relates to water flow characteristics and root extension. They considered that measurements of hydraulic conductivity (preferably *in situ*) and penetrometer resistance would be better indices of the soil condition.

A number of researchers have worked with permanent traffic lanes and wide frame vehicles in an attempt to minimize the soil compaction problem. Taylor (1983) reported that controlled traffic research began for the purpose of increasing crop yields by reducing traffic-induced soil compaction in the crop zone. However, he further explained that permanent traffic lanes have many additional benefits: improved tractive efficiency, improved mobility, and better timeliness of operations. Carter (1985) reported on a 5-year study in which wheel traffic was eliminated from normal furrow placement for irrigated row crops. No wheels were allowed within 0.75 m of plant rows, as measured from the edge of the tire to the row. Water infiltration, volume of soil for root exploration, and root density were increased. Tillage energy was reduced. The nontrafficked plots had substantially less high-penetration-resistance subsoil barriers. The trafficked plots had layers with greater than 1.3 MPa penetrometer resistance. However, over the 5 years of the study, cotton lint yields were only 2% better in the nontrafficked plots. Hadas et al (1990) studied forage wheat yields using conventional and wide-frame tractors with wheels of the same size. Distinct yield reductions caused by soil compaction in the conventional traffic plots were statistically quantified. They found a linear relation between wheat dry matter yields and soil bulk density. Soil bulk densities were generally lower under the wide-frame power system. Carter et al (1989) compared the following three treatments for cotton (*Gossypium hirsutum* L.) production: (1) broadcast system, a conventional management system with no restraint on tractor or equipment wheel placement, (2) precision tillage system, an adaptation of the broadcast system in which the last primary tillage was subsoiling under the intended drill row accompanied by bedding, and (3) zone tillage system, in which wheels were restrained to permanent paths and tillage was varied by specific zones to enhance water infiltration and root exploration. The study indicated that tillage was required under any surface area where wheels were operated to return the soil to a low mechanical impedance and to maintain a reasonable infiltration rate. For the zone system with no traffic or tillage the soil was changed to a stable state not achievable with trafficked systems. If roots had the capacity to penetrate at a resistance of 1 MPa, then 69.3% of the soil was available for exploration with the zone system. Major traffic increased the penetration resistance by a factor of 1.5 to 3.0 over nontrafficked, nontilled soils. Rechel et al (1990) found that wheel traffic did not alter fine root development (FRD) in alfalfa (*Medicago sativa* L.) until early in the second season of production. In this second year FRD was reduced by traffic in the upper 0.45-m soil layers, and with extreme traffic, FRD was reduced to 1.8 m. The plants' ability to extract water and nutrients was reduced by traffic. Hood et al (1991) have developed equipment and methods for implementing controlled traffic with conventional tractors in small grain — row crop production systems. With these methods, small grain drill rows are established in a controlled pattern to serve as guide rows for later production operations. The row crops, soybeans or cotton, are then interseeded in the small grain before it is harvested. Various drill and row spacings have been investigated in this on-going research. Tractor, implement, and harvester wheel spacings of 1.93 and 2.44 m have been successfully utilized with this system.

Soane et al (1982) gave some excellent recommendations for planning crop production systems. They recommended that we devote effort to avoid subsoil compaction. They called attention to widespread evidence that such compaction may persist for many years even with deep freezing. Compaction from wheel traffic has generally an adverse influence on all stages of crop growth. However, in some situations, crop responses to compaction are beneficial. They recommended fewer wheel passes, reduced loads and inflation pressures, confining traffic to pre-arranged strips, and zero tillage where possible.

Tillage Effects on Soil Conditions Affecting Root Growth

Surface soils from long-term no-till and conventional tillage plots at seven US locations were characterized by Doran (1980). The seven locations were situated in the states of Kentucky, Minnesota, West Virginia, Nebraska, and Oregon. Differences in soil microbial populations and enzyme activities with no-till compared to plowing were related to soil water, organic carbon, and nitrogen levels and pH. Water appeared to be the primary factor influencing microbial populations. Soil pH, C, and N may be as much a result of microbial activity as they are regulators of it. Higher C, N, and water of the surface soil under no-till are reflected by higher microbial populations and enzyme activities as compared to conventional tillage. These relationships are reversed at the 0.075- to 0.150-m depth with the conventional systems having higher microbial and enzyme activities, apparently the result of placement of crop residue at these depths with plowing and the higher soil moisture at this depth than on the tilled surface.

In a special publication by the Agronomy Society of America (Van Doren et al, 1982), the following mutual statement by the Presidents of the Agronomy Society of America and the Soil Science Society of America appeared: "There is an increasing awareness in the USA and in the world that much of the current level of agricultural production is being achieved at the expense of our nonrenewable soil resources".

Further discussion in this report emphasizes that research and farmers' experience indicate that tillage is responsible for a major part of soil structure deterioration. Adverse effects of tillage on soil structure are well-established: oxidation of organic matter by exposure at the surface, mechanical dispersion by puddling through the compaction and shearing action of implements, and by rainfall impact on bare soil (Larson and Osborne, 1982).

Fitzgerald et al (1968) conducted a study in South Dakota in which they mechanically injured roots of corn (*Zea mays* L.) to simulate the feeding of root worm larvae. They removed approximately 1, 25, 50, and 75% of the roots by cutting with a sharpened spade on one, two, or three dates. Some varieties were able to tolerate considerable root damage because they produced new roots quickly in response to damage. Corn plants could withstand severe root damage and still produce yields that were 85 to 90% of those produced by undamaged plants, provided moisture and fertility were favorable and injury occurred relatively early in the life of the plant. Later injury means that the plant has less time to recover. This data might be useful in making decisions concerning cultivation for the control of weeds. Reicosky et al (1977) stated that the Southeast could probably benefit more from conservation tillage than any region of the US, since the primary advantage of no-till is the conservation of soil water. However, in developing the conservation tillage systems, the nature of soil problems should be considered. Hardpans have been a problem. These frequently are traffic pans, but some are genetic in nature. Pans may slow root growth to 6 to 8 mm/day compared to 50 to 70 mm/day in loose soils. Time for roots to penetrate may result in water stress at particular times, yet roots may look normal at the end of the season. In addition to increasing rooting depth and thus the amount of water available to plants, chiseling also increased oxygen content of the soil during extremely wet periods. Chiseled soil permitted water to infiltrate and percolate to greater depths, thus decreasing the degree of saturation in the upper root zone. Chiseling was not effective if machinery subsequently recompacted the soil.

Spoor and Godwin (1978) gave some practical facts concerning deep tillage. They said that each deep tillage tine has a maximum useful working depth. Below this depth, soil compaction rather than loosening occurs and more force is required to pull the tool. The optimum depth is dependent on tine geometry and soil conditions. They stated that increases in soil disturbance with lower specific resistance can be achieved by attaching wings or sweeps and through the use of shallow tines working ahead of deep tines. At shallow working depths the soil is displaced forward, sidewise, and upwards, failing along well-defined rupture planes which radiate from just above the tine tip to the surface at angles of approximately 45° to the horizontal. At a certain depth, known as the critical depth, soil at the tine base begins to flow forward and to the side, creating compaction at depth.

Kamprath et al (1979) found that by breaking the tillage pan by chisel plowing or subsoiling, they could achieve increased top growth and leaf area of soybeans (*Glycine max* L.) at full bloom. Percentage of roots below 0.3 m was increased by chisel plowing and subsoiling, when rainfall was below normal during the growth period. Yield increases from chisel plowing and subsoiling were attributed to increased root proliferation below the tillage pan and greater utilization of subsoil moisture.

Rouse and Stone (1981) increased depth of tillage from 0.23 to 0.45 m and found that potatoes, broad beans, summer cabbage, and red beet extracted less water from the 0.00- to 0.30-m depth and extracted more water from the 0.3- to 0.7-m depth than with shallow tillage. They indicated that improved root distribution accounted for the beneficial effects on water use.

Bowen (1981) recommended the following for evading root impedance by compact soil: (1) control traffic, (2) control water content at critical times to optimize root growth, (3) use plants to shatter pans, (4) alter soil organic matter content, polyelectrolytes, earthworm management, and lime addition.

Whitely and Dexter (1982) worked with linseed (*Linum usitatissimum*), pea (*Pisum sativum*), rape (*Brassica napus*), safflower (*Carthamus tinctorius*), soybeans (*Glycine max* L.), sunflower (*Helianthus annuus* L.) and wheat (*Triticum aestivum* L.) to investigate the effects of different degrees of soil disturbance beneath planted rows. They had five soil treatments: (1) broadcast tine-tilled to a depth of 0.120 m, (2) nontilled with seeds sown in 0.03-m slots, (3) nontilled with seeds sown in 0.030 m deep holes, (4) nontilled with seeds sown in 0.003 m slots with cracks extending from base of slots downward to 0.12 m, and (5) same as treatment 4 except slots extended to a depth of 0.3 m. Treatments 2 and 3 restricted seminal root growth relative to treatment l. The degree of restriction was dependent on soil penetrometer resistance before planting. Growth of lateral roots was also less in treatments 2 and 3 than in treatment 1. The reductions in early growth for treatments 1 and 2 were associated with reduced dry matter and yield of the crops. However, the presence of cracks in treatments 4 and 5 provided a zone for unrestricted root growth and dry matter production and yields were comparable to treatment 1. Suitability for growth in nontilled soil decreases in the order: wheat > pea > rape > linseed > safflower and sunflower. The authors concluded that it may be worthwhile to develop and test a tillage machine that produces cracks below the seed row.

Elkins et al (1983) reported that subsoiling has a high energy requirement, has short-term beneficial effects, and leaves an undesirable mix of soil horizons. A deep tillage system was thus developed that required less energy, provided long-term amelioration of plow pans, and did not mix soil horizons. A narrow, vertical slit was cut through the plow pan beneath the row at planting. The slit eliminated soil strength and low soil oxygen as deterrants to roots growing through the pan. Once the slit was filled with organic matter from decaying roots, it became a lasting feature of the soil profile. Energy savings over conventional subsoiling were 12 to 43%. Soybean yields were 7.5 to 18.9% greater for slit-planted treatment in 1982.

Box and Langdale (1984) showed significant increases in corn (*Zea mays* L.) yields associated with in-row subsoiling in the coastal plains of Georgia. Campbell et al (1984) described three special experiences in working with conservation tillage for soybeans (*Glycine max* L.):

1. Full-season soybean under conservation tillage produced yields equal to or better than full-season, full-tillage because suppressed biomass production with the conservation tillage conserved soil water and favored growth during the reproductive phase.
2. Late-season soybean yields behind wheat favored the conservation tillage practice of in-row subsoil planting. However, planting in burned-off wheat stubble produced highest yields.

3. In a dry spring, a growing cover crop accelerated water use, which resulted in lower soybean yields under conservation tillage.

Karlen et al (1991) also documented a situation in which a cover crop transpired much of the soil water built up through the winter. Low soil water in the cover crop treatment resulted in abnormally high soil penetration resistance and higher energy requirements for subsoiling.

Chaudhary et al (1985) conducted a study on a loamy sand soil in which the soil bulk density distribution did not show a distinct root-limiting soil zone. They imposed treatments of subsoiling (SS) to 0.4 m, deep mold board plowing (MB) to 0.2 m, and a hand deep digging (DD) to 0.45 m. These treatments were compared with conventional tillage and without irrigation. Tillage operations slightly decreased bulk density at all working depths. SS and DD decreased soil penetrometer resistance in the 0.2- to. 04-m layer to one-tenth that of the control. SS and DD induced deeper and greater rooting and increased profile water use compared with conventional tillage. SS, MB, and DD increased plant height by 0.30 to 0.35 m, yielded 80 to 100% more stover and 70 to 350% more grain. This soil had a high penetrometer resistance at a relatively low bulk density. High strength without high bulk density could be ascribed to the rough surface of the sand particles. Loosening of the soil and disturbance of particle orientation brought about by subsoiling promoted root penetration. These findings suggest that bulk density may be influenced by soil-particle surface roughness and that surface roughness of soil, as well as soil wetness and soil aeration, may affect root morphology.

Bennie and Botha (1986) found that deep ripping and controlled traffic led to a significant increase in rooting depth, rooting density in the subsoil, water use efficiency and a yield increase of 30% for corn (*Zea mays* L.) and 19% for wheat (*Triticum aestivum*). Busscher and Sojka (1987) acknowledged that there was some disagreement regarding cone index readings for limiting root growth. However, most of the existing literature currently indicated that root growth was restricted beyond 2.0 MPa, measured with a flat-tipped penetrometer. This value corresponds to 2.5 to 3.0 MPa for the 13-mm, 30-degree cone tip. These researchers found higher yields of soybeans (*Glycine max* L.) in 1984 (1534 kg/ha compared to 1332 kg/ha) and corn (*Zea mays* L.) in 1985 (6620 kg/ha compared to 6360 kg/ha) for conservation tillage treatments. Conservation tillage treatments had a more uniform distribution of soil strength. NeSmith et al (1987) studied soil compaction in double-cropped wheat (*Triticum aestivum* L.) and soybean (*Glycine max* L.). They had the following treatments: no tillage, disking, and moldboard plow plus disk. Fall tillage prior to wheat sowing was accomplished by disking. The disked and no-till treatments had a compacted layer at 0.15 to 0.25 m attributed in part to fall disking. At 30 days after planting, when soil moisture was low, soil penetration resistance rose to 4 to 6 MPa at the 0.2-m depth. This occurred in both the disked and no-till treatments. The moldboard plow treatment had a penetration resistance of 1.0 MPa at the 0.2 m depth. For this soil condition they recommended moldboard plowing to alleviate the compaction problem.

Dickey et al (1991) conducted an 8-year study to evaluate six tillage and planting systems used for soybean (*Glycine max* L.) and grain sorghum (*Sorghum vulgare* L.). They evaluated no-till, no-till with cultivation, disked, double-disked, chiseled, and plowed treatments. For the first 3 years of measurement there were no significant differences in soybean yields. For grain sorghum, no-till had lower yields in the first year, but there were no significant differences in yields in the next 2 years. For the last five years of the study, no-till tended to have greatest yields for both crops. Soil organic matter was highest for continuous no-till treatment and lowest for the plow treatment. Soil resistance to penetration as measured by a cone penetrometer was not appreciably different for the tillage systems evaluated. Of the factors measured, residue cover exhibited the largest difference among the tillage and planting systems. More residue was associated with less tillage and resulted in larger yields.

Conclusions

The reader of the aforementioned books and articles and similar materials on soil-plant dynamics will be amazed at the elegant work that has been completed and surprised at the subject matter that is not yet completely understood.

There are opportunities for improvements at the level of basic as well as applied research. The mystery of root signals, which could be of much importance to producers and basic scientists, has not yet been solved.

There is no doubt that soil compaction is a serious problem. Yet some researchers continue to find circumstances in which limiting plant vegetative growth through restriction of roots has resulted in higher yields. It is possible that a way may be found to limit vegetative growth through the use of plant growth regulators while allowing roots to explore subsoils for maximum water availability during the fruiting phase of plant production. Understanding the plant responses to physical soil parameters will enable us to better respond to conditions of stress such as drought, root impedance, etc.

References

Advances in Soil Dynamics

Chapter 1*

Agricultural Engineering Abstracts. 1989-92. Commonwealth Bureau of Agriculture. Wallingford, Oxon, U.K. (2)

Agricultural Research Service. 1965. The National Tillage Machinery Laboratory, Auburn, Ala. USDA-ARS, ARS 42-9-2. (6)

Anon. 1934. The 1934 McCormick Medal Award. *Agric. Eng*. 15:5:175. (v)

Anon. 1985a. *Proc. International Conference on Soil Dynamics* Vol. 1-5. Auburn, Ala. (v)

________. 1985b. *Proc. International Conference on Soil Dynamics* Vol. 1:155. Auburn, Ala. (v)

Araya, K. and K. Kawanishi. 1984. Soil failure by introducing air under pressure. *Transactions of the ASAE* 27(5):1292-1297. (6)

Armbruster, K. and H. D. Kutzbach. 1989. Development of a single wheel tester for measurements of angled driven wheels. ISTVS *Proc. 4th European Conference* Vol. 1:8-14. Waginengen, The Netherlands. (21-23 March). (17)

ASAE, 1977. *Similitude of Soil-Machine Systems*. 1977. ASAE Publication No. 3-77. St. Joseph, Mich.: ASAE. (18)

ASAE Standards, 39th Ed. 1992. S313.2 Soil cone penetrometer. St. Joseph, Mich.: ASAE. (16,18)

Bailey, A. C. and E. C. Burt. 1988. The effects of tire dynamic load on soil stress state. ASAE Paper No. 88-1511. St. Joseph, Mich.: ASAE. (16)

Balovnev, V. I. 1969. *Physical Modeling of Soil Cutting*, p. 160. Moscow: Machine Construction Publishing House. (Technical Translation TT 81-58108. Translated for the Science and Education Administration, USDA and the National Science Foundation, Washington D.C. 1981). (Translation available at NSDL.) (18)

Batchelder, D. G., J. G. Porterfield, T. S. Chisholm and G. L. McLaughlin. 1970. A continuous linear soil bin. ASAE Paper No. 70-121. St. Joseph, Mich.: ASAE. (11,12)

Black, C. A., ed. 1965. *Methods of Soil Analysis*, 114-125, 383-390. Madison, Wis.: American Society of Agronomy. (16)

Burt, E. C., C. A. Reaves, A. C. Bailey and W. D. Pickering. 1980. A machine for testing tractor tires in soil bins. *Transactions of the ASAE* 23(3):546-547, 552. (18)

Burt, E. C., R. K. Wood and A. C. Bailey. 1987. A three-dimensional system for measuring tire deformation and contact stresses. *Transactions of the ASAE* 30(2):324-327. (17)

Chancellor, W. J. and R. H. Schmidt. 1962. A study of soil deformation beneath surface loads. *Transactions of the* ASAE 4:224-225. (7)

Durant, D. M., J. V. Perumpral and C. S. Desai. 1981. Soil bin test facility for soil-tillage tool interaction studies. *Soil Tillage Res*. 1(3):289-298. (6)

Dwyer, M. J., J. A. Okello and F. B. Cottrell. 1990. A comparison of the tractive performance of a rubber track and a tractor driving wheel tyre. *Proc. 10th International Conference of the ISTVS* Vol 1:289-299. Kobe, Japan. (20-24 Aug). (17)

Ellen, H. 1984. Tillage effects and specific energy requirements of rotary tillage. *Soil Tillage Res*. 4:471-484. (9)

---

* Numbers in parenthesis following each reference are the pages on which the reference is cited.

Feller, R., C. Charalambous, S. Orlowski, D. Wolf and A. Yavnai. 1971. *Effect of knife angles and velocities on cutting roots and rhizomes in the soil.* Bed Dagan, Israel: The Volcani Institute of Agricultural Research and The Agricultural Engineering Institute. (Final report submitted to the USDA). (6,9)

Freitag, D. R., R. L Schafer and R. D. Wismer. 1970. Similitude studies of soil-machine systems. *Transactions of the ASAE* 13(2):201-213. (18)

Freitag, D. R. 1971. A soil bin study of wheels for the lunar roving vehicle. ASAE Paper No. 71-131. St. Joseph, Mich.: ASAE. (18)

Fielke, J. M. and S. D. Pendry. 1986. SAIT tillage test track. In *Conference on Agricultural Engineering*, Adelaide, Australia: The Institution of Engineers. (2,8)

Gill, W. R. 1968. Influence of compaction hardening of soil on penetration resistance. *Transactions of the ASAE* 11(6):741-745. (7)

________. 1969. Soil deformation by simple tools. *Transactions of the ASAE* 12(2):234-239. (6,7,16)

________. 1985. A history of soil dynamics and the National Tillage Machinery Laboratory. In *Proc. International Conference on Soil Dynamics* Vol. 1:162-178. Auburn, Ala. (June). (v)

________. 1990. *A history of the USDA National Tillage Machinery Laboratory 1935-1985.* Auburn, Ala.: W. R. Gill. (2)

Gill, W. R. and R. C. Clark. 1985. History of the National Tillage Machinery Laboratory. In *Proc., International Conference on Soil Dynamics* Vol. 1:37-56. Auburn, Ala. (June). (1)

Gill, W. R. and G. E. VandenBerg. 1967. *Soil Dynamics in Tillage and Traction.* Agriculture Handbook No. 316. USDA-Agricultural Research Service, Washington, D.C. (vii).

Godwin, R. J., G. Spoor and J. Kilgour. 1980. The design and operation of a simple low cost soil bin. *J. Agric. Eng. Res.* 25:99-104. (6)

Goryachkin, V. P. 1968/1898. The graphical theory of the plow (1898). *Collected Works in Three Volumes* Vol. II:6-55. Moscow: Kolos Publishing House, Republished 1968. (Translation No. TT 71-50087 by the USDA and the National Science Foundation, Washington, D.C., 1971). (1)

Hadas, A. and I. Shmulevich. 1990. Spectral analysis of cone penetrometer data for detecting spatial arrangement of soil clods. *Soil Tillage Res* 18(1):47-62. (16)

Jin, W. J., J. X. Nan, S. X. Liu and L. F. Guo. 1986. The developing of experimental equipment of soil bin for the working device of earth moving machine. In *Proc. First Asian-Pacific Conference of International Society for Terrain-Vehicle Systems*, 926-936. Beijing: China Academic Publishers. (4-8 Aug). (18)

Jones, W. B. Another agricultural engineering milestone. 1941. *Agricultural Engineering* 22:12:441-442. (v)

Karczewski, T. 1986. Research in soil bins. In *ISTVS 3rd European Conference. Proc. Off the Road Vehicles and Machinery in Agriculture and Forestry.* 13-25, Warsaw, Poland. (Sept). (2)

Koolen, A. J. and H. Kuipers. 1983. *Agricultural Soil Mechanics.* Berlin: Springer-Verlag. (vii)

Küehne, G. 1914. *Untersuchungen über den Zugwiderstand eines Pflugwerkzeugmodelles bie verschiedenen Arbeitsbedingungen und ihre Anwendung auf praktische Verhältnisse.* p. 52. Berlin Druck von Gebr. Unger. (2)

Lyne, P. W. L., E. C. Burt and J. D. Jarrell. 1983. Computer control for the NTML single-wheel tester. ASAE Paper No. 83-1555. St. Joseph, Mich.: ASAE. (18)

Martin, E. R. and N. L. Buck. 1987. A computer-controlled soil bin for soil dynamics research. ASAE Paper No. 87-1013. St. Joseph, Mich.: ASAE. (2,5,6)

McRae, J. L., C. D. Powell and R. D. Wismer. 1965. Test facilities and techniques. Technical Report No. 3-666: 69. Vicksburg, Miss.: U. S. Waterways Experiment Station, Corps of Engineers. (2,8,9,10,12,13)

Moechnig, B. W. and D. L. Hoag. 1979. Dynamic parameters of artificial soils. ASAE Paper No. 79-1042. St. Joseph, Mich.: ASAE. (11)

Morgan, M. T., R. K. Wood and R. G. Holmes. 1991. Dielectric moisture measurement of soil cores. ASAE Paper No. 91-1528. St. Joseph, Mich.: ASAE. (16)

Nartov, P. S. and I. I. Shapiro. 1971. Use of similitude methods to evaluate the design of soil bins. In *Mechanization and Electrification of Soviet Agriculture* 5:51-52. (Translated by W. R. Gill, National Soil Dynamics Laboratory, Auburn, Ala. Translation No. NTML-WRG-235). (18)

Nedorezov, I. A. and V. G. Moiseenko. 1986. Study of underwater working of soil. *Construction and Highway Machines* 4:23-24. (Translated by W. R. Gill, National Soil Dynamics Laboratory, Auburn, Ala.). (6)
Nichols, M. L. 1925. The sliding of metal over soil. *Agric. Eng.* 6(4):80-84. (7)
________. 1929. Methods of research in soil dynamics as applied to implement design. Bulletin 229:27. Alabama Experiment Station of the Alabama Polytechnic Institute. (1,6)
Nichols, M. L. and L. D. Baver. 1930. An interpretation of the physical properties of soil affecting tillage and implement design by means of the Atterberg Consistency Constants. In *International Congress for Soil Sciences,* Leningrad. 6:175-188. (1)
Onwuala, A. P. and K. C. Watts. 1989. Development of a soil bin test facility. ASAE Paper No. 89-1106. St. Joseph, Mich.: ASAE. (2)
Raper, R. L., L. E. Asmussen and J. B. Powell. 1990. Sensing hard pan depth with ground-penetrating radar. *Transactions of the ASAE* 33(1):41-46. (16)
Reaves, C. A. 1966. Artificial soils simulate natural soils in tillage studies. Part II. Similitude of plane chisels in artificial soils. *Transactions of the ASAE* 9(2):147-150. (11)
Reidy, F. and I. F. Reed. 1966. Traction in submerged soil. *Transactions of the ASAE* 9(4):464-467. (16)
Revut, I. B. and A. A. Rode. 1969. *Experimental Methods of Studying Soil Structure.* Leningrad: Kolos Publishing House. (Translation No. TT 75-52008, USDA and the National Science Foundation, Washington, D.C., 1981). (16)
SAE. 1990. The National Soil Dynamics Laboratory, USDA-ARS—International Historic Mechanical Engineering Landmark—Historic Landmark of Agricultural Engineering. SAE - ASAE joint brochure. Auburn, Ala. (19 Oct). (v,1)
Schafer, E. D. 1993. Saving the good earth: Mark Lovell Nichols, soil dynamics, the pioneering of agricultural engineering. Ph.D. diss., Auburn University. (1)
Schafer, R. L., C. W. Bockhop and W. G. Lovely. 1968. Model-prototype studies of tillage implements. *Transactions of the ASAE* 11(5):661-664. (2)
Schafer, R. L., W. R. Gill and C. A. Reaves. 1979. Experiences with lubricated plows. *Transactions of the ASAE* 22(1):7-12. (17)
Sitkei, G. 1967. *Soil Mechanics Problems of Agricultural Machines*. Budapest: Academic Press. (Translation No. TT 75-58047, USDA and the National Science Foundation, Washington, D.C., 1976). (vii)
Smith, B. E., R. L. Schafer and R. L. Raper. 1993. Knowledge-based approach for determining soil bin preparation procedure. ASAE Paper No. 93-3042. St. Joseph, Mich.: ASAE. (13)
Soehne, W. H. 1985. Soil dynamics research at the National Tillage Machinery Laboratory: An international perspective. In *Proc. International Conference on Soil Dynamics* Vol. 1:98-128. Auburn, Ala. (June). (2,3)
Stafford, J. V. 1979. A versatile high-speed soil tank for studying soil and implement interactions. *J. Agric. Eng. Res.* 24:57-66. (6,9,10)
Studman, C. J. and J. E. Field. 1975. Motion of a stone embedded in a non-cohesive soil disturbed by a moving tine. *J. Terramechan.* 12(3/4):131-147. (6)
Sun, T. C., J. D. Jarrell and R. L. Schafer. 1986. Computer instrumentation for the NSDL soil bin penetrometer car. *Comput. Electronics Agric.* 1:281-288. (16,17)
Tajima, K., K. Tamaki and T. Kobayashi. 1992. Measurements of properties of granular material by a tine-sensor. *J. Japan. Soc. Agric. Mach.* 54:6:13-29. (13)
Tollner, E. W. 1993. Measurement of density and water content in soils with x-ray linescan and x-ray computed tomography. ASAE Paper No. 93-1086. St. Joseph, Mich.: ASAE. (16)
Turnage, G. E. 1970. Resistance of fine grained soils to high-speed penetration. Technical Report No. 3-652. Measuring soil properties in vehicle mobility research. Report 5-47. Vicksburg, Miss.: U. S. Army Engineer Waterways Experiment Station. Mobility and Environmental Systems Laboratory. (9)
Upadhyaya, S. K., J. Mehlschau, D. Wulfsohn and J. L. Glancey. 1986. Development of a unique single wheel traction testing machine. *Transactions of the ASAE* 29(5):1243-1246. (17)
Vincent, E. T. 1961. Land Locomotion in the United States Universities. In *Proc. International Conference on the Mechanics of Soil-Vehicle Systems.* 929. Turin, Italy. (12-16 July). (v)

Wang, R., H. Guo, C. Ay, R. T. Schuler, S. Gunmasekaran and K. J. Shinners. 1991. Ultrasonic method to evaluate soil moisture and compaction. ASAE Paper No. 91-1522. St. Joseph, Mich.: ASAE. (16)

Wells, L. G. and J. D. Buckles. 1987. PC-controlled soil/tire tester. ASAE Paper No. 87-1014. St. Joseph, Mich.: ASAE. (18)

Wilkins, D. E., W. J. Conley and P. A. Adrian. 1979. Soil bin for studying planting equipment and systems, AAT-W-4/Feb. U. S. Department of Agriculture, Science and Education Administration, Advances in Technology, Berkeley, Calif. (6,10)

Wismer, R. D. 1984. Soil bin facilities: Characteristics and utilization. In *Proc. 8th International Conference, International Society for Terrain-Vehicle Systems* Vol. III:1201-1216. Cambridge, England. (6-10 Aug). (10)

Wismer, R. D. and M. W. Forth. 1969. Soil dynamics research and Deere & Company. In *Proc. 3rd International Conference of the International Society for Terrain-Vehicle Systems* Vol. III:107-138. Tagungsort, Germany. (9-12 July). (2,6,9,10)

Wismer, R. D., H. J. Luth and W. W. Brixis. 1978. Field testing of soil-machine systems. In *Proc. 6th International Conference of the International Society for Terrain-Vehicle Systems* Vol. III:1211-1227. Vienna, Austria. (Aug). (17)

Yu, Q. and G. Xu. 1990. A new method for measuring shape and size of tire-soil contact zone. In *10th International Conference of the ISTVS* 1:171-175. Kobe, Japan. (Aug). (17)

Zelenin, H. N., V. I. Balovnev and I. P. Kerov. 1975. *Machines for Moving the Earth*. Moscow: Machine Construction Publishing House. (Translated by the USDA and the National Science Foundation, Washington, D.C., 1985). (vii)

Chapter 2*

Abernathy, G. H., M. D. Cannon, L. M. Carter and W. J. Chancellor. 1975. Tillage systems for cotton—A comparison in the western region. Bulletin 870. Berkeley, Calif.: University of California. (34)

Ali, O. S. and E. McKyes. 1979. Effects on soil thrust of lug angle, length and soil consistency. *Transactions of the ASAE* 22(6):1294-1298, 1304. (122,124)

Allen, W. H. and J. I. Sewell. 1973. Remote sensing of fallow soil moisture by photography and infrared line scanner. *Transactions of the ASAE* 16(4):700-706. (73)

Amir, I., G. S. V. Raghavan, E. McKyes and R. S. Broughton. 1976. Soil compaction as a function of contact pressure and soil moisture content. *Can. Agric. Eng.* 18(1):54-57. (190,192)

Araya, K. and K. Kawanishi. 1984. Soil failure by introducing air under pressure. *Transactions of the ASAE* 27(5):1292-1297. (115)

Archer, J. R. and M. J. Marks, eds. 1985. *Techniques for Measuring Soil Physical Properties*. ADAS Reference Book 441. London: Her Majesty's Stationery Office. (22,130)

Aref, K. E. F. 1973. Investigations of shear strength characteristics of an unsaturated soil with regard to stress state, loading rate and hysteresis effect. Ph.D. thesis, University of California, Davis. (169,172,175)

Aref, K. E., W. J. Chancellor and D. R. Nielsen. 1975. Dynamic shear strength properties of unsaturated soils. *Transactions of the ASAE* 18(5):818-823. (169,212,213,229,231)

ASAE Standards, 31st Ed. 1984. ASAE S 313.1. Soil cone penetrometer. St. Joseph, Mich.: ASAE. (124)

ASTM. 1989. ASTM D421-85. Standard practice for dry preparation of soil samples for particle-size analysis and determination of soil constants. Vol. 04.08. Soil and Rock, Building Stones; Geotextiles, 84-85. Philadelphia, Pa.: Annual Book of ASTM Standards. (22)

---

* Numbers in parentheses following each refrerence are the pages on which the reference is cited.

_______. 1989. ASTM D422-63 (Reapproved 1972). Standard method for particle-size analysis and determination of soils. Vol. 04.08 Soil and Rock, Building Stones; Geotextiles, 86-92. Philadelphia, Pa.: Annual Book of ASTM Standards. (22)

_______. 1992. ASTM D854-91. Standard test method for specific gravity of soils. Vol. 04.08 Soil and Rock, Building Stones; Geosynthetics, 177-180. Philadelphia, Pa.: Annual Book of ASTM Standards. (27)

Awadhwal, N. K. and C. P. Singh. 1985. Mechanical and viscoelastic characteristics of a puddled soil. In *Proc., International Conference on Soil Dynamics* 3:471-480. Auburn, Ala.: National Soil Dynamics Laboratory. (212,213)

Ayers, P. D. 1987. Moisture and density effects on soil shear strength parameters for coarse grained soils. *Transactions of the ASAE* 30(5):1282-1287. (95,136)

Ayers, P. D. and H. D. Bowen. 1987. Predicting soil density using cone penetration resistance and moisture properties. *Transactions of the ASAE* 30(5):1331-1336. (126,136,138)

_______. 1988. Laboratory investigation of nuclear density gage operation. *Transactions of the ASAE* 31(3):658-661. (33,35)

Bailey, A. C. 1971. Compaction and shear in compacted soils. *Transactions of the ASAE* 14(2):201-205. (196)

_______. 1973. Shear and plastic flow in unsaturated clay. *Transactions of the ASAE* 16(2):218-221, 226. (196)

Bailey, A. C. and C. E. Johnson. 1989. A soil compaction model for cylindrical stress states. *Transactions of the ASAE* 32(3):822-825. (197.198)

Bailey, A. C. and G. E. VandenBerg. 1968. Yielding by compaction and shear in unsaturated soil. *Transactions of the ASAE* 11(3):307-311, 317. (195)

Bailey, A. C. and J. A. Weber. 1965. Comparison of methods of measuring soil shear strength using artificial soils. *Transactions of the ASAE* 8(2):153-156, 160. (95,136,138,183)

Bailey, A. C., C. E. Johnson and R. L. Schafer. 1984. Hydrostatic compaction of agricultural soils. *Transactions of the ASAE* 27(4):952-955. (176,177)

_______. 1986. A model for agricultural soil compaction. *J. Agric. Eng. Res.* 33(4):257-262. (177,178)

Barshad, I. 1965. Thermal analysis techniques for mineral identification and mineralogical composition. In *Methods of Soil Analysis, Part 1: Physical and Mineralogical Properties, Including Statistics of Measurement and Sampling*, 699-742. Madison, Wis.: American Society of Agronomy. (25)

Bateman, H. P., M. P. Naik and R. R. Yoerger. 1965. Energy required to pulverize soil at different degrees of compaction. *J. Agric. Eng. Res.* 10(2):132-141. (146,218)

Bausch, W. 1983. Soil moisture model for conjunctive use with microwave radiometers. ASAE Paper No. 83-2152. St. Joseph, Mich.: ASAE. (37)

Baver, L. D. and W. H. Gardner. 1972. *Soil Physics*, 4th Ed. New York: Wiley & Sons. (22,27,30)

Bekker, M. G. 1960. *Off-the-road Locomotion; Research and Development in Terramechanics.* Ann Arbor: University of Michigan Press. (246)

Bekker, M. G. and Z. Janosi. 1958. Rolling resistance of pneumatic tires in soft soils. In *A Soil Value System for Land Locomotion Mechanics*, 50-54. Report No. 5. Centerline, Mich.: Dept. of the Army, Ordinance Tank-Automotive Command, Research and Development Division, Land Locomotion Branch. (248)

Bethlahmy, N. and P. J. Zwerman. 1959. The use of electrical resistivity apparatus in soil investigations. *Transactions of the ASAE* 12(1):68-70. (60,63,64)

Bigsby, F. W. and C. W. Bockhop. 1964. Effects of an air slide on soil engaging tools: Results from a model tool in soil boxes. ASAE Paper No. 64-106. St. Joseph, Mich.: ASAE. (115,148,149)

Blackwell, P. S. and B. D. Soane. 1978. Deformable spherical devices to measure stresses within field soils. *J. Terramechan.* 15(4):207-222. (236,237)

Blake, G. R. 1965a. Particle density. In *Methods of Soil Analysis, Part 1: Physical and Mineralogical Properties, Including Statistics of Measurement and Sampling*, ed. C. A. Black, 371-373. Madison, Wis.: American Society of Agronomy. (27)

________. 1965b. Bulk density. In *Methods of Soil Analysis, Part 1: Physical and Mineralogical Properties, Including Statistics of Measurement and Sampling*, ed. C. A. Black, 374-390. Madison, Wis.: American Society of Agronomy. (33)

Bodman, G. B. and G. K. Constantin. 1965. Influence of particle-size distribution in soil compaction. *Hilgardia* 36(15):567-591. (199,200)

Boedicker, J. J. and H. D. Bowen. 1976. Air permeability from a moving vehicle—feasibility study. ASAE Paper No. 76-1514. St. Joseph, Mich.: ASAE. (52)

Bolling, I. 1987. Bodenverdichtung und Triebkraftverhalten bei Reifen—Neue Mess—und Rechenmethoden—. (Soil compaction and tractive force characteristics of tires—new methods of measurement and computation.) Forschungsbericht Agrartechnik des Arbeitskreises Forschung und Lehre der Max-Eyth-Gesellschaft (MEG). Dissertation. Munich. (238,248,252,253)

Bonneau, M. and B. Souchier. 1982. *Constituents and Properties of Soils*. London: Academic Press. (30)

Bowen, H. D. 1966. Measurement of edaphic factors for determining planter specifications. *Transactions of the ASAE* 9(5):725-735. (51)

________. 1985. Air permeability measurement. In *Proc. International Conference on Soil Dynamics* 3:481-489. Auburn, Ala.: National Soil Dynamics Laboratory. (56)

Bowen, H. D. and P. Liang. 1988. Interpreting air permeability readings in growth area. ASAE Paper No. 88-1629. St. Joseph, Mich.: ASAE. (51,52,53,54,55)

Bowles, J. E. 1978. *Engineering Properties of Soils and their Measurement*, 2nd Ed. New York: McGraw Hill. (27,35)

Brandon, J. R., T. T. Weitzel, J. V. Perumpral and F. E. Woeste. 1986. Shear rate effects on strength parameters of agricultural soils. ASAE Paper No. 86-1044. St. Joseph, Mich.: ASAE. (212)

Brooks, R. H. and R. C. Reeve. 1959. Measurement of air and water permeability of soils. *Transactions of the ASAE* 2(1):125-126, 128. (51,54)

Bruce, R. R. and R. J. Luxmore. 1986. Water retention: Field methods. In *Methods of Soil Analysis, Part 1: Physical and Mineralogical Methods*, 2nd Ed., ed. A. Klute, 663-686. Madison, Wis.: American Society of Agronomy. (41)

Brutsaert, W. 1964. The propagation of elastic waves in unconsolidated, unsaturated granular mediums. *J. Geophy. Res.* 69(2):243-257. (55)

Butterfield, R. and K. Z. Andrawes. 1972. On the angles of friction between sand and plane surfaces. *J. Terramechan.* 8(4):15-23. (107,109,110)

Butts, C. L., J. W. Mishoe and J. W. Jones. 1989. Validating a coupled heat and mass transfer model for soil. ASAE Paper No. 89-7026. St. Joseph, Mich.: ASAE. (83)

Cady, J. G. 1965. Petrographic microscope techniques. In *Methods of Soil Analysis, Part 1: Physical and Mineralogical Properties, Including Statistics of Measurement Sampling*, ed. C. A. Black, 604-631. Madison, Wis.: American Society of Agronomy. (28)

Campbell, G. S. and G. W. Gee. 1986. Water potential: Miscellaneous methods. In *Methods of Soil Analysis, Part 1: Physical and Mineralogical Methods*, 2nd Ed., ed. A. Klute, 619-633. Madison, Wis.: American Society of Agronomy. (44)

Carter, L. M. 1970. Automatic controls for a cotton planter for the irrigated west. USDA-ARS 42-173. Shafter, Calif.: USDA-ARS Cotton Research Station. (60)

Chancellor, W. J. 1966. Combined hypotheses for anticipating soil strains beneath surface impressions. *Transactions of the ASAE* 9(6):887-892, 895. (232,236,253,254)

________. 1971. Effects of compaction on soil strength. In *Compaction of Agricultural Soils*, eds. K. K. Barnes, W. M. Carleton, H. M. Taylor, R. I. Throckmorton and G. E. VandenBerg, 190-212. St. Joseph, Mich.: ASAE. (95,104,105)

________. 1977. *Compaction of soil by agricultural equipment*. Bulletin No. 1881, University of California, Division of Agricultural Sciences, Richmond, Calif. (22,24,178,179,188,189,198, 199)

Chancellor, W. J. and A. Y. Korayem. 1965. Mechanical energy balance for a volume element of soil during strain. *Transactions of the ASAE* 8(3):426-430, 436. (155,156,158,159,160,184,186, 194,201,204,207,220,227,228)

Chancellor, W. J. and J. A. Vomocil. 1970. Relation of moisture content to failure strengths of seven agricultural soils. *Transactions of the ASAE* 13(1):9-13, 17. (93,94,103)

________. 1985. Stress and energy characteristics of agricultural soils during deformation and failure. In *Proc. International Conference on Soil Dynamics* 2:225-240. Auburn, Ala.: National Soil Dynamics Laboratory. (223)

Chancellor, W. J., J. A. Vomocil and K. S. Aref. 1969. Energy disposition in compression of three agricultural soils. *Transactions of the ASAE* 12(4):524-528, 532. (155,156,157,158,229,230)

Chapman, H. D. 1965. Cation exchange capacity. In *Methods of Soil Analysis, Part 2: Chemical and Microbiological Properties*, ed. C. A. Black, 891-901. Madison, Wis.: American Society of Agronomy. (25)

Chen, L. - S. 1948. An investigation of stress-strain and strength characteristics of cohesionless soils by triaxial compression tests. In *Proc. 2nd International Conference on Soil Mechanics and Foundation Engineering* 5:35. International Society for Soil Mechanics and Foundation Engineering. Office of the Secretary General, University Engineering Dept., Trumpington St., Cambridge, U.K. (168)

Chen, Y., J. T. Tarchitzky, J. Brouwer, J. Morin and A. Benin. 1980. Scanning electron microscope observations of soil crusts and their formation. *Soil Sci.* 130(1):49-55. (51)

Chesness, J. L., E. E. Ruiz and C. Cobb, Jr. 1972. Quantitative description of soil compaction in peach orchards utilizing a portable penetrometer. *Transactions of the ASAE* 15(2):217-219. (126)

Chi, L. D., S. Tessier and C. Laguë. 1993a. Finite element prediction of soil compaction induced by various running gears. *Transactions of the ASAE* 36(3):629-636. (160,170,175)

________. 1993b. Finite element modeling of soil compaction by liquid manure spreaders. *Transactions of the ASAE* 36(3):637-644. (158)

Clyma, H. E. and D. L. Larson. 1991. Evaluating the effectiveness of electro-osmosis in reducing tillage draft force. ASAE Paper No. 91-3533. St. Joseph, Mich.: ASAE. (115)

Cohron, G. T. 1962. The soil sheargraph. ASAE Paper No. 62-133. St. Joseph, Mich.: ASAE. (136)

Cohron, G. T. 1963. Soil sheargraph. *Agric. Eng.* 44(10):554-556. (135,136)

Costello, T. A. and H. J. Brand, Jr. 1989. Thermal diffusivity of soil by nonlinear regression analysis of soil temperature data. *Transactions of the ASAE* 32(4):1281-1286. (73,74)

Czako, T. 1958. Prediction of rigid wheel sinkage. In *A Soil Value System for Land Locomotion Mechanics*. Report No. 5:46-49. Centerline, Mich.: Dept. of the Army, Ordnance Tank-Automotive Command, Research and Development Division, Land Locomotion Branch. (248)

Czako, T. and M. G. Bekker. 1958. *Determination of Vehicle Sinkage Parameters by Means of Rigid Wheels*, Part 1: Report No. 33. Centerline, Mich.: Department of the Army, Ordnance Tank-Automotive Command, Research Division, Land Locomotion Laboratory. (248)

Danielson, R. E. and P. L. Sutherland. 1986. Porosity. In *Methods of Soil Analyses, Part I: Physical and Mineralogical Methods,* 2nd Ed., ed. A. Klute, 443-461, Madison, Wis.: American Society of Agronomy. (47)

DeRoock, B. and A. W. Cooper. 1967. Relation between propagation velocity of mechanical waves through soil and soil strength. *Transactions of the ASAE* 10(4):471-474. (57)

Desai, C. S. and H. J. Siriwardane. 1984. *Constitutive Laws for Engineering Materials with Emphasis on Geologic Materials*. Englewood Cliffs, N.J.: Prentice-Hall, Inc. (177,185)

DeVries, D. A. and N. H. Afgan. 1975. Heat transfer in soils. In *Heat and Mass Transfer in the Biosphere, Part 1: Transfer processes in the plant environment,* eds. D. A. DeVries and N. H. Afgan, 1-28. Washington, D.C.: Scripta Book Co. (76,80,81,82)

Dexter, A. R. 1981. Soil shear strengths measured with different levels of uniaxial stress acting in a direction tangential to the shear plane. *J. Terramechan.* 18(4):195-199. (97,98)

Domzal, H. 1970. Preliminary studies of the influence of moisture on physio-mechanical properties of some soils with regard to estimation of optimum working conditions of implements. *Polish J. Soil Sci.* 3(1):61-70. (140,143)

Duncan, J. M. 1980. Hyperbolic stress-strain relationship. In *Proc. Workshop on Limit Equilibrium, Plasticity and Generalized Stress-Strain in Geotechnical Engineering*, eds. R. K. Yong and H. K. Ko, 443-466. New York: ASCE. (169,233)

Duncan, J. M. and C.-Y. Chang. 1970. Nonlinear analysis of stress and strain in soils. In *Proc. American Society of Civil Engineers, J. Soil Mechan. Foundations Div.* (Sept):1629-1653. (232,233)

Dunlap, W. H. and J. A. Weber. 1971. Compaction of an unsaturated soil under a general state of stress. *Transactions of the ASAE* 14(4):601-607, 611. (88,89,90)

Dunlap, W. H., G. E. VandenBerg and J. G. Hendrick. 1966. Comparison of soil shear values obtained with devices of different geometrical shapes. *Transactions of the ASAE* 9(6):896-900. (95,136,137)

Edwards, D. M. and E. J. Monke. 1968. Electrokinetic studies of porous media systems. *Transactions of the ASAE* 11(3):412-415. (60)

Erbach, D. C. 1987. Measurement of soil bulk density and moisture. *Transactions of the ASAE* 30(4):922-931. (33,35,37,44)

Eshbach, O. W. 1952. *Handbook of Engineering Fundamentals*, 2nd Ed. New York: Wiley & Sons. (54,74,176)

Everts, C. J. and R. S. Kanwar. 1989. Quantifying macropores for modeling preferential flow. ASAE Paper No. 89-2162. St. Joseph, Mich.: ASAE. (51)

Farrell, D. A., E. L. Greacen and W. E. Larson. 1967. The effect of water content on axial strain in a loam soil under tension and compression. In *Proc. Soil Sci. Soc. Am.* 31:445-450. (98)

Foley, A. G., P. J. Lawton, A. W. Barker and V. A. McLees. 1984. The use of alumina ceramic to reduce wear of soil-engaging components. *J. Agric. Eng. Res.* 30(1):37-46. (118)

Foley, A. G. and V. A. McLees. 1986. A comparison of the wear of ceramic tipped and conventional precision seed drill coulters. *J. Agric. Eng. Res.* 35(2):97-113. (118)

Fox, W. R. and C. W. Bockhop. 1965. Characteristics of a Teflon-covered simple tillage tool. *Transactions of the ASAE* 8(2):227-229. (111)

Freeland, R. S. 1989. Review of soil moisture sensing using soil electrical conductivity. *Transactions of the ASAE* 32(6):2190-2194. (38,60,62,63)

Freitag, D. R. 1965. *A dimensional analysis of the performance of pneumatic tires on soft soils.* Technical Report No. 3-688. Vicksburg, Miss.: U. S. Army Engineer Waterways Experiment Station, Corps of Engineers. (24,124,126)

Freitag, D. R. 1968. Penetration tests for soil measurement. *Transactions of the ASAE* 11(6):750-753. (126)

Froelich, O. K. 1934. *Druckverteilung im Baugrunde, mit Besonderer Berucksichtung der Plastischen Erscheinungen.* (Pressure distribution in soils under structures, with special consideration of plastic phenomena). Vienna: J. Springer. (235)

Gameda, S., G. S. V. Raghavan and E. McKyes. 1988. Soil penetrometry for compaction modelling. ASAE Paper No. 88-1019. St. Joseph, Mich.: ASAE. (126)

________. 1989. Correlations between constitutive properties and soil strength parameters. ASAE Paper No. 89-1099. St. Joseph, Mich.: ASAE. (126)

Gameda, S., G. S. V. Raghavan, E. McKyes and R. Theriault. 1987. Single and dual probes for soil density measurement. *Transactions of the ASAE* 30(4):932-934, 944. (33,35,36,72)

Gameda, S., G. S. V. Raghavan, R. Theriault and E. McKyes. 1984. High axle load compaction effect on stresses and subsoil density. ASAE Paper No. 84-1547. St. Joseph, Mich.: ASAE. (235)

Gardner, W. H. 1986. Water content. In *Methods of Soil Analysis, Part I. Physical and Mineralogical Methods*, 2nd Ed., ed. A. Klute, 493-544. Madison, Wis.: American Society of Agronomy. (37)

Gerard, C. J. 1965. The influence of soil moisture, soil texture, drying conditions and exchangeable cations on soil strength. In *Proc. Soil Sci. Soc. Am.* 29(6):641-645. (27)

Gerlach, A. 1953. Physikalische untersuchungen über die zurischen den bodenteilchen wirkenden kräfte (Physical investigation of forces acting between soil particles). *Grundlagen der Landtechnik* 5:81-86. (60,65,114,115)

Ghildyal, B. P. and R. P. Tripathi. 1987. *Soil Physics*. New York: Wiley & Sons. (27,76,77,79,80)

Gibas, D. M., R. L. Raper, A. C. Bailey and C. E. Johnson. 1991. Cubical pneumatic cushion triaxial soil test unit. ASAE Paper No. 91-1530. St. Joseph, Mich.: ASAE. (88,90)

Gibbs, H. J. 1966. Research on electroreclamation of saline-alkali soils. *Transactions of the ASAE* 9(2):164-169. (60,63,65)

Gill, W. R. 1959. The effects of drying on the mechanical strength of Lloyd clay. In *Proc. Soil Sci. Soc. Am.* 23(4):255-257. (105)

________. 1968. Influence of compaction hardening of soil on penetration resistance. *Transactions of the ASAE* 11(6):741-745. (128)

Gill, W. R. and C. A. Reaves. 1957. Relationships of Atterberg limits and cation-exchange capacity to some physical properties of soil. In *Proc. Soil Sci. Soc. Am.* 21(5):491-494. (27)

Gill, W. R. and G. E. VandenBerg. 1968. *Soil Dynamics in Tillage and Traction.* Agriculture Handbook, No. 316. Agricultural Research Service, U. S. Department of Agriculture. (84,94, 95,98,106,129,135,146,148,149,154,246)

Gill, W. R. and W. I. McCreery. 1960. Relation of size of cut to tillage tool efficiency. *Agric. Eng.* 41(6):372-374, 381. (141,145,218)

Greacen, E. L. 1960. Aggregate strength and soil consistence. In *7th International Congress of Soil Science Proc.* 1:256-264. (198)

Greene, W. D. and W. B. Stuart. 1984. Core and nuclear methods compared for bulk density and moisture content. ASAE Paper No. 84-1040. St. Joseph, Mich.: ASAE. (35)

Grisso, R. D., C. E. Johnson and A. C. Bailey. 1987a. Soil compaction by continuous deviator stress. *Transactions of the ASAE* 30(5):1293-1301. (156,162,179,180,186,196)

________. 1987b. The influence of stress path on distortion during soil compaction. *Transactions of the ASAE* 30(5):1302-1307. (156,162)

Grisso, R. D., C. E. Johnson, A. C. Bailey and T. A. Nichols. 1984. Influences of soil sample geometry on hydrostatic compaction. *Transactions of the ASAE* 27(6):1650-1653. (177)

Grover, B. L. 1955. Simplified air permeameters for soil in place. *Soil Sci. Soc. Am. Proc.* 19(4):414-418. (51,52)

Gupta, C. P. and A. C. Pandya. 1967. Behavior of soil under dynamic loading: its application to tillage implements. *Transactions of the ASAE* 10(3):352-358. (56)

Gupta, S. C., A. Hadas, W. B. Voorhees, D. Wolf, W. E. Larson and E. C. Schneider. 1985a. Field testing of a soil compaction model. In *Proc. International Conference on Soil Dynamics* 5:979-994. Auburn, Ala.: National Soil Dynamics Laboratory. (191)

Gupta, S. C., A. Hades, W. B. Voorhees, D. Wolf, W. E. Larson and E. C. Schneider. 1985b. *Development of guides for estimating the ease of compaction of world soils.* Research Report, St. Paul, Minn.: Binational Agricultural Development Fund, and USDA-ARS. (236)

Gustafson, R. J., S. L. Green and T. M. Brennan. 1987. Survey of distribution system grounding and neutral-to-earth voltages in Minnesota. ASAE Paper No. 87-3040. St. Joseph, Mich.: ASAE. (60)

Gutwein, R. J., E. J. Monke and D. B. Beasly. 1986. Remote sensing of soil water content. ASAE Paper No. 86-2004. St. Joseph, Mich.: ASAE. (38)

Hallikainen, M. T., F. T. Ulaby, M. C. Dobson, M. A. El-Rayes and L. K. Wu. 1985. Microwave dielectric behavior of wet soil, Part 1: Empirical models and experimental observations. *IEEE Trans. Geosci. Remote Sensing* GE-23(1):25-34. (60,62,70)

Hanamoto, B. and E. Hegedus. 1958. Techniques of soil measurement. In *A Soil Value System for Land Locomotion Mechanics.* Report No. 5:34-42. Centerline, Mich.: Department of the Army, Ordnance Tank Automotive Command, Research and Development Division, Land Locomotion Branch. (248,249)

Hassan, A. El-Domiaty. 1968. *Stress-strain characteristics of a saturated clay soil at various rates of strain.* Ph.D. thesis. Davis, Calif.: University of California. (183,184)

Hassan, A. El-Domiaty and W. J. Chancellor. 1970. Stress-strain characteristics of a saturated clay soil at various rates of strain. *Transactions of the ASAE* 13(5):685-689. (92,93,97,209)

Hassan, O. S. A. 1983. Study on the relationships between the cone penetrometer and the soil cutting resistance of the Caribbean soils. ASAE Paper No. 83-1039. St. Joseph, Mich.: ASAE. (124)

Hayhoe, H. N., A. R. Mack, E. J. Brach and D. Blanchin. 1986. Evaluation of an electrical frost probe. *J. Agric. Eng. Res.* 33(4):281-287. (59)

Head, K. H. 1986. *Manual of Soil Laboratory Testing. Vol. 3: Effective Stress Tests.* London: Pentech Press. (86,91,92,154,165,166)
Hegedus, E. 1965. Plate sinkage study by means of dimensional analysis. *J. Terramechan.* 2(2):25-32. (251)
Hegedus, E. and R. A. Liston. 1966. Recent investigations of vertical load-deformation characteristics of soils. In Technical Report No. 9560, Research Report No. 6:2-20. Warren, Mich.: Land Locomotion laboratory, ATAC Components Research and Development Laboratories, U. S. Army Tank-Automotive Center. (251)
Hendrick, J. G. and A. C. Bailey. 1982. Determining components of soil-metal sliding resistance. *Transactions of the ASAE* 25(4):845-849. (106,107)
Hendrick, J. G. and G. E. VandenBerg. 1961. Strength and energy relations of a dynamically loaded clay soil. *Transactions of the ASAE* 4(1):31-32, 36. (220,221,222)
Hendrick, J. G. and W. F. Buchele. 1963. Tillage energy of a vibrating tillage tool. *Transactions of the ASAE* 6(3):213-216. (218)
Hillel, D. 1980. *Fundamentals of Soil Physics.* New York: Academic Press. (24,25,26,30,40,42, 43,45,75,98,118,120,130)
Hirschi, M. C. and I. D. Moore. 1980. Estimating soil hydraulic properties from soil texture. ASAE Paper No. 80-2523. St. Joseph, Mich.: ASAE. (23)
Hoffmann, O. H. 1975. Neuere grundlagen der mechanik körniger haufwerke (New fundamentals in the mechanics of granular media). *Grundlagen der Landtechnik* 25(2):48-58. (97)
Hough, B. K. 1957. *Basic Soils Engineering.* New York: Ronald Press Co. (29)
Ito, N. 1974. A study on the slip sinkage of the driven wheel for farm use vehicle. (Tsu, Japan: Mie University) *Bull. Faculty Agric.* 46 (Jan):121-147. (245)
Jackson, T. J., P. O'Neill, J. Wang and J. Shule. 1984. Evaluation of a pushbroom microwave radiometer aircraft soil moisture remote sensing system. ASAE Paper No. 84-2516. St. Joseph, Mich.: ASAE. (37,60)
Jackson, R. D. and S. A. Taylor. 1986. Thermal conductivity and diffusivity. In *Methods of Soil Analysis, Part 1: Physical and Minerological Methods,* 2nd Ed., ed. A. Klute, 945-956. Madison, Wis.: American Society of Agronomy. (77)
Jakobsen, B. F. and A. R. Dexter. 1989. Prediction of soil compaction under pneumatic tires. *J. Terramechan.* 26(2):107-119. (236)
Jamison, V. C. 1953. Changes in air-water relationships due to structural improvement of soils. *Soil Sci.* 76:143-151. (103)
Janosi, Z. and B. Hanamoto. 1961. The analytic determination of drawbar pull as a function of slip for tracked vehicles in deformable soils. Paper No. 41. *1st International Conference on the Mechanics of Soil-Vehicle Systems*, Turin, Italy. (June). (245)
Janosi, Z. 1962. Theoretical analysis of the performance of tracks and wheels operating on deformable soils. *Transactions of the ASAE* 5(2):133-134, 146. (242,243)
Ji, C.-Y., M.-N. Chen and J.-Z. Pan. 1986. Approach and instrumentation for predicting sinkage of wetland vehicle based on rheological characteristics of paddy soils. *Trans. Chinese Soc. Agric. Mach.* 17(1):21-31. (English version appears as Pan, J.-Z. and C.-Y. Ji. 1987. Prediction of sinkage for wetland vehicles, *J. Terramechan.* 24(2):159-168.). (208)
Johnson, C. E. and A. C. Bailey, 1990. A shearing strain model for cylindrical stress states. ASAE Paper No. 90-1085. St. Joseph, Mich.: ASAE. (203,204,205,207)
Johnson, C. E. and E. C. Burt. 1986. Theoretical soil stress state under tires. ASAE Paper No. 86-1059. St. Joseph, Mich.: ASAE. (238)
Johnson, C. E., A. C. Bailey, T. A. Nichols and R. D. Grisso. 1984. Soil behavior under repeated hydrostatic loading. ASAE Paper No. 84-1548. St. Joseph, Mich.: ASAE. (176,177,178)
Johnson, C. E., R. D. Grisso, T. A. Nichols and A. C. Bailey. 1987. Shear measurement for agricultural soils—A review. *Transactions of the ASAE* 30(4):935-938. (95)
Johnson, C. E., L. L. Jensen, R. L. Schafer and A. C. Bailey. 1978. Some soil-tool analogs. ASAE Paper No. 78-1037. St. Joseph, Mich.: ASAE. (128)
Jumikis, A. R. 1962. *Soil Mechanics.* Princeton, N.J.: D. Van Nostrand Co., Inc. (189,191)

Kalachev, V. Y. 1974. *Stickiness of clay soils* (Lipkost' glinistynkh grunov). Dissertation, Moscow State University. (Translated by W. R. Gill. Available from USDA, National Agricultural Library, Beltsville, Md. Report No. NTML-WRG-656-NAL. Translation No. 22812). (144,145)

Kano, Y., W. F. McClure and R. W. Skaggs. 1985. A near infrared reflectance soil moisture meter. *Transactions of the ASAE* 28(6):1852-1855. (37)

Kanwar, R. S. 1986. Effect of tillage systems on the variability of soil water tensions and soil water content. *Transactions of the ASAE* 32(2):605-610. (46,47)

Karol, R. H. 1955. *Engineering Properties of Soils*. New York: Prentice Hall. (202)

Kemper, W. D. and W. S. Chepil. 1965. Size distribution of aggregates. In *Methods of Soil Analysis, Part 1: Physical and Mineralogical Properties, Including Statistics of Measurement and Sampling*, ed. C. A. Black, 499-510. Madison, Wis.: American Society of Agronomy. (30)

Kenny, J. F. and K. E. Saxton. 1988. Tillage impacts on thermal and hydraulic characteristics of Palouse silt loam. ASAE Paper No. 88-2005. St. Joseph, Mich.: ASAE. (83)

Kenny, T. C. 1959. Discussion. *Proc. ASCE*. Vol. 85, No. SM3 pp. 67-79. (122)

Kézdi, A. 1974. *Handbook of Soil Mechanics, Vol. 1. Soil Physics*. Amsterdam: Elsevier. (22,25, 26,27,28,29,41,75,77,78,80,81,93,95,97,98,99,101,119,120,121,122,150,154,155,180,195,207)

________. 1979. *Soil Physics: Selected Topics, Developments in Geotechnical Engineering*, 25. Amsterdam: Elsevier. (22,23,48,95,96,201,203)

Khan, M. H. 1993. Anisotropic elastic parameters for unsaturated soils. ASAE Paper No. 93-1538. St. Joseph, Mich.: ASAE. (164,173,176)

Khan, M. H. and D. L. Hoag. 1978. Three-dimensional stress-strain relationships of unsaturated soils. ASAE Paper No. 78-1536. St. Joseph, Mich.: ASAE. (88,91)

Kirkham, D. 1946. Field method for determination of air permeability of soil in its undisturbed state. *Soil Sci. Soc. Am. Proc.* 11:93-99. (51)

Kirkham, D. and W. L. Powers. 1972. *Advanced Soil Physics*. New York: Wiley-Interscience. (38,39,40)

Kitani, O. 1965. Fundamental studies on tillage machinery V(III), soil failure tests under various methods of force application. *J. Soc. Agric. Mach.* (Japan) 27(2):98-104. (218,221)

Kitani, O. and S. P. E. Persson. 1967. Stress-strain relationships for soil with variable lateral strain. *Transactions of the ASAE* 10(6):738-741, 745. (156,161,162,185,191)

Kitani, O. 1975. Studies of soil failure criterion under tensile stress. (Mie University, Tsu, Japan) *Bull. Faculty Agric.* 49:195-201. (101)

________. 1978. Einige grundlagen für eine "pneumatische" bodenbearbeitung (Some fundamentals for pneumatic soil cultivation). *Grundlagen der Landtechnik* 28(5):204-207. (115)

Kitani, O., T. Okamoto and S. Yonekawa. 1985. Double blade tillage system and soil dynamics in tensile stress zone. In *Proc. International Conference on Soil Dynamics* 2:282-297. Auburn, Ala.: National Soil Dynamics Laboratory. (148,149)

Kittrick, J. A. 1965. Electron microscope techniques and electron-diffraction techniques for mineral identification. In *Methods of Soil Analysis, Part 1: Physical and Mineralogical Properties, Including Statistics of Measurement and Sampling*, ed, C. A. Black, 632-670. Madison, Wis.: American Society of Agronomy. (25)

Klute, A. 1986. Water retention: Laboratory methods. In *Methods of Soil Analysis, Part 1: Physical and Minerological Methods*, 2nd Ed., ed. A. Klute, 635-662. Madison, Wis.: American Society of Agronomy. (42)

Klute, A. and D. K. Cassell. 1986. Water potential: Tensionmetry. In *Methods of Soil Analysis, Part 1: Physical and Minerological Methods*, 2nd Ed., ed. A. Klute, 563-596. Madison, Wis.: American Society of Agronomy. (44)

Knight, S. J. and D. R. Freitag. 1962. Measurement of soil trafficability characteristics. *Transactions of the ASAE* 5(2):121-124, 132. (124,126,132)

Ko, H.-Y. and R. F. Scott. 1967. A new soil testing apparatus. *Geotechnique* 17:40-57. (88)

Kocher, M. F. and J. D. Summers. 1988. Wave propagation theory for evaluating dynamic soil stress-stain models. *Transactions of the ASAE* 31(3):683-691, 694. (57,166)

Kogure, K., Y. Ohira and H. Yamaguchi. 1983. Prediction of sinkage and motion resistance of a tracked vehicle using plate penetration test. *J. Terramechan.* 20(3/4):121-128. (252)

Koolen, A. J. and H. Kuipers. 1983. *Agricultural Soil Mechanics*. Berlin: Springer-Verlag. (30,32, 74,106,110,111,112,113,129,193,195,201,202,207)

Koolen, A. T. and P. Vaandrager. 1984. Relationships between soil mechanical properties. *J. Agric. Eng. Res.* 29(4):313-319. (127,188,190)

Koorevaar, P., G. Menelik and C. Dirksen. 1983. *Elements of Soil Physics*. (Developments in soil science - 13). Amsterdam: Elsevier. (26,27,46,77)

Krishna, R. and C. P. Gupta. 1972. Dynamic behavior of soil in compression by propagation of stress waves with a tillage tool. *Transactions of the ASAE* 15(6):1031-1034. (57,60)

Kuipers, H. and B. Kroesbergen. 1966. The significance of moisture content, pore space, method of sample preparation and type of shear annulus used on laboratory torsional shear testing of soils. *J. Terramechan.* 3(4):17-28. (95)

Kunz, G. L. 1971. *Granular viscous parameter investigation*. M.S. thesis, University of Illinois, Urbana. (209,210)

Ladd, C. C. 1964. Stress-strain modulus of clay from undrained triaxial tests. *Proc. ASCE*, Vol. 90, No. SM5 (Sept.). (169)

Lambe, T. W. and R. V. Whitman. 1969. *Soil Mechanics*. New York: Wiley & Sons. (28,30,97, 120,121,123,132,151,152,154,155,167,168,169,174,179,181)

Lambert, V. M. and M. J. McFarland. 1987. Land surface temperature estimation over the Northern Great Plains using dual polarized passive microwave data from the Nimbus 7. ASAE Paper No. 87-4041. St. Joseph. Mich.: ASAE. (73)

Larson, W. E. and J. R. Gilley. 1976. Soil-climate-crop considerations for recycling organic wastes. *Transactions of the ASAE* 19(1):85-89, 96. (74)

Lawton, P. J. and A. G. Foley. 1986. Alumina tipped spring tine points – field assessments. *J. Agric. Eng. Res.* 34(4):343-355. (118)

Leviticus, L. I. 1973. Investigation into properties of the soil-wheel interface, Part II: Results of tests with a rotating cone in sand. *Transactions of the ASAE* 16(1):52-57. (128)

Ligon, J. T. 1969. Evaluation of the gamma transmission method for determining soil water balance and evapo-transpiration. *Transactions of the ASAE* 12(1):121-126, 129. (72,73)

Lu, N.-Z., Y.-Q. Qian and J.-Z. Pan. 1982. Rheological characteristics of paddy-field soils in China. *Trans. Chinese Soc. Agric. Mach.* 13(2):43-54. (215,216)

Luth, H. J. and R. D. Wismer. 1971. Performance of plane soil cutting blades in sand. *Transactions of the ASAE* 14(2):255-259, 262. (127)

Marshall, T. J. and J. W. Holmes. 1988. *Soil Physics*, 2nd Ed. Cambridge: University Press. (47,48,71)

Matthes, R. K. and H. D. Bowen. 1968. Steady-state heat and moisture transfer in an unsaturated soil. ASAE Paper No. 68-534. St. Joseph, Mich.: ASAE. (83)

Matzkanin, G. A., W. L. Rollwitz, J. D. King and R. F. Paetzold. 1984. Principles of nuclear magnetic resonance for agricultural operations. Agricultural electronics - 1983 and beyond 1:309-318. Field equipment, irrigation and drainage. St. Joseph, Mich.: ASAE. (73)

McKyes, E. 1985. *Soil Cutting and Tillage*. Amsterdam: Elsevier. (94,195,196,199,200,247)

Mein, R. G. and C. L. Larson. 1973. Modeling infiltration during a steady rain. *Water Resources Res.* 9(2):384-394. (23)

Merva, G. E. 1975. *Physioengineering Principles*. Westport, Conn.: AVI Publishing Co. (83)

Meyer, M. P. and S. J. Knight. 1961. *Trafficability of soils, soil classification*. Technical Memorandum No. 3-240. Vicksburg, Miss.: U. S. Army Engineers, Waterways Experiment Station, Corps of Engineers. (22,23,127,133,134)

Mil'tsev, A. I. 1966. Prylinanie i trenie pochny po metallam i plastmassm (*Sticking and friction of soil on metals and plastics*). (Translated by W. R. Gill from *Studies of working tools of agricultural machines. Reports of Conference of Young Scientists*, 3-14. Moscow: Viskhom. Available NTIS, Springfield, Va. Report No. NTML-WRG-419, PB-232140-T). (138,141)

Mink, L. A., W. H. Carter and M. M. Mayeux. 1964. Effects of an air slide on soil engaging tools-results from ammonia knives in artificial soil. ASAE Paper No. 64-105. St. Joseph, Mich.: ASAE. (115)

Mitchell, B. W. 1988. Applications of remote infrared temperature measurements to environmental control systems. *Transactions of the ASAE* 31(6):1864-1868. (73)

Moore, M. A. and V. A. McLees. 1980. Effect of speed on wear of steels and a copper by bonded abrasive and soils. *J. Agric. Eng. Res.* 25(1):37-45. (117,118)
Morgan, M. T., R. G. Holmes and M. J. Lichtensteiger. 1988. Air permeability measurement system for soil cores. ASAE Paper No. 88-1632, St. Joseph, Mich.: ASAE. (51,52,53,56)
Morrow, N. R. and C. C. Harris. 1953. Capillary equilibrium in porous materials. *Soc. Petrol. Eng. J.* (March):15-24. (102)
Murphy, N. R. 1967. Discussion of "plate sinkage by means of dimensional analysis". *J. Terramechan.* 4(2):59-64. (252)
Nau, K. R. 1987. *Air permeability: A measure of soil compaction.* M.S. thesis, The Ohio State University, Columbus, Ohio. (53)
Neal, M. S. 1966. Friction and adhesion between soil and rubber. *J. Agric. Eng. Res.* 11(2):108-112. (Also see: Correspondence, *J. Agric. Eng. Res.* 1967. 12(1):83-87.) (109)
Nelson, N. W. and L. E. Sommers. 1982. Total carbon, organic carbon and organic matter. In *Methods of Soil Analysis, Part 2*, ed. A. L. Page, 539-578. Madison, Wis.: American Society of Agronomy. (29,30)
Nelson, S. O. 1983. Density dependence of dielectric properties of particulate materials. *Transactions of the ASAE* 26(6):1823-1826, 1829. (59,60)
NeSmith, D. S., W. L. Hargrave, E. W. Tollner and D. E. Radcliffe. 1986. A comparison of three soil surface moisture and bulk density sampling techniques. *Transactions of the ASAE* 29(5):1297-1299. (33,108)
Nichols, M. L. 1931. The dynamic properties of soil II: Soil and metal friction. *Agric. Eng.* 12(8):321-324. (112)
Nikolaeva, I. N. and P. U. Bakhtin. 1975. Lipkost' temno-kashtanovykh tyazhelosuglinistykh i supeschanykk pochr kustanaiskoi oblasti pri vertical' nom otryrei tangetsial' nom sdvige (*Stickiness of dark chestnut heavy loam soils of the kustani oblast under conditions of vertical tearing and of tangential shear*). (Translated by W. R. Gill from *Soil Science* 4:68-78, Moscow. Beltsville, Md.: USDA National Agricultural Library. Report No. NTML-WRG-617, NAL Accession No. 22918). (138,139,140,142)
NTML. 1961. *Soil-Tool Relationships*, Annual report. Auburn, Ala.: National Tillage Machinery Laboratory. (218)
O'Callaghan, J. R., K. M. Farrelly and P. J. McCullen. 1965. Limitations of the torsion shear test. *J. Agric. Eng. Res.* 10(2):114-117. (135)
Ohu, J. O., G. S. V. Raghavan and E. McKyes. 1985a. The shear strength of compacted soils with varying organic matter contents. ASAE Paper No. 85-1039. St. Joseph, Mich.: ASAE. (30)
________. 1985b. Peatmoss effect on the physical and hydraulic characteristics of compacted soils. *Transactions of the ASAE* 28(2):420-424. (30)
________. 1988. Cone index prediction of compacted soils. *Transactions of the ASAE* 31(2):306-310. (126,131)
Ohu, J. O., G. S. V. Raghavan, E. McKyes and G. Mehuys. 1986. Shear strength prediction of compacted soils with varying added organic matter contents. *Transactions of the ASAE* 29(2):351-355, 360. (126,131)
Oida, A. and T. Tanaka. 1981. Analysis of viscoelastic deformation of soil by means of finite element method. In *Proc. 7th International Conference of the International Society of Terrain Vehicle Systems* 3:1473-1492. Calgary, Alberta. (214)
Oskoui, K. E., D. H. Rackham and B. D. Whitney. 1982. The determination of plough draught, Part II, The measurement and prediction of plough draught for two mouldboard shapes in three soil series. *J. Terramechan.* 9(3):153-164. (128)
Osman, M. S. 1964. The measurement of soil shear strength. *J. Terramechan.* 1(3):54-60. (95)
Pall, R. and N. N. Mohsenin. 1980a. A soil air pycnometer for determination of porosity and particle density. *Transactions of the ASAE* 23(3):735-741. (27)
Pall, R. and N. N. Mohsenin. 1980b. Permeability of porous media as a function of porosity and particle size distribution. *Transactions of the ASAE* 23(3):742-745. (23,27)
Pan, J.-Z. 1984. Effect of load conditions and rheological parameters on sinkage of tracked vehicles in paddy fields. *Trans. Chinese Soc. Agric. Mach.* 15(4):7-12. (215,216)

Pan, J.-Z., Y.-Q. Qian, M.-N. Chen, C.-Y. Ji and Q.-S. Liu. 1983. Rheological characteristics of paddy-field soils in China (4). *Trans. Chinese Soc. Agric. Mach.* 14(3):34-44. (215,217)

Panwar, J. S. and J. C. Siemens. 1972. Shear strength and energy of soil failure related to density and moisture content. *Transactions of the ASAE* 15(3):423-427. (220,222)

Parchomchuk, P. and W. W. Wallender. 1986. Electromagnetic sensing of subsurface soil moisture. ASAE Paper No. 86-2005. St. Joseph, Mich.: ASAE. (38,60,64,70,71)

Pavlics, F. 1958. Instruments for the measurement of physical soil values. In *A Soil Value System for Land Locomotion Mechanics*. Report No. 5:14-24. Centerline, Mich.: Department of the Army, Ordnance Tank-Automotive Command, Research and Development Division, Land Locomotion Branch. (239,240,241,247)

Persson, S. P. E. and B.-S. Chang. 1966. Viscous properties of a slippery soil surface. ASAE Paper No. 66-145. St. Joseph, Mich.: ASAE. (210,211)

Perumpral, J. V. 1987. Cone penetrometer applications—A review. *Transactions of the ASAE* 30(4):939-944. (123)

Pichon, J. D. and G. W. Steinbruegge. 1965. Propagation of audio-frequency magnetic fields through soil. *Transactions of the ASAE* 8(2):264-266. (70)

Pikul, J. L., Jr., J. F. Juzel and D. E. Wilkins. 1988. Measurement of tillage induced soil macroporosity. ASAE Paper No. 88-1641. St. Joseph, Mich.: ASAE. (47,50,51)

Proctor, R. R. 1948. The relationship between foot pounds per cubic foot of compactive effort and the shear strength of compacted soils. In *Proc. 2nd International Conference on Soil Mechanics and Foundation Engineering* 5:219-223. International Society for Soil Mechanics and Foundation Engineering Office of the Secretary General, University Engineering Depart.. Cambridge, U.K. (231,232)

Qian, Y.-Q., Z.-J. Lu and J.-Z. Pan. 1982. Study of rheological characteristics for paddy-field soil of China (2) Investigation on thixotropic behavior for clayey paddy-field soil. *Trans. Chinese Soc. Agric. Mach.* 13(3):9-15. (134)

Raghavan, G. S. V. and E. McKyes. 1977. Laboratory study to determine the effect of slip-generated shear on soil compaction. *Can. Agric. Eng.* 19(1):40-42. (198)

Ram, R. B. and C. P. Gupta. 1972. Relationship between rheological coefficients and soil parameters in compression test. *Transactions of the ASAE* 15(6):1054-58. (214,215)

Raper, R. L. and D. C. Erbach. 1988a. Prediction of soil stresses using the finite element method. ASAE Paper No. 88-1017. St. Joseph, Mich.: ASAE. (164,232,233)

________. 1988b. Effect of variable linear elastic parameters on finite element prediction of soil compaction. ASAE Paper No. 88-1640. St. Joseph, Mich.: ASAE. (164,232,233)

Rawlins, R. L. and G. S. Campbell. 1986. Water potential: Thermocouple psychrometry. In *Methods of Soil Analysis, Part 1: Physical and Mineralogical Methods*, 2nd Ed., ed. A. Klute, 597-618. Madison, Wis.: American Society of Agronomy. (44)

Redmund, C. P., D. D. Schulte and J. Skopp. 1988. Effect of compaction on thermal contact conductance at heat exchanger surfaces in clay soils. ASAE Paper No. 88-3049. St. Joseph, Mich.: ASAE. (83)

Redmund, C. P. and D. D. Schulte. 1989. Interaction between soil drying, shrinkage, and thermal contact conductance. ASAE Paper No. 89-4062. St. Joseph, Mich.: ASAE. (83)

Reece, A. R. 1964. *Problems of soil-vehicle mechanics*. Land Locomotion Laboratory Report No. 8470 (LL97). Warren, Mich.: U.S. Army Tank-Automotive Center. (248)

Rich, C. A. 1965. Elemental analysis by flame photometry. In *Methods of Soil Analysis, Part 2, Chemical and Microbiological Properties*, ed. C. A. Black, 849-865. Madison, Wis.: American Society of Agronomy. (26)

Richardson, R. C. D. 1967. The wear of metallic materials by soil - Practical phenomena. *J. Agric. Eng. Res.* 12(1):22-39. (116,117)

Rickman, R. W. 1971. Sonic radiation for soil mechanical property measurement. *Transactions of the ASAE* 14(6):1126-1128. (166,167)

Riggle, F. R. and D. C. Slack. 1980. Rapid determination of soil water characteristic by thermocouple psychrometry. *Transactions of the ASAE* 23(1):99-103. (37)

Robbins, D. H., Jr., J. G. Hendrick and C. E. Johnson. 1987a. Instrumentation for soil-material sliding resistance research. ASAE Paper No. 87-1011. St. Joseph, Mich.: ASAE. (108,110,111)

Robbins, D. H., Jr., C. E. Johnson and R. L. Schafer. 1987b. Modeling soil-metal sliding resistance. ASAE Paper No. 87-1580. St. Joseph, Mich.: ASAE. (108,110)
Robbins, D. H., R. L. Schafer and C. E. Johnson. 1988. Minimizing tillage energy due to sliding resistance. ASAE Paper No. 88-1628. St. Joseph, Mich.: ASAE. (108,113)
Rohani, B. and G. Y. Baladi. 1981. Correlation of cone index with soil properties. In *Cone Penetration Testing and Experience*, eds. G. M. Norris and H. D. Holtz, 128-144. New York: American Society of Civil Engineers. (126,181)
Roscoe, K. H., A. N. Schofield and C. P. Wroth. 1958. On the yielding of soils. *Geotechnique* 8(1):22-53. (194)
Roscoe, K. H. 1970. The influence of strains in soil mechanics. Tenth Rankine Lecture. *Geotechnique* 20:129-170. (95)
Rowe, P. W., L. Barden and I. K. Lee. 1964. Energy components during the triaxial cell and direct shear tests. *Geotechnique* 14:247-261. (228)
Rowe, R. S. and E. Hegedus. 1959. *Drag coefficients of locomotion over viscous soils*, Part II. Report No. 54. Centerline, Mich.: Land Locomotion Laboratory, Research Division, Ordnance-Tank-Automotive Command. (210)
Rush, E. S. 1968. Trafficability tests with a two-wheel-drive industrial tractor. *Transactions of the ASAE* 11(6):778-782. (126,133)
Salokhe, V. M. and D. Gee-Clough. 1988. Coating of cage wheel lugs to reduce soil adhesion. *J. Agric. Eng. Res.* 41(3):201-210. (111)
Schaefer, S. W., J. H. Bischoff, D. P. Froehlich and D. W. DeBoerr. 1989. Effects of exchangeable soil sodium on implement draft. *Transactions of the ASAE* 32(3):812-816. (27)
Schafer, R. L., C. W. Bockhop and W. G. Lovely. 1963. Vane and torsion techniques for measuring soil shear. *Transactions of the ASAE* 6(1):57-60. (65,130,136)
Schafer, R. L., C. A. Reaves and D. F. Young. 1969. An interpretation of distortion in the similitude of certain soil-machine systems. *Transactions of the ASAE* 12(1):145-149. (124,125,128)
Schafer, R. L., W. R. Gill and C. A. Reaves. 1975. Lubrication of soil-metal interfaces. *Transactions of the ASAE* 18(5):848-851. (114)
________. 1977. Lubricated plows vs. sticky soils. *Agric. Eng.* 58(10):34-38. (114)
Schmidt, R. H. 1963. Calculated hydraulic conductivity as a measure of soil compaction. *Transactions of the ASAE* 6(3):177-181. (47)
Schmugge, T. 1983. Remote sensing of soil moisture with microwave radiometers. *Transactions of the ASAE* 26(3)748-753. (71)
Schreier, H. 1977. Quantitative predictions of chemical soil conditions from multispectral airborne, ground, and laboratory measurements. In *Proc. 4th Canadian Symposium on Remote Sensing*, 106-112. Quebec City, Quebec. (31)
Shonk, J. L. and L. D. Gaultney. 1989. Development of a real-time soil color sensor. ASAE Paper No. 89-1553. St. Joseph, Mich.: ASAE. (31,64,71)
Sirois, D. L., B. J. Stokes and C. L. Rawlins. 1989. Cone penetrometers - How do they measure up? ASAE Paper No. 89-7067. St. Joseph, Mich.: ASAE. (126)
Sitkei, G. 1985. Basic regularities of soil clod breakup at the seedbed preparation. In *Proc. International Conference on Soil Dynamics* 2:364-376. Auburn, Ala.: National Soil Dynamics Laboratory. (146,147,148)
Sitkei, G., J. Csermely and L. Fenyvesi. 1992. General regularities of compaction for viscoelastic agricultural materials. Paper No. 9203. In *13th International Conference on Agricultural Engineering*, Uppsala, Sweden. (1-4 June). (217)
Skempton, A. W. 1953. The colloidal activity of clays. In *Proc. 3rd International Conference on Soil Mechanics and Foundation Engineering* 1:57. International Society of Soil Mechanics and Foundation Engineering Office of the Secretary General, University Engineering Dept. Trumpington St., Cambridge, U.K. (121).
Sloss, D. 1966. *Wheeled bevameter tests*. Technical Report No. 9560, Research Report No. 6. Warren, Mich.: Land Locomotion Laboratory, ATAC Components Research and Development Laboratories. U. S. Army, Tank-Automotive Center. (248)

Smirnov, M. A. and M. S. Gorbunov. 1966. *Study of the stress in the soil under the powered and non-powered traction devices of agricultural tractors* (in Russian). Chelyabinski, Russia: Chelyabinsk Institute of Mechanization and Electrification. Trudy Issue (20):134-141. (Translated by W. R. Gill. Report No. NTML-WRG-423, PB-232-384-T. Springfield, Va.: NTIS). (237,238)

Smith, E. M., T. H. Taylor and S. W. Smith. 1967. Soil moisture measurement using gamma transmission techniques. *Transactions of the ASAE* 10(2):205-208. (72)

Snyder, V. A. and R. D. Miller. 1985. Tensile strength of unsaturated soils. *Soil Sci. Soc. Am. J.* 49(1):50-65. (102)

Soane, B. D. 1968. A gamma-ray transmission method for the measurement of soil density in field tillage studies. *J. Agric. Eng. Res.* 13(4):340-349. (35)

Soehne, W. H. 1953. Reibung und kohäsion bei ackerbodens (Friction and cohesion in arable soils). *Grundlagen der Landtechnik* 5:64-80. (107,108,109,110,112,113)

Soehne, W. H. 1956. Einige grundlagen für eine landtechnische bodenmechanik (Basic considerations of soil mechanics as applied to agricultural engineering). *Grundlagen der Landtechnik* 7:11-27. (Translation No. 53 by W. E. Klinner. Bedford, UK: NIAE, Silsoe). (146)

Soehne, W. H. 1958. Fundamentals of pressure distribution and soil compaction under tractor tires. *Agric. Eng.* 39(5):276-281, 290. (188,198,233,234,235)

________. 1965. Mechanics of the system: Off-Road Vehicles - Soil and Soil Working Tools. Meisenstr. 6:D8032, Graefelfing, Germany. (223,226)

Soiltest, Inc. 1973. Speedy moisture tester (AASHO Designation T-217-67). Evanston, Ill.: Soiltest, Inc. (37)

Sommer, C. 1976. Über die verdictungsempfindlichkeit von ackerböden (The susceptibility of agricultural soils to compaction). *Grundlagen der Landtechnik* 26(1):14-23. (47,49)

Sowers, G. B. and G. F. Sowers. 1961. *Introductory Soil Mechanics and Foundations*, 2nd Ed. New York: MacMillian. (28,29,84,85,86,88,93)

Sowers, G. F. 1965. Consistency. In *Methods on Soil Analysis, Part 1: Physical and Mineralogical Properties, Including Statistics of Measurement and Sampling*, ed. C. A. Black, 391-399. Madison, Wis.: American Society of Agronomy. (118,119)

Spangler, M. G. 1960. *Soil Engineering*, 2nd Ed. Scranton, Pa.: International Textbook Co. (24,155)

Spektor, M., N. Solomon and S. Malkin. 1985. Measurement of frictional interactions at soil-tool interfaces. *J. Terramechan.* 22(2):73-80. (107)

Stafford, J. V. and D. W. Tanner. 1976. *The frictional characteristics of steel sliding on soil.* NIAE Departmental Note No. DN/T/728/1162. Silsoe, UK: National Institute of Agricultural Engineering. (109,114)

Stefanelli, G. 1968. Soil cutting resistance with a wire. Paper No. 75. In *Trans. 9th International Congress of Soil Science* 1:739-750. Adelaide, Australia. (148,149)

Sture, S. and C. S. Desai. 1979. Fluid cushion truly triaxial or multiaxial testing device. *Geotech. Test. J.* (ASTM) 2(1):20-33. (88)

Sudduth, K. A., J. W. Hummel and R. C. Funk. 1989. NIR soil organic matter sensor. ASAE Paper No. 89-1035. St. Joseph, Mich.: ASAE. (31,71)

Surbrook, T. C., N. D. Reese and C. Jensen. 1982. Grounding electrode to earth resistance and earth voltage gradient measurements. ASAE Paper No. 82-3507. St. Joseph, Mich.: ASAE. (60)

Tagg, G. F. 1964. *Earth Resistance.* New York: Pitman Pub. Corp. (62)

Tavernetti, J. R. 1935. Characteristics of the resistance type soil sterilizer. *Agric. Eng.* 16(7):271-274. (60,62,63,64)

Taylor, J. H. and G. E. VandenBerg. 1966. Role of displacement in a simple traction system. *Transactions of the ASAE* 9(1):10-13. (95,187,188,242,243,245)

Taylor, S. A. and G. L. Ashcroft. 1972. *Physical Edaphology – The Physics of Irrigated and Nonirrigated Soils.* San Francisco: W. H. Freeman & Co. (46,59)

Taylor, S. A. and R. D. Jackson. 1986a. Temperature. In *Methods of Soil Analysis, Part 1: Physical and Mineralogical Methods*, 2nd Ed., ed. A. Klute, 927-940. Madison, Wis.: American Society of Agronomy.(73)

Taylor, S. A. and R. D. Jackson. 1986b. Heat capacity and specific heat. In *Methods of Soil Analysis, Part 1: Physical and Minerological Methods*, 2nd Ed., ed. A. Klute, 941-944. Madison, Wis.: American Society of Agronomy. (76)

Terzaghi, K. 1942. *Theoretical Soil Mechanics*. New York: Wiley. (246)

Tijink, F. G. J. and A. J. Koolen. 1985. Prediction of rolling resistance and soil compaction using cone, shear vane, and a falling weight. In *Proc. International Conference on Soil Dynamics* 4:800-813. Auburn, Ala.: National Soil Dynamics Laboratory. (128,131)

Tollner, E. W. 1983. Performance of a new soil impedometer in selected southeastern soils. ASAE Paper No. 83-1050. St. Joseph, Mich.: ASAE. (128)

Tollner, E. W. and B. P. Verma. 1984. Modified cone penetrometer for measuring soil mechanical impedance. *Transactions of the ASAE* 27(2):331-336. (128)

________. 1987a. X-ray CT for soil moisture analysis. ASAE Paper No. 87-1555. St. Joseph, Mich.: ASAE. (38,72)

________. 1987b. Apparent thermal conductivity of organic potting mixes. *Transactions of the ASAE* 30(2):509-513. (76,77,78,82)

Tollner, E. W. and W. L. Rollwitz. 1988. Nuclear magnetic resonance for moisture analysis of meals and soils. *Transactions of the ASAE* 31(5):1608-1615. (73)

Topp, G. C., J. L. Davis and A. P. Annan. 1980. Electromagnetic determination of soil water content: Measurement in coaxial transmission lines. *Water Res. Res.* 16(3):574-582. (60,61)

Trask, P. D. and H. Klehn, Jr. 1958. Pressure-sinkage tests of mixtures of kaolin and illite with clastic silt. Report Series 116, Issue 4. Berkeley, Calif.: Wave Research Laboratory, Institute of Engineering Research, University of California. (248,250,251)

Truman, C. C., H. F. Perkins, L. E. Asmussen and H. D. Allison. 1988. Using ground-penetrating radar to investigate variability in selected soil properties. *J. Soil Water Conserv.* 43(4):341-345. (59,60,70)

U. S. Dept. of the Army. 1959. Soils trafficability, Technical Bulletin TB ENG 37. Washington, D.C.: Headquarters, Dept. of the Army. (126,132)

Upadhyaya, S. K. 1989. *Development of a portable instrument to measure soil properties relevant to traction*. Research report. Davis, Calif.: Agricultural Engineering Department, University of California. (244,251)

Upadhyaya, S. K., L. J. Kemble, N. E. Collins and T. H. Williams. 1982. Cone index prediction equations for Delaware soils. ASAE Paper No. 82-1542. St. Joseph, Mich.: ASAE. (126,174)

Upadhyaya, S. K., D. Wulfsohn and G. Jubbal. 1989. Traction prediction equations for radial ply tires. *J. Terramech.* 26(2):149-175. (126)

Upadhyaya, S. K., D. Wulfsohn and J. Mehlschau. 1990. An instrumented device to obtain traction related parameters. ASAE Paper No. 90-1097. St. Joseph, Mich.: ASAE. (251)

________. 1993. An instrumented device to obtain traction related parameters. *J. Terramech.* 30(1):1-20. (244)

Upadhyaya, S. K., K. Zeier, R. Southard, W. J. Chancellor and S. Ahmed. 1988. Infiltration in crusted soils. ASAE Paper No. 88-2003. St. Joseph, Mich.: ASAE. (51)

VandenBerg, G. E. 1966. Triaxial measurements of shear strain and compaction in unsaturated soil. *Transactions of the ASAE* 9(4):460-467. (199)

Vomocil, J. A. 1954. *In situ* measurement of soil bulk density. *Agric. Eng.* 35(9):651-654. (35)

________. 1965. Porosity. In *Methods of Soil Analysis, Part 1: Physical and Mineralogical Properties, Including Statistics of Measurement and Sampling*, ed. C. A. Black, 299-314. American Society of Agronomy, Madison, Wis. (43,50)

Vomocil, J. A. and W. J. Chancellor. 1969. Energy requirements for breaking soil samples. *Transactions of the ASAE* 12(3):375-383, 388. (98,100,104,105,143,145,146,148,149,168,170, 171,172,173,174,219,220)

Vomocil, J. A. and W. J. Flocker. 1961. Effect of soil compaction on storage and movements of soil, air and water. *Transactions of the ASAE* 4(2):242-246. (51)

Vomocil, J. A., L. J. Waldron and W. J. Chancellor. 1961. Soil tensile strength by centrifugation. *Soil Sci. Soc. Am. Proc.* 25(3):176-180. (105)

Vomocil, J. A., W. J. Chancellor and K. S. Aref. 1968. Relation of moisture content to tensile-failure strength of glass bead systems. *Transactions of the ASAE* 11(5):616-618, 625. (101,102)

Wallender, W. W., G. L. Sackman, K. Kone and M. S. Kaminaka. 1985. Soil moisture measurement by microwave forward scattering. *Transactions of the ASAE* 28(4):1206-1211. (37,60)

Wang, G. and G. C. Zoerb. 1988. Indirect determination of tractor tractive efficiency. ASAE Paper No. 88-1517. St. Joseph, Mich.: ASAE. (127)

Warner, G. S. and J. L. Nieber. 1988. CT scanning of macropores in soil columns. ASAE Paper No. 88-2632. St. Joseph, Mich.: ASAE. (51)

Weber, F., 1932. *Untersuchungen über den Einfluss des elektrischen Stromes auf Zugkraftbedarf beim Pflügen.* (Investigation of the influence of electric currents on the draft force requirements for plowing). Dissertation, Technischen Hochscule, Munich. (60,65,114,115)

Wechsler, A. E., 1966. Development of thermal conductivity probes for soils and insulations. U. S. Army Cold Regions Research and Engineering Laboratory, Hanover, NH. Technical Report 182. (80)

Wheeler, P. A. and G. L. Duncan. 1984. Measuring soil moisture electromagnetically. *Agric. Eng.* 65(9):12-15. (60,65,66,73)

Whittig, L. D. 1965. X-ray diffraction techniques for mineralogical composition. In *Methods of Soil Analysis, Part 1: Physical and Mineralogical Properties, Including Statistics of Measurement and Sampling,* ed. C. A. Black, 671-698. Madison, Wis.: American Society of Agronomy. (25)

Wills, B. M. D. 1963. The measurement of soil shear strength and deformation moduli and a comparison of the actual and theoretical performance of a family of rigid tracks. *J. Agric. Eng. Res.* 8(2):115-131. (95,242,243,244)

Wismer, R. D. and H. J. Luth. 1972. Performance of plane soil cutting blades in clay. *Transactions of the ASAE* 15(2):211-216. (127)

________. 1974. Off-road traction prediction for wheeled vehicles. *Transactions of the ASAE* 17(1):8-10, 14. (126)

Withey, M. O. and J. Aston. 1950. *Johnson's Materials of Construction*, 8th Ed. New York: Wiley & Sons. (28,87)

Womac, A. R., F. D. Tompkins and E. C. Drumm. 1988. Resonant column testing of agricultural soils. *Transactions of the ASAE* 31(5):1326-1332. (58,182)

Womac, A. R., F. D. Tompkins, E. C. Drumm, R. S. Freeland and J. B. Wilkerson. 1987. Measuring dynamic response of soil subjected to impact loading. ASAE Paper No. 87-1010. St. Joseph, Mich.: ASAE. (126)

Wong, J. R. and T. J. Schmugge. 1980. An empirical model for the complex dielectric permittivity of soils as a function of water content. *IEEE Trans. Geosci. Remote Sensing* GE-18(4):288-295. (60)

Wood, D. M. 1990. *Soil Behaviour and Critical State Soil Mechanics*. Cambridge, England: Cambridge University Press. (177,185,233)

Wray, W. K. 1986. *Measuring Engineering Properties of Soil*. Englewood Cliffs, N.J.: Prentice-Hall. (27)

Yao, Y. and D. Zeng. 1988. Investigation on the relationship between sliding speed and soil-metal friction. *Trans. Chinese Soc. Agric. Mach.* 19(4):33-40. (108,109,113)

Yong, R. N. and E. A. Fattah. 1976. Prediction of wheel-soil interaction and performance using the finite element method. *J. Terramechan.* 13(4):227-240. (232)

Yong, R. N., E. A. Fattah and N. Skiadas. 1984. *Vehicle Traction Mechanics*. Developments in Agricultural Engineering (3). Amsterdam: Elsevier. (239,240,242,243,246)

Zadneprovski, R. P. 1975. Vliyanie davleniya vremeni kontaka i temporary na ageziyu gruntov k rabochim organam (in Russian) (*The influence of pressure time of contract and temperature on the adhesion of soils to working tools*). Mining, Construction and Highway Machines, No. 19. (Translated by W. R. Gill. 23-31. Kiev, Teknika. Available from USDA National Agricultural Library, Beltsville, Md. Report date 16 April 1985). (138,142,143)

Zareian, S. 1989. Application of electro-osmosis to reclaim saline soils. *Agric. Mech. Asia, Africa, Latin America* 20(3):9-11, 14. (60,63,65)

Zhang, J. X., Z. Z. Sang and L. R. Gao. 1986. Adhesion and friction between soils and solids. *Trans. Chinese Soc. Agric. Mach.* 17(1):32-40. (109,112)

Zhang, N., K. H. Kromer, P. L. Hien, K. L. Dreesen and B. Linde. 1989. Measuring soil moisture content and density using a polarized laser light. ASAE Paper No. 89-3031. St. Joseph, Mich.: ASAE. (71)

Zhou, P. A. 1986a. Research on effect of soil particle abrasive on wear behaviors of materials in agricultural machinery. *Trans. Chinese Soc. Agric. Mach.* 17(3):53-62. (116,117)

________. 1986b. Research on effect of soil particle abrasives on wear behaviors of materials in agricultural machinery. *Trans. Chinese Soc. Agric. Mach.* 17(4):61-68. (117)

Chapter 3*

Abdalla, A. M., D. R. P. Hettiaratchi and A. R. Reece. 1969. The mechanics of root growth in granular media. *J. Agric. Eng. Res.* 14:236-248. (261)

Arndt, W. 1965a. Nature of mechanical impedance to seedlings by soil surface seals. *Aust. J. Soil Res.* 3:45-54. (256)

________. 1965b. The impedance of soil seals and the forces of emerging seedlings. *Aust. J. Soil Res.* 3:55-68. (256)

Baldovinos, de la Pena. G. 1953. Growth of the root tip. In *Growth and Differentiation in Plants,* ed. W. E. Loomis. Ames, Iowa: Iowa State College Press. (258,259)

Bamford, S. J., C. J. Parker and M. K. J. Carr. 1991. Effects of soil physical conditions on root growth and water use of barley grown in containers. *Soil Tillage Res.* 21:309-323. (266)

Barley, K. P. 1962. The effects of mechanical stress on the growth of roots. *J. Exp. Bot.* 13:95-110. (257,258)

Barley, K. P., D. A. Farrell, and E. L. Greacen. 1965. The influence of soil strength on the penetration of a loam by plant roots. *Aust. J. Soil Res.* 3:69-79. (260)

Barley, K. P. and E. L. Greacen. 1967. Mechanical resistance as a soil factor influencing the growth of roots and underground shoots. *Adv. Agron.* 19:2-43. (258)

Ben-Porath, A. and D. N. Baker. 1990. Taproot restriction effects on growth, earliness and dry weight partitioning of cotton. *Crop Sci.* 30:809-814. (270)

Bennie, A. T. P. 1991. Growth and mechanical impedance. In *Plant Roots: The Hidden Half,* eds. Y. Waisel, A. Ashel and U. Katkafi, 393-414. New York: Marcel Dekker, Inc. (271)

Bennie, A. T. P. and F. J. P. Botha. 1986. Effect of deep tillage and controlled traffic on root growth, water use efficiency and yield of irrigated maize and wheat. *Soil Tillage Res.* 7:85-95. (279)

Birdsall, M. and R. D. MacLeod. 1990. Early growth of the root system in Allium cepa. *Can. J. Bot.* 68:747-753. (269)

Blackwell, P. S. and B. D. Soane. 1981. A method of predicting bulk density changes in field soils resulting from compaction by agricultural traffic. *J. Soil Sci.* 32:51-65. (273)

Blancher, R. W., C. R. Edmonds and J. M. Bradford. 1978. Root growth in cores formed from fragipan and B2 horizons of Hobson soil. *Soil Sci. Soc. Am. J.* 42:437-440. (266)

Blum, A., J. W. Johnson, E. L. Ramseur and E. W. Tollner. 1991. The effect of a drying top soil and a possible non-hydraulic root signal on wheat growth and yield. *J. Exp. Bot.* 42:1225-1231. (270)

Boone, F. R. and B. W. Veen. 1982. The influence of mechanical resistance and phosphate supply on morphology and function of maize roots. *Neth. J. Agric. Sci.* 30:179-192. (261)

Bowen, H. D. 1966. Measurement of edaphic factors for determining planter specifications. *Transactions of the ASAE* 9:725-735. (260)

________. 1981. Alleviating mechanical impedance. In *Modifying the Root Environment,* eds. G. F. Arkin and H. M Taylor, 21-57. St. Joseph, Mich.: ASAE. (255,256,278)

Box, J. E. and G. W. Langdale. 1984. The effects of in-row subsoil tillage on corn yields in the southeastern Coastal Plains of the *US. Soil Tillage Res.* 4:67-78. (278)

* Numbers in parenthesis following each reference are the pages on which the reference is cited.

Bozoa, R. C. and L. R. Oliver, 1990. Competitive mechanisms of common cocklebur (*Xanthium strumarinus* L.) and soybean (*Glycine max* L.) during seedling growth. *Weed Science* 38:344-350 (271)

Brock, T. G. and P. B. Kaufman. 1991. Growth regulators: An account of hormones and growth regulation. In *Plant Physiology, A Treatise,* 10:277-340. San Diego, Calif.: Academic Press. (261)

Brown, D. A. 1984. Characterizing root growth and distribution. *Ark. Farm Res.* 33:3. (271)

Busscher, W. J. and R. E. Sojka. 1987. Enhancement of subsoiling effect on soil strength by conservation tillage. *Transactions of the ASAE* 30(4):888-892 (279)

Campbell, R. B., R. E. Sojka and D. L. Karlen. 1984. Conservation tillage for soybeans in the U.S. southeastern Coastal Plains. *Soil Tillage Res.* 4:531-541. (278)

Campbell, R. B., D. C. Reicosky and C. W. Doty. 1974. Physical properties and tillage of paleudults in the southeastern Coastal Plains. *J. Soil Water Conserv.* 29:220-224. (267)

Cannell, R. Q. 1977. Soil aeration and compaction in relation to root growth and soil management. *Appl. Biol.* 2:1-86. (263)

Carmi, A., J. D. Hesketh, W. T. Enos and D. B. Peters. 1983. Interrelationships between shoot growth and photosynthesis, as affected by root growth restriction. *Photosynthetica* (Prague) 17:240-245. (270)

Carpenter, T. G., N. R. Fawesey and R. C. Reeder. 1985. Theoretical effects of wheel loads on subsoil stresses. *Soil Tillage Res.* 6:179-192. (273)

Carr, M. K. V. and S. Dodds. 1983. Some effects of soil compaction on root growth and water use of lettuce. *Exp. Agric.* 19:117-130. (272)

Carter, L. M. 1985. Wheel traffic is costly. *Transactions of the ASAE* 28(2):430-434. (275)

Carter, L. M., B. D. Meek and E. A. Rechel. 1989. Cone index and cotton zone production systems. ASAE Paper No. 89-1542. St. Joseph, Mich.: ASAE. (275)

Cassel, D. K. 1982. Tillage effects on soil bulk density and mechanical impedance. In *Predicting Tillage Effects on Soil Physical Properties and Processes*, eds. D. M. Van Doren, R. R. Allmaras, D. R. Linden and F. D. Whisler. Madison, Wis.: Agronomy Society of America. (266)

Chaudhary, M. R., P. R. Gajri, S. S. Prihar and R. Khera. 1985. Effect of deep tillage on soil physical properties and maize yields on coarse textured soils. *Soil Tillage Res.* 6:31-44. (279)

Collis-George, N. and P. Yoganathan. 1985. The effect of soil strength on germination and emergence of wheat (*Triticum aestivum* L.). I. Low shear strength conditions. *Aust. J. Soil Res.* 23:577-587. (260)

Cornish, P. S., H. B. So and J. R. McWilliam. 1984. Effects of bulk density and water regime on root growth and uptake of phosphorus by rye grass. *Aust. J. Agric. Res.* 35:631-644. (272)

Dexter, A. R. and J. S. Hewitt. 1978. The deflection of plant roots. *J. Agric. Eng. Res.* 23:17-22. (258)

Dexter, A. R. 1986. Model experiments on the behavior of roots at the interface between a tilled seedbed and a compacted sub-soil: l. Effect of seedbed aggregate size and sub-soil strength on wheat roots. *Plant Soil* 95:123-134. (264)

________. 1987. Mechanics of root growth. *Plant Soil.* 98:303-312. (264,265,271)

Dickey, E. C., P. J. Jasa and R. D. Grisso. 1991. Long term tillage treatment used in a soybean/grain sorghum rotation. ASAE Paper No. 91-1005. St. Joseph, Mich.: ASAE. (279)

Doran, J. W. 1980. Soil microbial and biochemical changes associated with reduced tillage. *Soil Sci. Soc. Am. J.* 44:765-771. (276)

Drew, L. O. and W. F. Buchele. 1962. Emergence force of plants. ASAE Paper No. 62-641. St. Joseph, Mich.: ASAE. (256)

Drew, L. O., T. H. Garner and D. G. Dickson. 1971. Seedling thrust vs. soil strength. *Transactions of the ASAE* 14(2):315-318. (256)

Drew, M. C. 1979. Properties of roots which influence rates of absorption. In *The Soil-Root Interface*, eds. J. A. Harley and R. S. Russell, 21-38. New York: Academic Press. (260)

Eavis, B. W. and D. Payne. 1969. Soil physical conditions and root growth. In *Proceedings of 15th Easter School in Agricultural Science*, ed. W. J. Whittington, 315-338. London: Butterworth. (263)

Ehlers, W., V. Fopke, F. Hesse and W. Bohm. 1983. Penetration resistance and root growth of oats in tilled and untilled loam soil. *Soil Tillage Res.* 3:261-275. (265)
Elkins, C. B., D. L. Thurlow and J. G. Hendrick. 1983. Conservation tillage for long term amelioration of plow pan soils. *J. Soil Water Conserv.* 38:305-307. (278)
Erbach, D. C. 1987. Measurement of soil bulk density and moisture. *Transactions of the ASAE* 30(4):927-931. (266)
Fausey, N. R., and A. S. Dylla. 1984. Effects of wheel traffic along one side of corn and soybean rows. *Soil Tillage Res.* 4:147-154. (274)
Feldman, L. J. 1984. Regulation of root development. *Ann. Rev. Plant Phys.* 35:223-242. (261)
Fitter, A. H. 1991. Characteristics and functions of root systems. In *Plant Roots: The Hidden Half*, eds. Y. Waisel, A. Eshel and U. Katkafi, 3-25. New York: Marcel Dekker, Inc. (269)
Fitzgerald, P. J., E. E. Ortman and T. F. Branson. 1968. Evaluation of mechanical damage to roots of commercial varieties of corn (*Zea mays* L.). *Crop Sci.* 8:419-421. (277)
Foster, R. C., A. D. Rovira and T. W. Cook. 1983. *Ultra Structure of the Root-Soil Interface.* St. Paul, Minn.: American Phytopatholog Society. (262)
Gameda, S., G. S. V. Raghavan, R. Therwalt and E. McKyes. 1985. High axle load compaction and corn yield. *Transactions of the ASAE* 28(6):1759-1765. (274)
Garner, T. H. and H. D. Bowen. 1966. Plant mechanics in seedling emergence. *Transactions of the ASAE* 9(5):650-653. (256,258,259)
Gaultney, L., G. W. Krutz, G. W. Steinhardt and J. B. Liljedahl. 1982. Effect of subsoil compaction on corn yield. *Transactions of the ASAE* 25(3):563-569, 575. (273)
Gill, W. R. and G. H. Bolt. 1955. Pfeffer's studies of the root growth pressures exerted by plants. *Agron J.* 47:166-168. (255,256)
Goss, M. J., A. J. Wilson and A. W. Robards. 1977. Effects of mechanical impedance on root growth in barley *Hordeum vulgare* L. II. Effects on cell development in seminal roots. *J. Exp. Bot.* 28:1216-1227. (267)
Gowing, D. J. G., W. J. Davies and H. G. Jones. 1990. A positive root-sourced signal as an indicator of soil drying in apple, *malus* x *domestica borkh. J. Exp. Bot.* 41:1535-1540. (270)
Greacen, E. L., K. P. Barley and D. A. Farrell. 1968. The mechanics of root growth in soils with particular reference to the implications of root distribution. In *Proc. of 15th Easter School in Agriculture Science*, ed. W. J. Whittington, 256-268. London: Butterworth. (258)
Goyal, M. R., L. O. Drew, G. L. Nelson and T. J. Logan. 1980. Soybean seedling emergence force. *Transactions of the ASAE* 23(4):836-839. (257)
Hadas, A., I. Schmulevich, O. Hadas and D. Wolf. 1990. Forage wheat yields as affected by compaction and conventional vs. wide-frame tractor traffic patterns. *Transactions of the ASAE* 33(1):79-85. (275)
Hakansson, I., W. B. Voorhees, P. Elonen, G. H. S. Raghavan, B. Lowery, A. L. M. VanWijk, K. Rasmussen and H. Riley. 1987. Effect of high axle-load traffic on subsoil compaction and crop yield in humid regions with annual freezing. *Soil Tillage Res.* 10:259-268. (274)
Hale, M. G. and L. D. Moore. 1979. Factors affecting root exudation. II. 1970-1978. *Adv. Agron.* 31:93-123. (269)
Hardy, D. H., C. D. Raper, Jr. and G. S. Miner. 1990. Chemical restrictions of roots in ultisol subsoils lessened by long term management. *Soil Sci. Soc. Am. J.* 54:1657-1660. (273)
Hood, C. E., A. Khalilian, J. H. Palmer, T. H. Garner, T. R. Garrett and J. C. Hayes. 1991. Double-cropping interseeding system for wheat, soybeans and cotton. *Applied Engineering in Agriculture* 7(5):530-536. (276)
Itai, C. and H. Biunbaum. 1991. Synthesis of plant growth regulators by roots. In *Plant Roots: The Hidden Half*, eds. Y. Waisel, A. Ashe and U. Katkafi, 163-177. New York: Marcel Dekker, Inc. (261)
Jensen, E. H., J. R. Frelich and R. O. Gifford. 1972. Emergence force of forage seedlings. *Agron. Jour.* 64:635-639. (256)
Jones, C. A. 1983. Effect of soil texture on critical bulk densities for root growth. *Soil Sci. Soc. Am. J.* 47:1208-1211. (263)
Jones, C. A., W. L. Bland, J. T. Ritchie and J. R. Williams. 1991. Simulation of root growth. In *Modeling Plant and Soil Systems*, No. 31 in the series, eds. R. J. Hanks and J. T. Ritchie,

91-123. Madison, Wis.: Am. Soc. of Agronomy, Crop Science Society of America and Soil Science Society of America. (271)

Kamprath, E. J., D. K. Cassel, H. D. Gross and D. W. Dibb. 1979. Tillage effects on biomass production and moisture utilization by soybeans on Coastal Pain soils. *Agron. J.* 71:1001-1005. (277)

Karlen, D. L., W. J. Busscher, S. A. Hale, R. B. Dodd, E. E. Strickland and T. H. Garner. 1991. Drought condition energy requirement and subsoiling effectiveness for selected deep tillage implements. *Transactions of the ASAE* 34(5):1967-1972. (279)

Kirkham, M. B. 1987. Soil-oxygen and plant-root interactions: An electrical analog study. In *Plant and Soil: Interfaces and Interactions*, ed. A. vanDiest, 11-19. Boston: Kleuwer Academic Publishers. (268)

Kramer, P. J. 1988. Changing concepts regarding plant water relations. *Plant, Cell Environ.* 11:565-568. (270)

Lachno, D. R., R. S. Harrison-Murray and L. J. Andus. 1982. The effects of mechanical impedance to growth on the levels of ABA and IAA in root tips of *Zea Mays* L. *J. Exp. Bot.* 33:943-951. (261)

Larson, W. E and G. J. Osborne. 1982. Tillage accomplishments and potential. In *Predicting Tillage Effects on Soil Physical Properties and Processes*, eds. D. M. Van Doren, R. R. Allmaras, D. R. Linden and F. D. Whisler. Madison, Wis.: American Society of Agronomy. (277)

Letey, J. 1985. Relationship between soil physical properties and crop production. *Adv. Soil Sci.* 1:277-294. (262)

Lowery, B. and R. T. Schuler. 1991. Temporal effects of subsoil compaction on soil strength and plant growth. *Soil Sci. Soc. Am. J.* 55:216-223. (274)

Lowery, F. E., H. M. Taylor and M. G. Huck. 1970. Growth rate and yield of cotton as influenced by depth and bulk density of soil pans. Soil Sci. Soc. Am. Proc. 34:306-309. (272)

Lutz, J. F. 1952. Mechanical impedance. In *Soil Physical Conditions and Plant Growth,* ed. B. T. Shaw, 43-71. Madison, Wis.: American Society Agronomy. (256)

Lutz, A., J. Menge and N. O'Connell. 1986. Citrus root health; hardpans, claypans and other mechanical impedances. *Citrograph* 71:57-61. (268)

Martin, M. P. L. D. and R. W. Snaydon. 1982. Root and shoot interactions between barley and field beans when intercropped. *J. Appl. Ecol.* 19:263-272. (271)

Masle, J. and J. B. Passioura. 1987. The effect of soil strength on the growth of young wheat plants. *Aust. J. Plant Physiol.* 14:643-656. (272)

Masle, J., G. D. Farquhar and R. M. Gifford. 1990. Growth and carbon economy of wheat seedlings as affected by soil resistance to penetration and ambient partial pressure of CO_2. *Aust. J. of Plant Phys.* 17:465-487. (270)

Misra, R. C., A. R. Dexter and A. M. Alston. 1986. Maximum axial and radial growth pressures of plant roots. *Plant Soil* 95:315-326. (258)

Morton, C. T. and W. F. Buchele. 1960. Emergence energy of plant seedlings. *Agricultural Engineering* 41:428-431. (256)

Moss, G. I., K. C. Hall and M. B. Jackson. 1988. Ethylene and the responses of roots of maize (*Zea mays* L.) to physical impedance. *New Phytol.* 109:303-311. (261)

NeSmith, D. S., D. E. Radcliffe, W. L. Hargrove, R. L. Clark and E. W. Tollner. 1987. Soil compaction in double-cropped wheat and soybeans in an Ultisol. *Soil Sci. Soc. Am. J.* 51:183-186. (279)

Okada, K. and Y. Shimara. 1990. Reversible root tip rotation in *Arabidopsis* seedlings induced by obstacle-touching stimulus. *Science* 250:274-276. (260)

Okello, J. A. 1992. A review of soil strength measurement techniques for prediction of terrain vehicle performance. *J. Agric. Eng. Res.* 50:129-155. (263)

Parihar, S. S. and G. C. Aggarwal. 1975. A new technique for measuring emergence force of seedlings and some laboratory and field studies with corn (*Zea mays* L.) *Soil Sci.* 120:200-204. (257)

Person, H. 1978. Root dynamics in a young Scots pine stand in central Sweden. *Oikos* 30:508-519. (262)

Pfeffer, W. 1893. Druck und arbsitsleistung durch wachsende pflazen. *Abhandlungen der Koniglich Sachsischen Gesellschaft der Wissenschaften* 33:235-474. (255,256,258,271)
Powell, S. W. and F. P. Day Jr. 1991. Root production in four communities in the Great Dismal Swamp. *Amer. Jour. Bot.* 78:288-297. (262)
Prebble, R. E. 1970. Root penetration of smeared soil surfaces. *Exp. Agric.* 6:303-308. (260)
Raghavan, G. S. V. and E. McKyes. 1978. Statistical models for predicting compaction generated by off-road vehicular traffic on different soil types. *J. Terra. Mech.* 15:1-14. (274)
Raper, R. C. and D. C. Erbach. 1987. Bulk density measurement variability with core samplers. *Transactions of the ASAE* 30(4):878-881. (266)
Rechel, E. A., B. D. Meek, W. R. Detar and L. M. Carter. 1990. Fine root development of alfalfa as affected by wheel traffic. *Agron. J.* 82:618-622. (276)
Reicosky, D. C., D. K. Cassell, R. L. Blevins, W. R. Gill and G. C. Naderman. 1977. Conservation tillage in the southeast. *J. Soil Water Cons.* 32:13-19. (277)
Rendig, V. V. and H. M. Taylor. 1989. *Principles of Soil-Plant Inter-Relationships.* New York: McGraw-Hill Pub. Co. (255,269,270,271)
Richards, B. G. and E. L. Greacen. 1986. Mechanical stresses on an expanding cylindrical root analogue in granular media. *Aust. J. Soil Res.* 24:393-404. (261)
Rovira, A. D., R. C. Foster and J. F. Martin. 1979. Note on terminology: Origin, nature and nomenclature of the organic materials in the rhizosphere. In *The Soil-Root Interface*, eds. J. A. Harley and R. S. Russell, 1-4. New York: Academic Press. (260)
Rouse, H. R. and D. A. Stone. 1981. Deep cultivation of a sandy clay loam. II. Effects on soil hydraulic properties and on root growth, water extraction and water stress in 1977, especially of broad beans. *Soil Tillage Res.* 1:173-185. (277)
Russell, R. S. 1977. *Plant Root Systems: Their Function and Interaction with Soil.* London: McGraw-Hill. (255,256,260,262,263,271)
Russell, R. S. and M. J. Goss. 1974. Physical aspects of soil fertility. The response of roots to mechanical impedance. *Neth. J. Agric. Sci.* 22:305-318. (263)
Saini, G. R. and T. L. Chow. 1982. Effect of compact subsoil and water stress on shoot and root activity of corn (*Zea mays* L.) and alfalfa (*Medicago sativa* L.) in a growth chamber. *Plant Soil* 66:291-298. (267)
Sands, R., E. L. Greacen and C. J. Gerard. 1979. Compaction of sandy soils in radiata pine forests; A penetrometer study. *Aust. J. Soil Res.* 17:101-113. (267)
Schumaker, T. E. and A. J. M. Smucker. 1981. Mechanical impedance effects on oxygen uptake and porosity of dry bean roots. *Agron. J.* 73:51-55. (268)
Soane, B. D., J. W. Dickson and D. J. Campbell. 1982. Compaction by agricultural vehicles: A review. III. Incidence and control of compaction in crop production. *Soil Tillage Res.* 2:3-36. (276)
Sojka, R. E., W. J. Busscher, D. T. Gooden and W. H. Morrison. 1990. Subsoiling for sunflower production in the southeast Coastal Plains. *Soil Sci. Soc. Am. J.* 54(4):1107-1112. (268)
Spoor, G. and R. J. Godwin. 1978. An experimental investigation into the deep loosening of soil by rigid tines. *J. Ag. Engr. Res.* 23:243-258. (277)
Stolzy, L. H. and K. P. Barley. 1968. Mechanical resistance encountered by roots entering compact soils. *Soil Sci.* 105:297-301. (256,257)
Taylor, H. M. 1971. Root behavior as affected by soil structure and strength. In *The Plant Root and Its Environment,* ed. E. W. Carson, 270-291. Charlottesville, Va.: University of Virginia Press. (260)
Taylor, H. M. and H. R. Gardner. 1960. Use of wax substrates in root penetration studies. *Soil Sci Soc. Am. Proc.* 24:79-81. (257)
Taylor, H. M. and L. F. Ratliff. 1969. Root elongation rates of cotton and peanuts as a function of soil strength and soil water content. *Soil Sci.* 108:113-119. (256)
Taylor, H. M., L. F. Locke and J. E. Box. 1964. Pans in the southern Great Plains soils: III. Their effect on yield of cotton and grain sorghum. *Agron. J.* 56:542-545. (272)
Taylor, J. H. 1983. Benefits of permanent traffic lanes in a controlled traffic crop production system. *Soil Tillage Res.* 3:385-396. (275)

Threadgill, E. D. 1982. Residual tillage effects as determined by cone index. *Transactions of the ASAE* 25(4):859-963. (268)

Tollner, E. W. and B. P. Verma. 1984. Modified cone penetrometer for measuring soil mechanical impedance. *Transactions of the ASAE* 27(2):331-336. (263)

Torbert, H. A., J. H. Edwards and J. F. Pedersen. 1990. Fescues with large roots are drought tolerant. *Appl. Agric. Res.* 5:181-187. (273)

Tu, J. C. and C. S. Tan. 1991. Effect of soil compaction on growth, yield and root rots of white beans in clay loam and sandy loam soil. *Soil Biol. Biochem.* 23:233-238. (272)

Van Doren, D. M., R. R. Allmaras, D. R. Linden and F. D. Whisler. 1982. *Predicting Tillage Effects on Soil Physical Properties and Processes.* Madison, Wis.: Agronomy Society of America. (276)

VanLoon, C. D. and J. Bouma. 1978. A case study on the effect of soil compaction on potato growth in a loamy sand soil. 2. Potato plant responses. *Neth. J. Agr. Sci.* 26:421-429. (268)

Veen, B. W. 1982. The influence of mechanical impedance on the growth of maize roots. *Plant Soil* 66:101-109. (267,271)

Veihmeyer, F. J. and A. H. Hendrickson. 1948. Soil density and root penetration. *Soil Sci.* 65:487-493. (255)

Voorhees, W. B., W. W. Nelson and G. W. Randall. 1986. Extension and persistence of subsoil compaction caused by heavy axle loads. *Soil Sci. Soc. Am. J.* 50:428-433. (274,275)

Wanjura, D. F., E. B. Hudspeth, Jr. and I. W. Kirk. 1965. Measurement of pressures exerted by surface press wheels and their effect on cotton emergence. (Paper presented at Joint Meeting of Southeast and Southwest Regions ASAE) Dallas, 1-3 Feb. St. Joseph, Mich.: ASAE. (257)

Whiteley, G. M. and A. R. Dexter. 1982. Root development and growth of oil seed, wheat and pea crops on tilled and non-tilled soil. *Soil Tillage Res.* 2:379-393. (278)

Whiteley, G. M., J. S. Hewitt and A. R. Dexter. 1982. The buckling of plant roots. *Physiol. Plant* 54:333-342. (258)

Whiteley, G. M., W. H. Utomo and A. R. Dexter. 1981. A comparison of penetrometer pressures and the pressures exerted by roots. *Plant Soil* 61:351-364. (260)

Williams, W. A. 1963. The emergence force of forage legume seedlings and their response to temperature. *Crop Sci.* 3:472-474. (254)

Wolf, D., T. H. Garner and J. W. Davis. 1981. Tillage mechanical energy input and soil-crop response. *Transactions of the ASAE* 24(6):1412-1419. (266)

Zobel, R. W. 1989. Steady state control and investigation of root system morphology. In *Applications of Continuous and Steady-state Methods to Root Biology*, eds. J. G. Torrey and L. J. Winship, 165-182. Boston, Mass.: Kluwer Academic Publishers. (255)

Index